THE UNIVERSITY OF

Spectroscopic Measurement
An Introduction to the Fundamentals

Dedication

To Tracy, Jacob, and Kevin - the greatest gift I have ever received.

Spectroscopic Measurement
An Introduction to the Fundamentals

Mark A. Linne
Lund Institute of Technology
Lund, Sweden

ACADEMIC PRESS

An imprint of Elsevier Science

Amsterdam Boston London New York Oxford Paris
San Diego San Francisco Singapore Tokyo

Academic Press
An Elsevier Science Imprint
84 Theobald's Road, London WC1X 8RR, UK
http://www.academicpress.com

Academic Press
An Elsevier Science Imprint
525 B Street, Suite 1900, San Diego, California 92101-4495, USA
http://www.academicpress.com

ISBN 0-12-451071-X

Library of Congress Catalog Number: 2002100910

A catalogue record for this book is available from the British Library

Printed and bound in Great Britain by MPG Books Ltd, Bodmin, Cornwall
02 03 04 05 06 07 MP 9 8 7 6 5 4 3 2 1

Contents

Preface

In its first incarnation, this document was a much shorter article meant for publication in a review journal. The intended audience was (and is) the community of engineering thermosciences researchers who wish to apply a spectroscopic technique to a flowfield of interest, but who have not been formally trained in the fundamental topics that underlie spectroscopic measurement. At times I have been taken aside and asked very basic questions by people who were already using optical techniques, but who had found it necessary to learn about them "on the fly", almost as a footnote to their main research. Certain details of the field can indeed be confusing, but it is not an impenetrable subject. Unfortunately, there was no text or monograph I could recommend to the people who asked these questions. I had already asked several others working in the field if they would be willing to write such a book or monograph. These people were much better equipped to write one than I was. I had in mind a volume that would distill knowledge contained within roughly 40 sources, presenting the basic physics in a way that would make the subject accessible to practicing thermoscientists. This volume remained unwritten. Given the situation, I decided that a review article covering several common sources of confusion would prove useful.

The article encountered some resistance to publication in the journal to which I had submitted it. The reviewers said that the document should really be a book of some kind. One reviewer was especially negative: "My essential conclusion in going through this manuscript is that formal training in electromagnetism, quantum mechanics, and radiation should be prerequisites for researchers in this field". By "formal training", that reviewer meant one or two years worth of graduate-level coursework. I strongly disagree with that assertion. It is clear that some experimental flowfield researchers undertake to perform optical diagnostics without knowing beforehand what is really required. Problems can be minimized, however, by developing a more in-depth understanding, by learning "trade secrets" from people experienced in the area (if they are willing to help), and by applying more care. Imagine, however, a well-established fluid mechanics researcher who finds himself or herself in need of a good spectroscopic diagnostic. I do not believe that they should have to enroll in two years worth of coursework in order to understand fully their measurement technique. Much of the material in those courses is not germane to optical diagnostics anyway, and topics that are important are often not included in standard courses. There must be other avenues made available by which these researchers can become proficient.

The laser diagnostics community has claimed that development of new diagnostics is necessary to advance flowfield research, and that these techniques can uncover *critical* information heretofore unavailable. If this is indeed the case, then it is incumbent upon the diagnostics community to make these techniques accessible to mainline flowfield researchers. The restrictions proposed by that reviewer should not be, and de facto are not, taken seriously. Many members of the community have made a serious effort to make diagnostics more accessible. The books by Eckbreth [1] and Demtröder [2] are excellent examples, as is the new book edited by Kohse-Höinghaus and Jeffries [3], which is a community effort by many diagnostics developers and users. Many other monographs and review articles perform similar functions (many are referenced in this text).

This book is intended to complement those publications by providing a foundation for readers who are confused by material in those sources, because they have not seen how the more basic expressions were originally developed. This is not a stand-alone document; it will not teach diagnostic techniques. It is meant to be used as a guide, to explain material in the sources that do teach techniques.

That same negative reviewer wrote "I am skeptical that many of the formulations (in the article) will be accessible to the average engineering student or practicing combustion scientist that has no other exposure to the topic". Researchers who have mastered the physics and mathematics of fluid mechanics certainly have the background required to develop a deep understanding of spectroscopy, if they have the time and energy required to do so. This book is intended to make the task more straightforward by distilling many sources into one, to provide a first step in the process.

While I struggled with a response to the reviewer comments, the article metamorphosed into a book, thanks to the encouragement of Professor Derek Dunn-Rankin at the University of California at Irvine, and Dr Emma Roberts and Dr Egbert van Wezenbeek at Academic Press. It was not an easy task, however, because it required doubling the size of the document. While much of the text has been proof-read by colleagues, I fear that errors may remain. It is very difficult to proof-read one's own writing. I can misread a statement that is incorrectly typewritten or phrased because I know what I meant to say. Too many times I have stumbled across such a misstatement. Readers who may catch something that requires my attention are encouraged to contact me about it.

The mathematical approaches used in this book are similar to those used in engineering: in fluid mechanics, radiation heat transfer, chemical kinetics, and electronic signals and systems. This book is written by a thermoscience engineer, with thermoscience engineers in mind, and so engineering approaches to conceptual development are used whenever possible. This is not done simply for my convenience. Bransford *et al.* [4] have shown that learning always occurs within the context of previously learned material. People learn more quickly and in greater depth when the instructional source matches their own context to some degree. Any text will in fact resonate with some better than with others, in part because it was written (either intentionally or unintentionally) from a specific background and with a specific approach. Here, a specific audience, one that would most commonly need this volume, has been intentionally chosen. Having said that, the topics presented in this book are general. It would be fantastic if nonengineers could also find something useful here.

Finally, the same reviewer who has provided the structure for this Preface wrote that the article's "utility would be enormously improved if the authors had used their dense formalism to solve even one simple problem of quantitative spectroscopy in combustion systems. Several such examples might prove a more useful tool for those lacking sufficient background education in the subject areas". First, I am sensitive to the charge regarding the density of the material, but I do not see how this material can be made less so; it simply is what it is. A review on the jacket of another book on molecular spectroscopy boasts that it avoids mathematics. This could be considered a less dense path, but I do not like mathematics-free survey books. To present a result without explaining how it can be developed is, in my view, of minimal utility. I have found errors (most often typographical mistakes) in even the most respected books, but I have been able to correct them and make use of the book if they have shown how the result was meant to be found. If an author simply retypes a formula from another source, I am not sure I can trust it. More importantly, spectroscopic literature uses many different formalisms, units, and conventions. I do not believe it is possible to extract useful information from the literature unless one has a deep understanding of the subject. Mathematics-free treatments do not offer such depth.

I have chosen not to include example problems. People who feel forced to learn a new subject too quickly tend to seek patterns and then reproduce the pattern, without knowing the details underlying the pattern they have implemented. It would be far too easy to apply an example from this book to another molecule or diagnostic, incorrectly extracting spectroscopic data from other sources, and make the very errors this book seeks to minimize. This is a subject that cannot be grasped casually.

The literature contains many good examples that can be understood with the help of texts like this one. For work on examples, readers can select their own molecule and diagnostic and study the various papers in the literature. By sorting the problem completely, in all of its detail, readers can learn by example and ensure that this part of their research is also well sorted.

Best of luck with your research.

Mark Linne
June, 2002

Acknowledgments

I am indebted to many people. First, I would like to thank two colleagues who helped me prepare this book in direct ways. Dr Tom Settersten graduated from my group and is now working at the Combustion Research Facility at Sandia National Labs in Livermore, California. He contributed material that is now built into Chapters 3, 10, and 14. I am grateful to him for enjoyable discussions, incisive questions, and for his well-written contributions. Professor Philip Varghese at the University of Texas at Austin diligently proof-read most of the book. His detailed corrections and suggestions have made significant and important improvements throughout the text. I am not sure how I can properly thank him for the huge investment he has made in this document.

Dr Donald Lucas at Lawrence Berkeley Laboratories and Dr Phillip Paul at Eksigent Technologies have also read large portions of the book and provided very useful suggestions. I am grateful for their help. Professor Derek Dunn-Rankin at the University of California at Irvine first suggested the idea of a book and encouraged me to explore the possibilities. Without his insight and help, this book would not exist.

I am grateful to Professor Robert Dickson of the Georgia Institute of Technology and Dr Akira Tonomura of the Hitachi Advanced Research Laboratory, who kindly granted their permission to reproduce Figures 5.5 and 7.1, respectively. These two remarkable experimental images go a long way towards conclusively demonstrating the actual behavior of primary particles. They are invaluable didactic tools.

Active research collaboration has contributed to this book indirectly, by raising new issues and questions, and by identifying new directions. Research collaborators include Professor Marcus Aldén and the group at the Lund Institute of Technology in Sweden; Dr Roger Farrow and Dr Phillip Paul at Eksigent Technologies; Dr Greg Fiechtner and Dr Tom Settersten at Sandia Labs; Dr James Gord at Wright Labs; Professor Katharina Kohse-Höinghaus, Dr Andreas Brockhinke and the rest of the group at the University of Bielefeld in Germany; Dr Stuart Snyder (formerly at the Idaho National Environmental and Engineering Lab, now at Montana State University in Billings); and Professor Philip Varghese at the University of Texas at Austin. I have been fortunate to interact with, and in the process learn from, each of these colleagues.

This book was written while I was a member of the faculty at the Colorado School of Mines (CSM). In fact, I started on this project while on a sabbatical that was partially supported by CSM. Faculty at the Colorado School of Mines, including Professors Robert Kee, Terry Parker, Laxminarayan Raja (now at the University of Texas), and Jean-Pierre Delplanque, have also contributed to this book in many ways. I have genuinely enjoyed working with them. Moreover, I have relied upon Professor Tom Furtak of the CSM Physics Department for excellent help on questions regarding optical electromagnetics.

Much of my understanding of spectroscopy has been shaped by the research I have been able to pursue, thanks to generous support from Dr James Gord at Wright Labs; Dr Milton Linevsky and Dr Farley Fisher at the National Science Foundation (Chemical and Transport Systems); Dr David Mann at the Army Research Office, Dr Stuart Snyder at the Idaho National Environmental and Engineering Lab; and Ray Witten, Mark Nall and John West at NASA. I am grateful for their sustained support.

Guidance of graduate students has forced me to learn at much deeper levels, and this is perhaps the most rewarding part of my job. I have genuinely enjoyed working with and learning from every graduate student in my group, including past students: Mr Richard MacKay, Mr Adam Rompage, Mr Jon Luff, Mr Jim Weiler, Mr Dan Vogel, Dr Tom Settersten, Mr John Fowler, Dr Chad Fisher, Dr Scott Spuler, and current students: Mr Tom Drouillard, Mr Tyler Hall, Ms Megan Paciaroni, Mr David Sedarsky, and Mr Ryan Swartzendruber.

Finally, I must acknowledge my PhD thesis advisor, Professor Ron Hanson at Stanford University. I first met Ron in 1976, shortly after completing my undergraduate studies. In March of that year I was in the San Francisco Bay Area, and, driven mostly by curiosity, I stopped at Stanford. Ron had just installed a very nice shock tube lab and so he took me to see it. To the side of the lab he showed me an experiment that he and a student (now Dr Pat Falcone) had set up. They were using a lead-salt diode laser to do absorption spectroscopy in the post-flame region of a laminar, flat flame. I knew just enough about spectroscopy to appreciate the importance of that experiment. In my world view, Ron had just turned combustion spectroscopy completely upside down, in truly ground breaking ways. Later on I realized that he was not alone in applying lasers to combustion spectroscopy, but the community was small in 1976 and that was my first encounter with it. For me, a novice, to stand in Ron's shock tube lab and consider the possibilities for lasers in combustion spectroscopy was a career-defining moment. I made a decision then to work in his group at Stanford. During my dissertation research, I did indeed have the opportunity to perform both laser absorption and laser induced fluorescence measurements of OH in flames. I am extremely grateful to Ron for that opportunity, and for his support during those graduate years. Especially now that I support and advise graduate students myself, I understand and appreciate the price he paid to support and advise me. In the years since I have left Stanford, Ron has freely provided advice and counsel when I have asked for it. Perhaps more importantly, he has always been encouraging. I count myself extremely fortunate to have stopped by Stanford on that day in 1976.

Nomenclature

Abbreviations

ASE	amplified spontaneous emission
cv	control volume
cw	continuous wave
CVD	chemical vapor deposition
DFWM	degenerate four-wave mixing
DME	density matrix equations
EMF	electro motive force
ERT	equation of radiative transfer
E&M	electricity and magnetism, or electromagnetism
FTIR	Fourier transform infrared spectroscopy
FWHM	full-width at half-maximum (of a distribution)
LIDAR	light detection and ranging
LIF	laser-induced fluorescence
LTE	local thermodynamic equilibrium
MKSA	meter-kilogram-second-ampere
ODE	ordinary differential equation
PDF	probability density function
PSD	power spectral density
SI	international system of units
STP	standard temperature and pressure
TEM	transverse electromagnetic

Variables

a	eigenvalue of operator $\hat{A}$, non-dimensional broadening parameter, or mean polarizability
A	surface area, Einstein coefficient for spontaneous emission, electromagnetic vector potential, generalized operator (quantum mechanics), symmetric top rotational constant, or denotes first diatomic molecular excited electronic level
b	component of angular momentum along coordinate z (in quantum mechanics)
b	Raman quantum amplitude coefficient
B	Einstein coefficient for stimulated absorption or

emission, magnetic induction, rotational constant, or denotes second diatomic molecular excited electronic level

c — speed of light in vacuum, expansion coefficient for one wave function in terms of another, or $c^2 =$ electronic angular momentum squared (in quantum mechanics)

$c.c.$ — complex conjugate

C — number of energy microstates in the statistical mechanics development

d — differential

d — diameter

D — electric displacement, centrifugal distortion correction in molecular spectroscopy

$\mathcal{D}$ — dissociation energy in Morse potentials

e — charge on one electron

E — energy, or electric field

$\mathcal{E}$ — radiant energy

f — velocity or energy distribution function, Boltzmann fraction, photon distribution function, oscillator strength, or generalized function of position and/or time

$f\#$ — optics f-number

F — force

F — velocity density, or ro-

tational term energy $[\text{cm}^{-1}]$

g — degeneracy, or normalized autocorrelation

g — represents the eigenfunctions of the spin momentum equation

G — vibrational term energy $[\text{cm}^{-1}]$

$\mathcal{G}$ — autocorrelation

h — Planck's constant

$\hbar$ — $h/2\pi$

H — magnetic field, or Hamiltonian operator

i — $\sqrt{-1}$

I — irradiance, electric current, moment of inertia, or nuclear spin quantum number

J — radiance, current density, or total rotational quantum number

$\mathcal{J}$ — nuclear rotational quantum number

k — Boltzmann's constant, wave vector, spring constant, or Rayleigh/Raman amplitude coefficient

$\mathcal{K}$ — component of electronic orbital angular momentum along symmetric top figure axis

l — length

ℓ — interaction length, or electron orbital angular momentum quantum number

L — dimensionless spectral distribution function (in ERT), distance, Lagrangian operator (Hamiltonian mechanics), or orbital angular momentum (quantum mechanics)

m lower energy level, mass, number of constraints, or atomic magnetic quantum number

M total mass, magnetization, or molecular magnetic quantum number

n surface normal

n upper energy level, index of refraction, number of particles, or principal quantum number

N total number (not italicized)

N number density (italicized), or $R+\Lambda$ in Hund's case b for molecules

$\mathcal{N}$ number of moles

O denotes $\Delta J = -2$ for molecular rotational transitions

p pressure, momentum, probability density function, or component of polarization

P pressure, material polarization, power, permutation operator, or designates $\Delta J = -1$ for molecular rotational transitions

$\mathcal{P}$ probability, or velocity

parameterized polarization

q charge

Q partition function, quenching rate, denotes $\Delta J = 0$ for molecular rotational transitions, or classical vibrational normal mode coordinate

r position, or radius

R rotation matrix

R dimensionless transverse profile (in ERT), position vector, Rydberg constant, radial portion of the wave function (in quantum mechanics), diatomic nuclear angular momentum, or denotes $\Delta J = +1$ for molecular rotational transitions

$\mathcal{R}$ ideal gas constant or radial wave function

s relative velocity

s number of degrees of freedom, or spin quantum number

S surface area, source function (blackbody), Poynting vector, electronic spin, Frank-Condon factor, Hönl-London factor, denotes $\Delta J = +2$ for molecular rotational transitions, power spectral density function, or signal

t time

T temperature, dimension-

	less temporal distribution function (in ERT), decay or dephasing time, kinetic energy (Hamiltonian and quantum mechanics), energy "term", or electronic term energy $[\text{cm}^{-1}]$	y	part of the second anharmonic vibrational correction factor
u	general wave function	Y	lineshape function, or θ- and ϕ-dependent wave function (in quantum mechanics)
v	velocity, or volume		
v	specific volume, volume, speed of light in a medium, or vibrational quantum number	Z	charge number
		α	atomic or molecular polarizability, or spin wave function
V	volume, or Voigt function		
V	voltage (in E&M), potential energy (Hamiltonian and quantum mechanics), or total volume	β	factor in statistical mechanics development $= (k_B T)^{-1}$, or spin wave function
w	speed	γ	decay rate for coherences, or anisotropy of polarizability ellipsoid
w	Gaussian beam waist, or generalized coordinates (classical Hamiltonian mechanics)		
		Γ	decay rate for a population difference
W	number of microstates, optical pumping rate, work, or energy	δ	impulse function, Kronecker delta, velocity-dependent detuning, or phase angle
x	non-dimensional detuning parameter		
x	part of the first anharmonic vibrational correction factor	Δ	change in a variable, uncertainty, or designates $\Lambda = 2$
x, y, z	Cartesian coordinates	ϵ	energy level (statistical mechanics), electric permittivity, or volume emission coefficient
X	denotes the ground-state diatomic molecular electronic level		
		ε	electromotive force (EMF), or real and imaginary components of the electric field amplitude
		θ	vertical angle
		Θ	θ-dependent wave func-

tion

κ volume absorption coefficient, dielectric constant, or relative permeability

λ optical wavelength

Λ component of electronic orbital angular momentum along diatomic molecular internuclear axis

μ magnetic permeability, dipole moment, reduced mass, or Bohr magneton

ν optical frequency (Hz, or 1/s), vibrational constant

$\breve{\nu}$ wavenumbers or Kayser (cm^{-1})

ξ random variable or process

π linearly polarized transition

Π denotes $\Lambda = 1$

ρ mass density, spectral energy density (ERT), depolarization factor, or density matrix

ϱ charge density in E&M

σ cross-section, circularly polarized transition, or slowly varying envelope function for coherence

Σ component of electronic spin angular momentum along diatomic molecular internuclear axis, or denotes $\Lambda = 0$

τ characteristic time, optical depth, or volume element in Hilbert space (quantum mechanics)

ϕ azimuthal angle, or electromagnetic scalar potential

φ wave function, or phase angle

Φ magnetic flux, or ϕ-dependent wave function

χ susceptibility, or diatomic molecular nuclear wave function (in quantum mechanics)

ψ position-dependent wave function (in quantum mechanics)

Ψ time- and position-dependent wave function (in quantum mechanics)

ω angular frequency (rad/s), or vibrational energy constant

Ω total number of energy states, solid angle, $\Lambda + \Sigma$ for Hund's case a for molecules, or Rabi frequency

Subscripts

A Avagadro

A-S anti-Stokes

B Boltzmann, Bohr

BE Bose-Einstein

cv control volume

D displacement (current), or Doppler

e	electron, or electronic	p	photon, or plasma
ed	electric dipole	r	rotational
em	electromagnetic	R	right-circular, or Raman
f	free (current or charge density)	sat	saturation
f	final	ss	steady-state
FD	Fermi-Dirac	S	Stokes
h	horizontal	TH	top-hat
H	homogeneous	v	vibrational, at vibrational quantum number v, or vertical
i	counting subscript, or initial		
I	nuclear rotational	v	velocity
j	indicator for individual atoms or molecules	$x, y,$ or z	Cartesian positional dependence
J	at rotational quantum number J	ν	spectral quantity
k	vibrational normal mode index	1	on T, indicates decay time for population difference
L	left-circular, or laser	2	on T, indicates decay time for coherence
m	lower energy level, magnetic, or localized microscopic	$\frac{1}{2}$	FWHM

Superscripts

n	upper energy level, general indicator of a quantum level, or signifies nuclear	I	imaginary component
		R	real component
		○	normalized quantity
n	nuclear	∗	complex conjugate, or equilibrium level
nm	transition between levels n and m	$(1), (2)..$	order of nonlinear susceptibility
N	nuclear	′	first derivative, real part of susceptibility, upper energy level, derived polarizability, or retarded time
NL	nonlinear		
○	located at a peak, in vacuum, amplitude (e.g. electric field), at rest (mass), indicates unperturbed, or equilibrium	″	second derivative, imaginary part of susceptibility, or lower energy level
osc	oscillator		

Other

~	complex variable
^	unit vector, or operator in quantum mechanics
¯	average
→	vector
˘	wave numbers
⟨|	Dirac bra
|⟩	Dirac ket
⟨⟩	quantum mechanical average
↔	allowed transition
↮	forbidden transition

Chapter 1

INTRODUCTION

1.1 Spectroscopic Techniques

Spectroscopy has proven to be a powerful tool for the investigation of flowfield structure. The earliest observations of flame structure, for example, were based upon emission of light from reaction zones (see, for example, Gaydon [5] or Mavrodineanu and Boiteux [6]). Prior to the advent of laser-based techniques, absorption of lamp light was used to measure both species concentration and temperature (e.g. via sodium line reversal [6]). Both thermal and chemiluminescent emission remain powerful and straightforward flowfield research tools, while lamp absorption has adapted more sophisticated signal processing to become Fourier transform infrared spectroscopy (FTIR) [7]. These techniques have been overshadowed somewhat by the development of laser-based measurements in flames [1, 2, 3, 8, 9, 10]. Lasers provide directional and coherent beams that have very high spectral brightness, thus enabling entirely new approaches [1]. Laser diagnostic techniques have provided information heretofore unavailable, making it possible to develop and prove sophisticated models for combustion. The same diagnostic and computational techniques have been applied to other important flowfields as well, from chemical vapor deposition (CVD) to atmospheric sensing (via light detection and ranging, LIDAR).

Optical techniques for measurement of species concentration and temperature are important for several reasons. First, they do not physically intrude into a flowfield and disturb it (via hydrodynamic, thermal, or catalytic processes [1]), as a physical probe would. Disturbance by physical probes can alter the chemical makeup or thermo/fluid variables one hopes to sense, usually in ways that introduce significant uncertainty. In contrast, optical techniques do not perturb the flow, if performed correctly. They can be used to detect reactive species that would not survive within most extraction systems (molecular beams are the exception). Optical diagnostics can often be performed within small sample volumes, making it possible to study the structure of nonuniform flowfields in detail. They can also provide excellent temporal resolution, making it possible to observe rapid transients. An important example of this is the fact that, prior to 1975, there were very few high-quality measurements of OH, despite its importance as an intermediate species in combustion chemistry. Since then, the measurement of OH concentration in flames via laser diagnostics has become routine [1, 2, 3, 9, 10]. Extension of these techniques to optical imaging of species concentrations [11] has provided critical insight for development of advanced models.

The purpose of this book is not to review optical diagnostic techniques themselves. At the time of this writing, laser diagnostic techniques have been covered well by Eckbreth [1], Demtröder [2], Kohse-Höinghaus [9], Kohse-Höinghaus and Jeffries [3], Measures [12] and by others who are mentioned in these sources. These authors provide an overview, present expressions in basic physics, review the current status of diagnostic techniques, and provide example results.

The purpose of this book is complementary; it compiles topics in the fundamental background material required for spectroscopic techniques. Optical diagnostics require attention to the generation of a signal, collection of the signal, and electronic sensing and signal processing. The first of these components, generation of the signal, depends upon exactly which diagnostic technique one intends to use. One class of diagnostic does not rely upon the material in this book. It includes any nonspectroscopic, purely optical measurement (e.g. particle image velocimetry). The material in this book emphasizes spectroscopic techniques. Spectroscopic signal generation depends upon the physics of the interaction (e.g. a linear or nonlinear technique, orientation of the optical beams, and so forth), and upon the spectrum of the signal and the response of the atom or molecule, which can be either resonant or nonresonant at the optical wavelength.

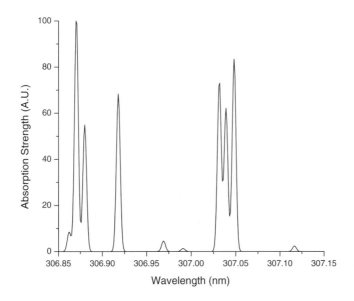

Figure 1.1: A portion of the rotational absorption spectrum of OH in the first electronic transition and fundamental vibrational band, taken from LIFBASE [13].

An example spectrum of a resonant response is represented by Figure 1.1, which contains a portion of the absorption spectrum (strength of optical absorption vs wavelength) for OH. The spectral locations (horizontal axis) of the features in the figure are determined by changes in the molecular energy during an optical interaction. In brief, the OH molecule is capable of absorbing light (or emitting it) at wavelengths in the vicinity of the peaks shown in Figure 1.1 (where a response is indicated), but not where there is no response. As with any atom or molecule, OH has specific, quantized energy levels (or states). The spectral locations indicate a transition between two of these levels (often labeled "n" for the higher energy level and "m" for the lower energy level). A change in molecular energy (represented by ΔE_{nm}) occurs during a transition between these two levels, and when an optical interaction is responsible, this change corresponds to absorption or emission of light at the optical frequency ν_{nm} (Hz or 1/s), given by

$$\Delta E_{nm} = h\nu_{nm} \tag{1.1}$$

where h is Planck's constant ($h = 6.626 \times 10^{-34}$ J-s).

The subject areas describing the spectral location of these transitions (what energy levels and transitions are allowed), and the specific jargon that describes states and transitions, are introduced in this book. They are also introduced (but with less background material) by Banwell and McCash [14], Eckbreth [1], and Demtröder [2]. The books by Herzberg provide a detailed explanation of the spectroscopy of atoms [15], diatomic molecules [16, 17], and polyatomic molecules [18, 19]. Unfortunately, each particular atom or molecule has its own specific patterns (coupling of spin, orbital and nuclear rotation cases; anharmonicities; symmetries; etc.). It is not possible to describe all of these in detail in this book, nor are they described fully in the other sources just mentioned. Typically, one must build a large file of journal articles on the spectroscopy of each species of interest. Fortunately, some groups (see, for example, Luque and Crosley [13]) provide such information in databases that are easy to use.

The heights of the peaks in Figure 1.1, together with the spectral lineshapes, are controlled by the response of an atom or molecule to radiation, a topic discussed throughout this book. Other sources on this topic include Banwell and McCash [14], small sections of Herzberg's books, the previously mentioned texts by Eckbreth, Demtröder, Measures [12], Allen and Eberly [20], Steinfeld [21], Boyd [22], Lalanne *et al.* [23], and Mukamel [24], and the laser texts by Siegman [25] and Verdeyen [26]. In addition, the articles by Hillborn [27] and by Lucht *et al.* [28] are quite useful.

Despite these many resources, there is no single location where the fundamental topics associated with spectroscopic response are distilled into one reference for use by flowfield researchers who need to apply optical diagnostics. There is certainly no such treatment for those who are new to the subject, and one can often encounter confusion. Sources of confusion include the differences between the many ways one can describe the atomic or molecular resonance response (e.g. absorption cross-section vs band strength), or the presence of $\sqrt{4\pi\epsilon_o}$ in some expressions and its absence in others. Without a presentation of the underlying physics and the context of a particular result, it becomes very difficult to use a result with confidence. The goal of this book, therefore, is to distill knowledge contained within roughly 40 other sources, presenting the basic physics in a way that will make the subject accessible to practicing thermoscientists.

1.2   Overview of the Book

Several fundamental topics are presented in this book, starting with a brief review of statistical mechanics. Chapter 2 is not exhaustive because most thermoscientists have taken a formal course on statistical thermodynamics.

Next, the equation of radiative transfer is discussed in detail. The standard rate equations for resonance response are developed, and the equation of radiative transfer is related to the Planck black-body distribution. Standard forms such as Beer's law are developed. Radiative transfer is then followed by the theory of electromagnetism. The concept of resonance response via the classical oscillator, or Lorentz atom, is developed. Both absorption and emission are modeled in both formalisms (equation of radiative transfer and electromagnetic theory). The book is intentionally structured in this way, because the two formalisms describe the same problem in parallel. The equation of radiative transfer deals with transport of optical power (irradiance) via the phenomena we wish to describe. It represents simple energy conservation. Electromagnetism deals with electromagnetic (in this case optical) waves, which are propagated via the wave equation. Irradiance is proportional to the electric field amplitude squared. The equation of radiative transfer and the wave equation are therefore linked by this power law. The equation of radiative transfer is simple to use, it readily provides expressions such as Beer's law for absorption, and it provides easy book-keeping for energy conservation using an engineering approach. At a fundamental level, however, the resonance interaction between light and an atom or molecule occurs between a traveling electromagnetic wave and a quantized polarization state. The two formalisms therefore complement each other.

Next, the classical formalism for Hamiltonian dynamics is presented. This is done because nonphysicists are often simply presented with the expressions of quantum mechanics with no background development. The Hamiltonian approach was developed to solve the classical dynamic problem for a particle in a potential field, in cases where it becomes difficult to account for Newtonian force vectors. This is exactly the situation associated with atoms and molecules, and for this reason the Hamiltonian is the appropriate operator to use in quantum mechanics. It is the author's view that a short introduction to classical Hamiltonian dynamics eases the entry into quantum mechanics.

Following that, key concepts in quantum mechanics are reviewed. The postulates of quantum mechanics are described and then formally presented.

Having built a foundation, the basics of atomic and molecular spectroscopy are presented in two companion chapters. A good deal of atomic spectroscopy is discussed, despite the fact that molecules are more commonly of interest in flowfield research, because atomic spectroscopy prepares a solid foundation for molecular discussions. General subjects such as the rigid rotator and corrections to the rigid rotator (e.g. centrifugal distortion) are developed.

The following chapter describes the various ways one can express the resonance response of atoms or molecules, and how they are interrelated. An entire chapter is devoted to this subject because it has generated so much confusion in the past. A formalism for using and interrelating the information that appears in the literature is provided, so that it can be applied effectively to a diagnostic problem.

To this point, spectral broadening will have been ignored, except that broadening is related to the decay and collision times. In the next chapter, therefore, the subject is discussed in significantly more detail.

Polarization of light and the material it interacts with is an important subject that is often overlooked. Polarization orientation, however, can make as much as a factor of 3 difference in certain diagnostics signals. For this reason, a chapter briefly reviews this topic.

Next, both Rayleigh and spontaneous Raman scattering are discussed. Rayleigh scattering is not a spectroscopic interaction, but it is closely related to Raman scattering and it is used routinely by spectroscopists. Raman scattering is a very common and useful nonresonant spectroscopic technique. For this reason we develop basic expressions on the topic and describe how it is used in flowfield research.

Finally, the density matrix equations are presented. These describe the dynamics of the quantum interaction, and they can be linked to the wave equation via the state polarization. Using appropriate assumptions, the density matrix equations can generate the same expressions as the classical result. The system of expressions is therefore self-consistent.

1.3 How to Use This Book

This is not a stand-alone volume on spectroscopic diagnostic techniques, nor is it a book for absolute beginners. This book provides portions of the background material that graduate students in Physical Chemistry or Chemical Physics receive when they are being trained as spectroscopists. Other researchers who need to apply a spectroscopic diagnostic to a flowfield are often not prepared in the same way, because the spectroscopic measurement constitutes only a portion of their flowfield research project. It is common in such a case to use printed sources that explain in a very direct way how a specific diagnostic is performed. Such an approach is certainly legitimate, but it does not provide all of the background material that is necessary if one hopes to go into further depth. This book is intended to provide the background material that makes it possible to achieve a deeper understanding. This book, therefore, should be used simultaneously with other sources on diagnostic techniques in order to develop a broad understanding.

One successful approach used in graduate coursework in the past has been to use both this material and the book by Eckbreth [1] or the book by Kohse-Höinghaus and Jeffries [3] which has just been published. In such a course, the material in this book is used to explain formalisms that appear in a diagnostics book (or article). Much of the material is presented in a lecture format, starting from basic concepts (from this book) and leading to a well-known diagnostic expression (found in the diagnostics book). Students are asked to perform several experiments during this course (e.g. emission, absorption, laser-induced fluorescence, etc.). The students are first given a preparatory assignment in which they must find the spectroscopic information for an assigned molecule or atom. They then create a model for the assigned diagnostic, including estimates for signal strength and detection limit. Both this book and the diagnostics text provide the necessary background. The class then meets to agree upon a group approach to the experiment, which is then performed in the lab. Each student then discusses results in a report, comparing experimental outcomes to expectations based upon their individual preparatory assignments.

Draft versions of this book have also been used by individual students outside of a formal course, during the performance of their research. Again, the combination of this text and more diagnostic-specific material has led to successful implementation of a measurement by students who were not formally trained in this area.

1.4 Concluding Remarks and Warnings

It is not possible to discuss every variation that one can encounter in this subject area within this text. For some reason, the topic seems to foster as many variants as there are authors. As just one example, some rotational line strengths may be normalized differently than others (requiring a different accounting of degeneracies). One should take warning that every source of spectroscopic information *must* be examined with extreme care and diligence, in order to produce a spectroscopic measurement that one can say with confidence is free from errors in interpretation. This is an unfortunate situation, but those are the facts. The goal of this book is to present the fundamentals that will allow one to proceed while using literature values, but it is not possible to explain what every individual author may have done in specific articles or databases. This text can be thought of, therefore, as a guide for other sources and for more advanced treatments.

Chapter 2

A BRIEF REVIEW OF STATISTICAL MECHANICS

2.1 Introduction

Statistical mechanics is a core subject in the thermosciences. It is used in areas such as advanced thermodynamics, gas dynamics, heat and mass transfer, and chemical kinetics. This chapter, therefore, will not be as detailed as the ones that follow, nor is it a complete treatment, because it is assumed that most readers are somewhat familiar with the subject. Readers who have experience using statistical mechanics can comfortably skip past this chapter.

In flowfield diagnostics, the Boltzmann distribution is used to describe populations of states (fraction of atoms or molecules in a particular energy state) that interact with radiation. As an example, each of the spectral features shown in Figure 1.1 provides evidence of coupling between two energy states. These features would not be discernable, however, if no OH molecules actually occupied the energy states that absorb the light. Local thermodynamic equilibrium (LTE) is commonly assumed for the collection of molecules under investigation prior to excitation by light. Statistical mechanics can therefore describe the populations of the absorbing levels that are responsible for the features in Figure 1.1. The topic of this chapter is therefore critical to the description of the overall spectral response.

An extreme example is provided by laser spectroscopy. Lasers have very high spectral brightness, meaning that a large amount of energy is contained within a very narrow spectral band. Often the laser bandwidth is narrower than the linewidths shown in Figure 1.1. This is one reason for the popularity of laser-based spectroscopy; a large amount of energy can be coupled directly into an atomic or molecular resonance. Unfortunately, narrow bandwidth means that the laser radiation can be absorbed only by atoms or molecules in one particular energy level out of many possible levels. The fraction of total population in just one level of a diatomic molecule can be as low as 10^{-3}, and it can be much lower for polyatomic molecules. Spectroscopic measurements are often aimed at species that exist in very low concentrations. The lowest measurable concentration, or "detection limit", then becomes the major figure of merit for a technique. This factor of 10^{-3} is thus a critical component of the diagnostic.

Much of what follows is adapted from portions of the books by Vincenti and Kruger [29] and by Garrod [30]. The classical text by Tolman [31] also provides a background on the topic. Many other well-written texts discuss statistical mechanics in a comprehensive way. The treatment provided here is by no means comprehensive because the purpose of this chapter is to review the basis for statistical mechanics very briefly, and then to present expressions that prove useful for diagnostics. Thermoscientists who are unfamiliar with statistical mechanics should consider going to more comprehensive sources such as references [29], [30], and [31].

The first section of this chapter introduces the Maxwellian velocity distribution, which is important in spectroscopy because it is used to describe Doppler broadening of spectral lines. Moreover, the Maxwellian distribution introduces the statistical mechanics formalism in a simple and intuitive way. Next, the Boltzmann energy distribution is developed. The presentation is simplified by taking advantage of the fact that many flowfields of interest can be considered ideal gases. Other short cuts are taken. Finally, equations that are useful for spectroscopic flowfield diagnostics are presented, using energy level expressions that will be developed in later chapters.

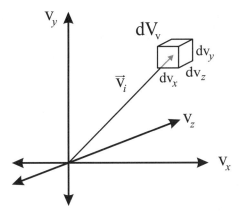

Figure 2.1: Single velocity value $\vec{v}_i$ in velocity space v_x, v_y, v_z with differential volume in velocity space $dV_v = dv_x \, dv_y \, dv_z$.

2.2 The Maxwellian Velocity Distribution

If the gas under investigation does not have appreciable bulk velocity (e.g. the Mach number is small), then the flowfield can be thought of as a collection of rapidly moving molecules with a relatively low (or zero) average velocity. The density of these molecules is sufficiently high for them routinely to collide with each other. Energy is transferred throughout the gas via these collisions, establishing an equilibrium distribution of energy among the constituents of the gas (LTE).

With respect to velocity, there are various values of speed and direction for the molecules that form the gas mixture. It is therefore necessary to develop a statistical distribution function that will produce values of interest, such as the average velocity. We begin by assuming a uniform mass distribution in physical (Euclidean) space, but a nonuniform velocity distribution in velocity space (see Figure 2.1). The gas would then be represented in Figure 2.1 by a cloud of points (one for each molecule) with a centroid at the average value for v.

Velocity distributions

Next, we define a local density in velocity space (number of points per unit volume in velocity space) as $F(\vec{v}_i)$. Then the number of points whose velocity coordinates lie within dV_v, located between $v_x \leftrightarrow$

v_x+dv_x, $v_y \leftrightarrow v_y+dv_y$, and $v_z \leftrightarrow v_z+dv_z$ is $F(\vec{v}_i)dV_v$. Each point in velocity space represents a molecule in physical space, and $F(\vec{v}_i)\ dV_v$ is then the number of molecules for which the velocity vectors land within dV_v, no matter where they may be located in physical space. A normalized distribution can be defined by

$$f(\vec{v}_i) \equiv \frac{F(\vec{v}_i)}{N} \tag{2.1}$$

where N is the total number of molecules in the gas. The probability that a molecule will have velocity within the range dV_v is $f(\vec{v}_i)\ dV_v$. The number of molecules whose velocity components lie within $v_x \leftrightarrow v_x+dv_x$, $v_y \leftrightarrow v_y+dv_y$, and $v_z \leftrightarrow v_z+dv_z$ is then

$$dN = Nf(\vec{v}_i)\ dV_v \tag{2.2}$$

From the properties of $f(\vec{v}_i)$, one can see that

$$\int_{-\infty}^{+\infty} Nf(\vec{v}_i)dV_v = N \tag{2.3}$$

and

$$\int_{-\infty}^{+\infty} f(\vec{v}_i)\ dV_v = 1 \tag{2.4}$$

because every molecule must have some value of velocity that falls within the bounds $-\infty$ to $+\infty$.

Average molecular pressure

Now consider an array of particles (atoms or molecules) that interact with a wall. For simplicity, we assume that all of the particles are identical (have the same mass m). Figure 2.2 depicts a small section of this wall as it undergoes collisions with incoming particles. Various particles strike the wall from different directions, and when they hit, v_y goes to zero just before the rebound. We will consider various classes of incoming vectors; the rebound will be dealt with afterwards. Each class will consist of vectors that are very close in magnitude and direction to the velocity class designator "$\vec{v}_i$". Consider the particles from one class that strike the wall within time interval dt, as shown in Figure 2.3. Those particles all lie within the slanting cylinder shown in the figure. Particles in class $\vec{v}_i$ that lie outside the slanting cylinder do

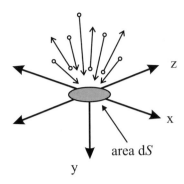

Figure 2.2: Small section of the wall with area dS. Particles in the gas randomly collide with the wall and rebound.

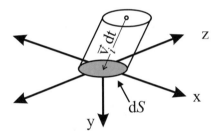

Figure 2.3: Molecules in velocity class $\vec{v}_i$ striking area dS within time dt (adapted with permission from Vincenti and Kruger [29]).

not strike the area dS, while all particles inside the slanting cylinder that reside within velocity class $\vec{v}_i$ must strike dS within time dt. The volume of this cylinder is v_{yi} dt dS.

The number of molecules in class $\vec{v}_i$ per unit volume (in physical space) is

$$dN = Nf(\vec{v}_i)\, dv_x\, dv_y\, dv_z \tag{2.5}$$

where N is the number density given by $N = \text{N}/\text{V}$ ($\#/\text{m}^3$) (using the volume in Euclidean space). The number of molecules in velocity class $\vec{v}_i$ that strike dS within dt is then given by dN (molecules/m^3)$\times$ volume of class $\vec{v}_i$ (m^3):

$$
\begin{aligned}
&\text{Number of molecules in } \vec{v}_i \text{ at } dS \text{ in } dt \\
={}& Nf(\vec{v}_i)\, dv_x\, dv_y\, dv_z\, v_{yi}\, dS\, dt \\
={}& Nv_{yi}f(\vec{v}_i)\, dv_x\, dv_y\, dv_z\, dS\, dt
\end{aligned}
\tag{2.6}
$$

The flux of molecules on a per unit area and per unit time basis is then

$$\text{Flux} = N\text{v}_{yi}f(\vec{\text{v}}_i)\, d\text{v}_x\, d\text{v}_y\, d\text{v}_z \tag{2.7}$$

To find the flux to surfaces in the other two directions, simply replace v_{yi} with v_{xi} or v_{zi}.

The pressure exerted on dS is developed by momentum exchange at the surface. By Newton's law, the force exerted is equal to the time rate of change in momentum. In the case just discussed, momentum goes from $m\text{v}_{yi}$ to zero during dt, so the momentum change is $m\text{v}_{yi}$. Therefore, the total normal momentum change for incoming particles per unit area and per unit time is

$$
\begin{aligned}
&\;\text{Total normal momentum change}\\
&= \frac{\Delta\text{momentum}}{\text{molecule}}\quad\frac{\#\,\text{molecules}}{\text{unit area unit time}}\\
&= (m\text{v}_{yi})\,(N\text{v}_{yi}f(\vec{\text{v}}_i)\, d\text{v}_x\, d\text{v}_y\, d\text{v}_z)\\
&= Nm\text{v}_{yi}^2 f(\vec{\text{v}}_i)\, d\text{v}_x\, d\text{v}_y\, d\text{v}_z \tag{2.8}
\end{aligned}
$$

This is a pressure component in the y direction exerted by the incoming molecules in class v_{yi}. Integrating over all incoming velocity classes gives

$$p_{in} = Nm \int_0^{+\infty} \int_{-\infty}^{+\infty} \int_{-\infty}^{+\infty} \text{v}_{yi}^2 f(\vec{\text{v}}_i)\, d\text{v}_x\, d\text{v}_z\, d\text{v}_y \tag{2.9}$$

where the first integral over $d\text{v}_y$ has limits from 0 to $+\infty$ because, for incoming particles, v_y is never negative (when $+y$ is defined as in Figure 2.3).

The same formalism can be used for the rebound molecules. We model this as a source term, as though there were a hole at dS and molecules from $y < 0$ enter through the hole:

$$p_{rebound} = Nm \int_{-\infty}^{0} \int_{-\infty}^{+\infty} \int_{-\infty}^{+\infty} \text{v}_{yi}^2 f(\vec{\text{v}}_i)\, d\text{v}_x\, d\text{v}_z\, d\text{v}_y \tag{2.10}$$

Note that, for a true rebound, $d\text{v}_y$ is negative. The total pressure exerted on dS is then the sum of these two:

$$
\begin{aligned}
P_{total} &= p_{in} + p_{rebound}\\
&= Nm \int_{-\infty}^{+\infty} \int_{-\infty}^{+\infty} \int_{-\infty}^{+\infty} \text{v}_{yi}^2 f(\vec{\text{v}}_i)\, d\text{v}_x\, d\text{v}_z\, d\text{v}_y
\end{aligned}
$$

$$= Nm \int_{-\infty}^{+\infty} v_{yi}^2 f(\vec{v}_i) \, dV_v$$

$$= Nm\overline{v_y^2} \tag{2.11}$$

where $\overline{v_y^2}$ denotes the average of v_y^2. Note also that N (# molecules/unit volume) $\times m$(mass/molecule) is the mass density ρ. This expression has been developed for just one particle mass. In order to describe mixtures of gases, one simply adds terms from each component, since we assume they do not interact. The sums pass through the integrals and Dalton's law of partial pressures is recovered.

Translational energy and the Boltzmann constant

Next, we assume that there are no gradients in temperature, pressure, mean velocity, and so forth. We therefore assume that the system is in equilibrium. Thus, $\vec{v}_i$ is totally random, no direction is preferred, and

$$\overline{v_x^2} = \overline{v_y^2} = \overline{v_z^2} \tag{2.12}$$

but

$$\overline{v^2} = \overline{v_x^2} + \overline{v_y^2} + \overline{v_z^2} = 3\overline{v_y^2} \tag{2.13}$$

The total pressure P is therefore reduced to

$$P = \frac{\rho}{3} \overline{v^2} \tag{2.14}$$

by using equations (2.11) and (2.13). Dividing by ρ gives

$$Pv = \frac{1}{3} \overline{v^2} \tag{2.15}$$

where v is the specific volume. Equation (2.15) was written for one species of collider. In what follows it will be simpler to write equation (2.15) as

$$PV = \frac{M}{3} \overline{v^2} \tag{2.16}$$

where M is now the total mass.

The average energy due to translational kinetic energy of this gas can be written as

$$\overline{E_{trans}} = \frac{M}{2} \overline{v^2} \tag{2.17}$$

where individual masses were pulled out of the average summation to produce M. This result can be related to equation (2.16) by:

$$PV = \frac{2}{3}\overline{E_{\text{trans}}} \tag{2.18}$$

If we now assume an ideal gas at equilibrium, the ideal gas law relates PV and temperature via

$$PV = \mathcal{N}\mathcal{R}T \tag{2.19}$$

where $\mathcal{N}$ is the number of moles, and $\mathcal{R}$ is thus written on a per-mole basis. Then

$$\overline{E_{\text{trans}}} = \frac{3}{2}\mathcal{N}\mathcal{R}T \tag{2.20}$$

This expression can be recast in terms of the energy per molecule $\overline{\epsilon_{\text{trans}}}$ by dividing equation (2.20) by the total number of particles N:

$$\overline{\epsilon_{\text{trans}}} \equiv \frac{\overline{E_{\text{trans}}}}{N} = \frac{3}{2}\frac{\mathcal{N}}{N}\mathcal{R}T \tag{2.21}$$

The ratio $N/\mathcal{N}$ is Avagadro's number N_A, and the ratio $\mathcal{R}/N_A$ defines the Boltzmann constant:

$$\boxed{k_B \equiv \frac{\mathcal{R}}{N_A}} \tag{2.22}$$

It is the single-molecule equivalent of the ideal gas constant, equal to 1.381×10^{-23} J/K-molecule. Equation (2.21) can then be written as

$$\overline{\epsilon_{\text{trans}}} = \frac{3}{2}k_B T \tag{2.23}$$

The Maxwellian distribution

An equilibrium velocity distribution can be cast in terms of collisions that remove and replenish molecules from velocity class $\vec{v}_i$. Molecules are continuously removed from this class, but others are also continuously inserted into it, and at equilibrium the rate of change in population in class $\vec{v}_i$ is zero.

Consider a collision between two particles as shown in Figure 2.4. In the figure, $\vec{v}_i$ represents the velocity class of particle A, while $\vec{w}_i$ represents the velocity class of particle B. The relative velocity vector

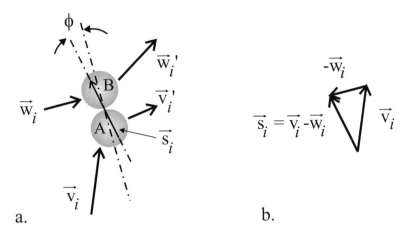

Figure 2.4: a. Geometry of a collision between two molecules A and B, and b. definition of relative vector $\vec{s}_i$ (adapted with permission from Vincenti and Kruger [29]).

$\vec{s}_i$ is defined in Figure 2.4b; $\vec{s}_i$ is the velocity of molecule A relative to B. Velocities before the collision are unprimed, while those after the collision are primed. Because the particles are different species, they have different masses m_A and m_B. If $\vec{v}_i$, $\vec{w}_i$, and ϕ are known, then $\vec{v}_i'$ and $\vec{w}_i'$ can be found via conservation of energy and momentum, together with geometric relations.

The inverse of the collision shown in Figure 2.4 is shown in Figure 2.5. This is not the reverse of the collision in Figure 2.4; rather it is the collision that returns molecule A to velocity class $\vec{v}_i$.

The number of molecules in velocity class $\vec{v}_i$ per unit volume per unit time is

$$N_A f_A(\vec{v}_i)\, \mathrm{dV_v} \qquad (2.24)$$

A similar expression could be written for velocity class $\vec{w}_i$.

The differential collision rate per unit volume per unit time can be developed via simple geometry. Imagine that molecule A is moving at relative velocity $\vec{s}_i$ within a field of stationary B molecules. Molecule A will sweep out a volume in space at a rate of

$$\vec{s}_i \sigma \qquad (2.25)$$

where σ is the collision cross-section (m^2) of molecule A. The cross-section is a function of $\vec{s}_i$ and the collision geometry (including angular dependence). For one molecule A, there will be a collision with a

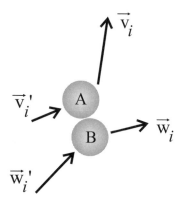

Figure 2.5: Inverse of the collision shown in Figure 2.4 (adapted with permission from Vincenti and Kruger [29]).

molecule B every time one B appears inside the swept volume. The rate of these collisions will then be the swept volume times the number density of molecules in velocity class $\vec{w}_i$. Then the total collision rate can be found by multiplying this number of collisions for one A molecule by the number of swept volumes, i.e. the number of molecules A in class $\vec{v}_i$:

$$N_A f_A(\vec{v}_i) N_B f_B(\vec{w}_i) s_i \sigma \, dV_w \, dV_v \qquad (2.26)$$

This is the number of collisions that deplete class $\vec{v}_i$ per unit volume per unit time.

For collisions that replenish class $\vec{v}_i$, one must simply consider the inverse collisions. The number of collisions that replenish class $\vec{v}_i$ per unit volume per unit time is then

$$N_A f_A(\vec{v}_i') N_B f_B(\vec{w}_i') s_i' \sigma \, dV_{w'} \, dV_{v'} \qquad (2.27)$$

As it turns out, $\sigma(s) = \sigma(s')$. This occurs because $|\,s\,| = |\,s'\,|$, an result of momentum and kinetic energy conservation in elastic collisions [29].

The rate of change in population in class $\vec{v}_i$ is the rate of replenishing collisions minus the rate of depleting collisions:

$$
\begin{aligned}
&\text{Rate of change of population in class } v_i \\
={}& N_A f_A(\vec{v}_i') N_B f_B(\vec{w}_i') s_i' \sigma \, dV_{w'} \, dV_{v'} \\
-{}& N_A f_A(\vec{v}_i) N_B f_B(\vec{w}_i) s_i \sigma \, dV_w \, dV_v \\
={}& N_A N_B \left[f_A(\vec{v}_i') f_B(\vec{w}_i') - f_A(\vec{v}_i) f_B(\vec{w}_i) \right] s_i \sigma \, dV_w \, dV_v \qquad (2.28)
\end{aligned}
$$

which is possible because $dV_w\ dV_v = dV_{w'}\ dV_{v'}$. This equality is developed via an application of Liouville's theorem in mechanics [29, 31].

For a gas in equilibrium, the rate of change in population in class $\vec{v}_i$ is zero:

$$N_A N_B \left[f_A(\vec{v}_i') f_B(\vec{w}_i') - f_A(\vec{v}_i) f_B(\vec{w}_i) \right] s_i \sigma\ dV_w = 0 \qquad (2.29)$$

Here we have ignored important issues, such as collision geometry for example. Vincenti and Kruger [29] discuss these issues in detail and arrive at the same outcome. One way for equation (2.29) to be satisfied is if

$$f_A(\vec{v}_i') f_B(\vec{w}_i') - f_A(\vec{v}_i) f_B(\vec{w}_i) = 0 \quad \text{or}$$
$$f_A(\vec{v}_i') f_B(\vec{w}_i') = f_A(\vec{v}_i) f_B(\vec{w}_i) \qquad (2.30)$$

It turns out that this is the only way equation (2.29) can vanish. This means that for a gas to be in equilibrium, each detailed process must be in equilibrium. This notion is called the principle of detailed balance. Following Vincenti and Kruger [29], we take the natural log of equation (2.30):

$$\ln[f(\vec{v}_i')] + \ln[f(\vec{w}_i')] = \ln[f(\vec{v}_i)] + \ln[f(\vec{w}_i)] \qquad (2.31)$$

where the subscripts on f have been removed because we assume that it has the same form for any species or velocity class.

Equation (2.31) states that there is a function $\ln[f]$ of the particle velocity which is the same for two particles before and after a collision. Both energy and momentum are conserved in a collision. For a particle of mass m, the energy can be written as

$$\epsilon = \frac{m}{2} \left(v_x^2 + v_y^2 + v_z^2 \right) \qquad (2.32)$$

and the components of momentum can be written as

$$
\begin{aligned}
p_x &= mv_x \\
p_y &= mv_y \\
p_z &= mv_z
\end{aligned} \qquad (2.33)
$$

The most general expression for $\ln[f]$ is a linear combination of energy and momentum terms:

$$\ln[f(\vec{v})] = -\beta \frac{m}{2} \left(v_x^2 + v_y^2 + v_z^2 \right) + \alpha_1 mv_x + \alpha_2 mv_y + \alpha_3 mv_z + \ln A \qquad (2.34)$$

where the minus in front of β anticipates the solution. Taking the exponential of both sides produces

$$f(\vec{v}_i) = A e^{-\beta \frac{m v^2}{2}} e^{\alpha_1 m v_x} e^{\alpha_2 m v_y} e^{\alpha_3 m v_z} \tag{2.35}$$

At equilibrium, velocity has no preferred direction; hence $f(v_x, v_y, v_z) = f(-v_x, -v_y, -v_z)$. This will hold only if $\alpha_1 = \alpha_2 = \alpha_3 = 0$. The velocity distribution is thus written as

$$f(\vec{v}_i) = A e^{-\beta \frac{m v^2}{2}} \tag{2.36}$$

and it becomes necessary to solve for A and β. A can be found by requiring a normalized distribution:

$$\int_{-\infty}^{+\infty} f(\vec{v}_i)\, dV_v = 1 \tag{2.37}$$

which leads to

$$A \int_{-\infty}^{+\infty} e^{-\beta \frac{m v_x^2}{2}} dV_{v_x} \int_{-\infty}^{+\infty} e^{-\beta \frac{m v_y^2}{2}} dV_{v_y} \int_{-\infty}^{+\infty} e^{-\beta \frac{m v_z^2}{2}} dV_{v_z} = 1 \tag{2.38}$$

One of these integrals is equal to

$$\int_{-\infty}^{+\infty} e^{-\beta \frac{m v_x^2}{2}} dV_{v_x} = \left(\frac{2\pi}{\beta m}\right)^{1/2} \tag{2.39}$$

so A is equal to

$$A = \left(\frac{\beta m}{2\pi}\right)^{3/2} \tag{2.40}$$

Next, to solve for β, we begin with equation (2.15):

$$Pv = \frac{1}{3}\overline{v^2} = RT = \frac{k_B}{m}T \tag{2.41}$$

where R is now on a per unit mass basis. The last relation indicates that $R = k_B/m$; because R is on a per unit mass basis, k_B is the gas constant for one particle, and m is the mass of one particle. The average velocity squared in one direction is then

$$\overline{v_x^2} = \frac{1}{3}\overline{v^2} = \frac{k_B}{m}T \tag{2.42}$$

From the properties of the distribution, $\overline{v_x^2}$ can also be written as

$$
\begin{aligned}
\overline{v_x^2} &= \int_{-\infty}^{+\infty} v_x^2 \, f(\vec{v}_i) \, dV_v \\
&= \int_{-\infty}^{+\infty} \int_{-\infty}^{+\infty} \int_{-\infty}^{+\infty} v_x^2 e^{-\beta \frac{m}{2}(v_x^2 + v_y^2 + v_z^2)} dV_x \, dV_y \, dV_z \\
&= \frac{1}{\beta m}
\end{aligned}
\tag{2.43}
$$

Equating relations (2.42) and (2.43) produces

$$
\beta = \frac{1}{k_B T}
\tag{2.44}
$$

This result then produces

$$
\boxed{f(\vec{v}_i) = \left(\frac{m}{2\pi k_B T}\right)^{3/2} e^{-\frac{m}{2k_B T}(v_x^2 + v_y^2 + v_z^2)} = \left(\frac{m}{2\pi k_B T}\right)^{3/2} e^{-\frac{mv^2}{2k_B T}}}
\tag{2.45}
$$

which is the Maxwellian velocity distribution. We will see it again when we discuss Doppler broadening in Chapter 11, and when we include Doppler broadening in the density matrix equations in Chapter 14.

2.3   The Boltzmann Energy Distribution

Here the focus is not on molecular motion and collisions; instead we will consider various possible energy microstates and the most probable arrangement of these microstates. There are many approaches to this problem. In this chapter we present a simplified, relatively intuitive approach that arrives at the useful (to the world of diagnostics) portions of statistical mechanics as quickly as possible. In part, it parallels a treatment presented by Vincenti and Kruger [29].

It is necessary at this point to recognize that atoms and molecules occupy discrete ("quantized") energy levels; they cannot occupy an arbitrary value within an energy continuum. This indisputable fact is developed further in Chapters 7, 8, and 9, where we introduce quantum mechanics and then apply it to atomic and molecular spectroscopy. For now, however, this notion will have to be accepted in order to develop Boltzmann statistics.

We begin by assuming that the gas is comprised of N identical particles (they could be atoms, molecules, or any other basic particle), each of which is in a quantized energy level denoted by $\epsilon_1, \epsilon_2, \epsilon_3$, etc. These energy levels are determined by the potential field within each atom or molecule (see, for example, Chapter 8 for the one-electron atom). It is assumed that these particles do not interact across the open spaces (vacuum) that separate them. We therefore assume an ideal gas, which works well for most flowfields of interest.

Microstates and macrostates

A microstate is defined as one possible way that the N particles can be distributed across the energy states. As an example, the first particle might be in ϵ_4, the second in ϵ_7 and so forth. We assume that each ϵ_i has an equal a-priori probability. A macrostate is a collection of microstates. Some macrostates are more probable than others because they have more microstates (each of which has the same probability).

N is a very large number in statistical mechanics. It is necessary, however, to ensure that the microstates can ultimately reproduce physical observation. This can be done using conservation expressions:

$$\sum_i N_i = N \qquad (2.46)$$

$$\sum_i N_i \epsilon_i = E \qquad (2.47)$$

where the N_i are the populations of each energy state ϵ_i. In short, the numbers of particles and the energy both must add up to known totals. In a more detailed treatment of statistical mechanics, "canonical ensembles" are used to apply other restrictions. In reality, E is often unknown and these other constraints provide more utility (see, for example, Garrod [30]).

Bosons and fermions

The distinction between bosons and fermions controls the way in which the various N_i are counted. As it turns out, in a relatively low-density ideal gas, the issue is not critical, but the discussion also introduces topics to be addressed in Chapters 7, 8, and 9.

As we will find in Chapter 7, the quantum mechanical wave function $\psi(r)$ (where r designates position) is a probability amplitude. When squared (using the complex conjugate ψ^*), we find the probability density. For example, the integral $\int \psi^* r \psi \, d\tau$ provides an expectation value for position (where $d\tau$ is a differential volume). The wave function has several important properties.

If ψ is unchanged when the coordinates of any pair of particles (e.g. in an atom) are switched, ψ is called a symmetric wave function. Systems with symmetric wave functions are called bosons. Examples include any particle made up of an even number of smaller particles, and photons. Bosons have integer values for their total angular momentum (in units of $\hbar \equiv h/2\pi$, where h is Planck's constant). For bosons, the various N_i are simply non-negative integers. Any number of bosons can occupy a given energy level ϵ_i. This has to do with the Pauli exclusion principle discussed in Chapter 7. Bosons follow Bose-Einstein statistics.

The picture just given is oversimplified. As one example, the wave function for a homonuclear diatomic molecule can be symmetric under the exchange of nuclei only if the number of neutrons and protons in each nucleus is even. Despite the simplifications, we can use this picture to accomplish our goal.

If ψ changes sign when the coordinates of any pair of particles are switched, ψ is called an antisymmetric wave function. Systems with antisymmetric wave functions, such as neutrons and protons, are called fermions. Fermions have half-integral values for their total angular momentum (in units of $\hbar$). No two fermions can occupy the same energy state. In such a case, $N_i = 0, 1$ exclusively. Fermions follow Fermi-Dirac statistics.

Mixed states (e.g. a homonuclear diatomic molecule with Bose-Einstein nuclei but Fermi-Dirac electrons) require more detailed accounting.

Energy accounting

The energy of a bound atomic system is discrete. A continuous probability distribution would thus be a very irregular function of ϵ. Instead, probability is found in this case by counting the microstates that contribute to each macrostate.

The spacing between energy states is very small, especially relative to the macroscopic total energy. It is therefore possible to group them into "quantum microcanonical ensembles" (called "groups" here for

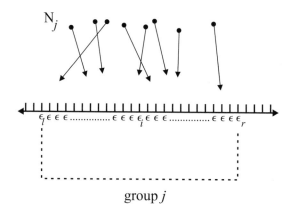

Figure 2.6: Depiction of C_j energy microstates (various values of ϵ_i) within group j, spanning from ϵ_l to ϵ_r. The N_j particles indicate a random distribution among the microstates within group j.

short) and simplify the counting process (see, for example, Figure 2.6). A large number of energy microstates (call this number C_j) can be grouped. All the states within that group will be assumed to have the same energy ϵ_j (using j as the index for groups). It is necessary to ensure, therefore, that $\mid \epsilon_l - \epsilon_r \mid$ (depicted in Fig. 2.6) is small relative to the total macroscopic energy. Vincenti and Kruger [29] show that this is the case.

The number of particles (not energy states) in group j will be designated N_j. Now C_j and ϵ_j are fixed, but N_j is variable within the following constraints:

$$\sum_j N_j = N \tag{2.48}$$

$$\sum_j N_j \epsilon_j = E \tag{2.49}$$

(now summing over all groups).

Boltzmann statistics

To count the number of microstates, we count the number of macrostates associated with each assignment of N_j and then count the number of possible N_j assignments. Vincenti and Kruger provide some details on the process. Here we simply quote the outcome.

For Bose-Einstein statistics, the number of ways that N_j indistinguishable particles can be assigned to C_j containers (all of them with $\epsilon = \epsilon_j$) within group j is

$$\frac{(N_j + C_j - 1)!}{(C_j - 1)!N_j!}$$

If we define $W(N_j)_{BE}$ as the number of microstates in a given arrangement of N_j particles, then

$$W(\{N_j\})_{BE} = \prod_j \frac{(N_j + C_j - 1)!}{(C_j - 1)!N_j!} \tag{2.50}$$

where $\{N_j\}$ denotes a set of N_j, and the operator $\prod_j$ indicates a product over all j.

For Fermi-Dirac statistics, the number of ways that N_j indistinguishable particles can be assigned to C_j containers (all of them with $\epsilon = \epsilon_j$) within group j is

$$\frac{C_j!}{(C_j - N_j)!N_j!}$$

and

$$W(\{N_j\})_{FD} = \prod_j \frac{C_j!}{(C_j - N_j)!N_j!} \tag{2.51}$$

These results are simply the outcome of statistical combinatorial analysis, subject to constraints on N_j.

Now take the sum over all the possible arrangements of $\{N_j\}$ (designated by Ω), consistent with constraints (2.48) and (2.49)

$$\Omega = \sum_j W(N_j) \tag{2.52}$$

The problem is how to find Ω? We assume that N is very large, which leads to

$$\Omega \approx W(N_j)_{\text{maximum}} \tag{2.53}$$

meaning that Ω is a sharply peaked function of N_j. This is proven in Garrod [30], while Vincenti and Kruger [29] demonstrate that it must be true at the limits we assume. As with the Maxwellian distribution, we maximize $\ln W$.

For Bose-Einstein statistics:

$$\ln(W_{BE}) = \sum_j [\ln(N_j + C_j - 1)! - \ln(C_j - 1)! - \ln N_j!] \qquad (2.54)$$

where the product has become a sum with the logarithm. For Fermi-Dirac statistics:

$$\ln(W_{FD}) = \sum_j [\ln(C_j)! - \ln(C_j - N_j)! - \ln N_j!] \qquad (2.55)$$

We now apply Stirling's formula, which states that, for large z, $\ln z! \cong z \ln z - z$. The outcome is

$$\ln(W) = \sum_j \left[\pm C_j \ln \left(1 \pm \frac{N_j}{C_j} \right) + N_j \ln \left(\frac{C_j}{N_j} \pm 1 \right) \right] \qquad (2.56)$$

where the $+$ in $\pm$ applies to Bose-Einstein statistics and the $-$ applies to Fermi-Dirac. Now $\ln(W)$ is maximized with respect to the N_j arrangements:

$$\sum_j \frac{\partial(\ln W)}{\partial N_j} \delta N_j = 0 \qquad (2.57)$$

which results in

$$\sum_j \left[\ln \left(\frac{C_j}{N_j} \pm 1 \right) \right] \delta N_j = 0 \qquad (2.58)$$

The restrictions on N_j generate matching restrictions on δN_j:

$$\sum_j \delta N_j = 0$$

$$\sum_j \epsilon_j \delta N_j = 0 \qquad (2.59)$$

A good way to find extrema subject to constraints is to use Lagrange's method. The outcome of such an analysis is that the most general value that $\ln[(C_j/N_j) \pm 1]$ can have in order to satisfy equations (2.58) and (2.59) is

$$\ln \left(\frac{C_j}{N_j} \pm 1 \right) = \alpha + \beta \epsilon_j \qquad (2.60)$$

or

$$\frac{N_j^*}{C_j} = \frac{1}{e^{\alpha + \beta \epsilon_j} \mp 1} \qquad (2.61)$$

where the asterisk on N_j^* denotes the N_j that maximize W. At relatively high temperatures, E is distributed over a wide array of states that are sparsely populated. The term N_j^*/C_j is thus very small, requiring that $e^{\alpha+\beta\epsilon_j}$ be large relative to 1. Equation (2.61) is then:

$$\frac{N_j^*}{C_j} = e^{-\alpha-\beta\epsilon_j} \qquad (2.62)$$

It is now necessary to solve for α and β. α can be found using the constraint on particle number [equation (2.48)]:

$$\sum_j N_j = N = \sum_j C_j e^{-\alpha} e^{-\beta\epsilon_j} \qquad (2.63)$$

α is not a function of j and it can be removed from the summation:

$$N = e^{-\alpha} \sum_j C_j e^{-\beta\epsilon_j} \qquad (2.64)$$

or

$$e^{-\alpha} = \frac{N}{\sum_j C_j e^{-\beta\epsilon_j}} \qquad (2.65)$$

Using this result in equation (2.62) yields

$$\frac{N_j^*}{N} = \frac{e^{-\beta\epsilon_j} C_j}{\sum_j C_j e^{-\beta\epsilon_j}} \qquad (2.66)$$

This expression gives the most probable fraction of particles (a probability distribution function) in group (energy level) j as a function of j (via C_j and ϵ_j). The function is indeed strongly peaked. If we assume thermal equilibrium, then the N_j^* represent naturally occurring equilibrium populations, at which point one might as well remove the asterisk. The summation in the denominator of equation (2.66) is called a "partition function" Q, defined by

$$Q \equiv \sum_j C_j e^{-\beta\epsilon_j} = C_1 e^{-\beta\epsilon_1} + C_2 e^{-\beta\epsilon_2} + \cdots \qquad (2.67)$$

Note that, based upon equation (2.66), the term $C_1 e^{-\beta\epsilon_1}$ is proportional to the population in level 1, N_1, the term $C_2 e^{-\beta\epsilon_2}$ is proportional to the population in level 2, N_2, and so forth. Q is called a partition function because it provides weighting for the distribution (partition-ing) of population among the various energy levels. Because equation (2.67) actually sums over all energy levels, it can be rewritten as

$$Q = \sum_i e^{-\beta \epsilon_i} \qquad (2.68)$$

where j has been replaced by i.

We have yet to find β. Equation (2.44) sets

$$\beta = \frac{1}{k_B T}$$

where β performs the same function in the development of the Maxwellian velocity distribution as it does here. These two formalisms represent the same problem. The Maxwellian expression [equation (2.45)] is the probability distribution function for kinetic energy, while equation (2.66) is the probability distribution function for energy in general. One could argue by comparison that equation (2.44) must hold here as well. Such an argument is certainly not rigorous, but we will accept it and move on. The goal of this chapter is to review quickly the portions of statistical mechanics that apply to optical diagnostics in flowfields. Much more rigorous treatments also arrive at equation (2.44). Readers who require additional depth are encouraged to consult the references for such treatments.

Now it is possible to write the full Boltzmann distribution. Before doing that, however, we make two changes in notation. First, we have used ϵ to represent the energy of a state. The use of ϵ provides a good reminder of the microstate formalism. Throughout spectroscopy, however, E is typically used to represent the energy of a quantum state. In order to conform to this convention, we now substitute E for ϵ. Next, C_j represented the total number of microstates in an energy level. In spectroscopy, however, when there is more than one state with the same energy it is called a degenerate energy level and the number of states with this energy is called the degeneracy g (a small integer). Spectroscopy deals with discrete energy states internal to the molecule (rotational, vibrational, and electronic). Translational energy, which is a continuum, does not normally affect spectra (except for Doppler broadening). C_j is therefore set equal to $C_{tr} \, g_j$, where C_{tr} (a large number) becomes part of the translational energy distribution and the various g_j are associated with internal energy states. With these definitions and results in hand, the Boltzmann fraction f_B in equation (2.66) can be rewritten as

$$f_B \equiv \frac{N_i}{N} = \frac{g_i e^{-\frac{E_i}{k_B T}}}{Q} \tag{2.69}$$

which is the recognized form for the Boltzmann distribution. This equation is used to calculate what fraction of the total is in a particular energy state. We often use the word "population" of a level to mean this fraction. Equation (2.69) thus provides a means by which to evaluate the distribution of the total population across the available energy levels, or quantum states.

2.4 Molecular Energy Distributions

Molecules have various internal energy storage modes, each with a different energy level. N will therefore be distributed among many possible translational, rotational, vibrational, and electronic levels, and the distribution can be described using equation (2.69). The Born-Oppenheimer approximation (see Chapter 9) states that the various energy modes can be treated as independent. The energy state of a molecule can then be decomposed by

$$E = E_{\text{trans}} + E_{\text{rot}} + E_{\text{vib}} + E_{\text{elec}} \tag{2.70}$$

The partition function is now given by

$$
\begin{aligned}
Q = & \sum_{\text{tr}} \sum_{\text{rot}} \sum_{\text{vib}} \sum_{\text{elec}} \left(C_{\text{tr}} e^{-\frac{E_{\text{trans}}}{k_B T}} \right) \left(g_{\text{rot}} e^{-\frac{E_{\text{rot}}}{k_B T}} \right) \\
& \times \left(g_{\text{vib}} e^{-\frac{E_{\text{vib}}}{k_B T}} \right) \left(g_{\text{el}} e^{-\frac{E_{\text{elec}}}{k_B T}} \right)
\end{aligned}
\tag{2.71}
$$

Consider, for example, the summation over vibrational modes alone. Vibrational energy levels are not degenerate, so $g_{\text{vib}} = 1$. We can therefore define a vibrational partition function by

$$Q_{\text{vib}} \equiv \sum_{\text{vib}} e^{-\frac{E_{\text{vib}}}{k_B T}} \tag{2.72}$$

and the total partition function is thus

$$Q = Q_{trans}Q_{rot}Q_{vib}Q_{elec} \tag{2.73}$$

This occurs because we assume that the various modes are independent of each other.

Imagine that we care to know the fraction of molecules in one of numerous possible rotational energy levels. Implicit in this statement is the notion that we will evaluate all translational, vibrational, and electronic modes, because we wish to focus on the fraction in a rotational level within all of the other modes. This fraction would be written as

$$
\begin{aligned}
f_{B,\,rot} &= \frac{\sum_{tr}\sum_{vib}\sum_{el}\left(C_{tr}e^{-\frac{E_{trans}}{k_B T}}\right)\left(g_{rot}e^{-\frac{E_{rot}}{k_B T}}\right)\left(g_{vib}e^{-\frac{E_{vib}}{k_B T}}\right)\left(g_{el}e^{-\frac{E_{elec}}{k_B T}}\right)}{Q} \\[2mm]
&= \left(\frac{g_{rot}e^{-\frac{E_{rot}}{k_B T}}}{Q_{rot}}\right)\left(\frac{\sum_{tr}g_{tr}e^{-\frac{E_{trans}}{k_B T}}}{Q_{trans}}\right)\left(\frac{\sum_{vib}g_{vib}e^{-\frac{E_{vib}}{k_B T}}}{Q_{vib}}\right)\left(\frac{\sum_{el}g_{el}e^{-\frac{E_{elec}}{k_B T}}}{Q_{elec}}\right) \\[2mm]
&= \left(\frac{g_{rot}e^{-\frac{E_{rot}}{k_B T}}}{Q_{rot}}\right)\frac{Q_{trans}}{Q_{trans}}\frac{Q_{vib}}{Q_{vib}}\frac{Q_{elec}}{Q_{elec}} \\[2mm]
&= \left(\frac{g_{rot}e^{-\frac{E_{rot}}{k_B T}}}{Q_{rot}}\right) \tag{2.7}
\end{aligned}
$$

All of the other fractions are equal to 1 because we have included every translational, vibrational, and electronic energy level.

The fraction of molecules is not normally evaluated in a particular translational energy level when using Boltzmann statistics. This is because we usually want to know about partitioning among internal energy modes. In the remaining discussion, we assume that translational motion has been summed over all states and that the fraction is now 1.

Electronic populations

Electronic levels are often degenerate. This can be caused by spin splitting, for example. Furthermore, as shown in Chapters 8 and 9, electronic energy levels are not normally written as E_{elec}, because quantum mechanics shows that electronic energy is specified by the principal quantum number $n = 1, 2, 3...$ The electronic partition function is therefore written as

$$Q_{\text{elec}} = \sum_n g_n e^{-\frac{E_n}{k_B T}}$$

(2.75)

where g_n is the electronic degeneracy. These energy levels are somewhat relative, we usually have the choice on where to set $E_n = 0$. We usually choose $E_1 = 0$. Sometimes, just one or two electronic levels occur before a molecule dissociates (or ionizes). In such a case:

$$Q_{\text{elec}} = g_1 + g_2 e^{-\frac{E_2}{k_B T}} + g_3 e^{-\frac{E_3}{k_B T}} + \dots$$

(2.76)

The fraction of atoms or molecules in one electronic level is:

$$f_{B,n} = \frac{g_n e^{-\frac{E_n}{k_B T}}}{Q_{\text{elec}}}$$

(2.77)

The values for the various electronic energy levels must be found for each atom or molecule of interest, because they are a function of the unique geometry of each species. They are typically derived from quantum calculations, spectroscopic measurement, or a combination of both. Often the fraction in the first electronic level is nearly 1, even at flame temperatures. This is because the jump in energy from E_1 to E_2 can be (but is not always) very large.

Vibrational populations

As mentioned above, vibrational states are not degenerate. The energy of a quantum harmonic oscillator (the simplest vibrational case) is given by equation (9.54):

$$E_{\text{vib}} = E_v = \left(v + \frac{1}{2}\right) h\nu_e, \quad \text{for } v = 0, 1, 2, \dots$$

where v is the vibrational quantum number and ν_e is a vibrational oscillation frequency (usually measured spectroscopically). We define

$$\Theta_v \equiv \frac{h\nu_e}{k_B}$$

(2.78)

The summation used in the vibrational partition function can then be compared with a series expansion for the following:

$$Q_{\text{vib}} = \frac{e^{-\frac{\Theta_v}{2T}}}{1 - e^{-\frac{\Theta_v}{T}}}$$

(2.79)

Table 2.1: Nuclear spin degeneracy (adapted from Long [32])

	O_2	H_2	N_2
Nuclear spin	0	1/2	1
Nuclear spin statistical weight g_I:			
· J odd	1	3	3
· J even	0	1	6

This expression is based upon the harmonic oscillator assumption. As discussed in Chapter 9, that assumption is not uniformly applicable. Anharmonicity effects can become important, especially at large v. In such a case, one must simply perform the summation

$$Q_{\text{vib}} = \sum_{\text{vib}} e^{-\frac{E_v}{k_B T}} \qquad (2.80)$$

where the values for E_v must be found using appropriate constants. The summation is simply carried out until higher-order terms make no effective difference to Q_{vib}.

The fraction of molecules in one vibrational level is

$$f_{B,v} = \frac{e^{-\frac{E_v}{k_B T}}}{Q_{\text{vib}}} \qquad (2.81)$$

Rotational populations

Rotational states have a magnetic degeneracy given by equation (9.112):

$$g_J = 2J + 1$$

where J is the rotational quantum number given by $J = 0, 1, 2, ...$ Molecules also have a nuclear spin degeneracy labeled by g_I. Heteronuclear diatomic molecules have a constant value of g_I; it divides out of the ratio that determines the Boltzmann fraction. For homonuclear molecules the nuclear spin degeneracy depends upon J, as shown in Table 2.1.

The energy of a quantum rigid rotator (the simplest rotational case) is given by equation (9.28):

$$E_{\text{rot}} = E_J = \frac{\hbar^2}{2I_e} J(J+1)$$

where $\hbar = h/2\pi$ and I_e is the nuclear moment of inertia (usually measured spectroscopically). We define

$$\Theta_r \equiv \frac{h^2}{8\pi^2 I_e k_B} \tag{2.82}$$

The summation used in the rotational partition function can then be compared with a series expansion for the following:

$$\boxed{Q_{\text{rot}} = \frac{T}{\sigma \Theta_r}} \tag{2.83}$$

where $\sigma = 1$ for a heteronuclear molecule and $\sigma = 2$ for a homonuclear molecule. This expression is based upon the rigid rotator assumption. As discussed in Chapter 9, that assumption is not uniformly applicable. Centrifugal distortion and vibrational coupling to the moment of inertia are usually very important. In such a case, one must simply perform the summation

$$\boxed{Q_{\text{rot}} = \sum_{\text{rot}} g_I (2J+1) e^{-\frac{E_J}{k_B T}}} \tag{2.84}$$

where the values for E_J must be found using appropriate constants. The summation is simply carried out until higher-order terms make no effective difference to Q_{rot}.

The fraction of molecules in one rotational level is

$$\boxed{f_{B,J} = g_I (2J+1) \frac{e^{-\frac{E_J}{k_B T}}}{Q_{\text{rot}}}} \tag{2.85}$$

It is necessary to remember that for a heteronuclear diatomic molecule, g_I divides out of the expression. In most review texts on resonant techniques it does not even appear, because homonuclear diatomic molecules do not absorb infrared light (they are not "IR active"). It normally does appear in treatments of Raman scattering, because homonuclear diatomics can be Raman active.

Coupled modes

If we wish to find the relative population in combined electronic, vibrational and rotational modes, we simply multiply the fraction in each mode. As an example, the Boltzmann fraction for one rotational level within one vibrational level (one ro/vibrational mode) would be

$$f_B(\text{combined ro/vibrational}) = \frac{g_I(2J+1)e^{-\frac{E_J}{k_B T}}}{Q_{\text{rot}}} \frac{e^{-\frac{E_v}{k_B T}}}{Q_{\text{vib}}} \qquad (2.86)$$

Such a case would assume that the electronic fraction is 1, or it might assume that all electronic levels are included in the calculation.

2.5   Conclusions

This chapter has simply provided a framework for the usage of statistical mechanics throughout the remainder of this book, and within other sources on diagnostics. Perhaps the most compelling reason to include such a chapter in an otherwise more detailed text is to call attention to the fact that population fractions cannot be neglected in spectroscopic calculations. Moreover, these fractions are actually quite useful. The Boltzmann fraction has a very strong temperature dependence. By probing two states (energy levels), one has measured the relative populations of the two states. Boltzmann statistics then make it possible to infer the temperature via this ratio. Kohse-Höinghaus [9] discusses the issues one must face in order to make such a temperature measurement accurately.

Chapter 3

THE EQUATION OF RADIATIVE TRANSFER

3.1 Introduction

The concepts underlying the equation of radiative transfer (ERT) are often presented in undergraduate courses on thermodynamics and heat and mass transfer. In order to accommodate expressions that describe resonance response of atoms and molecules, we simply modify the well-known unsteady energy equation. Inclusion of particle scattering is also straightforward. The critical difference between undergraduate heat transfer and this case is the fact that the radiation is emitted or absorbed by partially transmitting, distributed volumetric elements in a gas. Moreover, gas-phase spectra are usually fairly sparse and well defined, requiring a detailed description of the transitions between quantum states.

The ERT can be applied to several areas of study in combustion. Indeed, it constitutes the fundamental basis for radiation heat transfer from flames. This is a critical area of research for materials processing and for design of robust combustors, to give just two examples. Viskanta and Mengüç [33] have presented a complete review of the topic. Some of what follows will in fact parallel their presentation, but the goal of this book is to provide a basis for species measurements using the ERT. For this reason, the following development will diverge from theirs. Another related area is the measurement of particulate in flowfields via scattering and extinction (e.g. [12, 34, 35, 36, 37]).

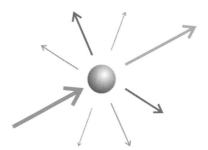

Figure 3.1: Schematic of the overall optical interaction.

Because particulate scattering is not spectroscopic in nature, it is not discussed in this book.

The ERT is used to describe the most common spectroscopic diagnostics, in order to relate the measured signal to the variable of interest (species concentration or temperature). The list of applications can include thermal or chemiluminescent emission from the flame, absorption of lamp or laser light, and laser-induced fluorescence (LIF). Indeed, the ERT is used for all of the linear techniques (linear in power) for which the optical interaction time is long relative to the molecular timescales.

Figure 3.1 contains a cartoon of the elements one must consider. The sphere represents a sample volume, or alternatively a distributed volumetric emitter. We typically assume that this volume is originally in local thermodynamic equilibrium (LTE). For combustion measurements, this is often a valid assumption. An example where LTE does not hold is for certain fuel fragments in the flame front. As one example, CH is chemically formed in an excited electronic state. It then chemiluminesces as it relaxes to thermal equilibrium, emitting light of the characteristic blue color (at roughly 400 nm) of a flame front. In low-pressure systems, LTE may not hold because collisional energy transfer is the primary mechanism leading to an equilibrium distribution, and collision rates may be too slow. In these cases, the validity of the LTE assumption should be checked. Temperatures in a combustion flowfield are sufficiently high for molecules to occupy higher-energy ro/vibrational levels. As the gas cools, these molecules relax into lower-lying levels, emitting thermal radiation in the infrared and visible parts of the electromagnetic spectrum.

Chemiluminescence and thermal radiation are both emitted isotropically from within a volume—there is no preferred direction of radiation as long as it emanates from a statistical ensemble of atoms and/or molecules (individual atoms and molecules can have a directional dependence).

In Figure 3.1, we also show light that can potentially enter the volume (from the lower left in the cartoon). It can come from any source—the neighboring volumetric emitter, a lamp, or a laser. That light can be absorbed, scattered, or simply transmitted through the volume. Scattering can occur at the same wavelength (elastic scattering), or be wavelength shifted (inelastic, e.g. by Raman scattering, see Chapter 13). Scattering can be directionally dependent, especially when the light is scattered from a particle that is large relative to the wavelength of the light [34, 35, 36, 37]. It is also possible for the light to be absorbed and then re-emitted as the atom or molecule relaxes back to LTE, which is fluorescence [1, 38].

Finally, the input light may be of sufficient energy to induce a nonlinear interaction. This will generate a nonlinear polarization state in the volume, and that new material state will then generate an output. The output will be controlled by coherent addition of the dipole emitter signals. This means that it will be highly directional and often (but not always) at a wavelength different from at least one of the inputs. Because the signal is constructed via generation of a nonlinear polarization state, followed by coherent addition from these dipoles, nonlinear optical interactions are described by electromagnetism. Nonlinear optical interactions are briefly introduced in Chapter 4.

In the remainder of this chapter, some definitions are provided, the ERT is derived and discussed, and then it is simplified in useful ways. We then discuss topics such as flame emissivity and optical depth. Finally, the regime in which the ERT reproduces Planck's law is described.

3.2 Some Definitions

Discussions on radiative transfer always seem to begin with definitions (see, for example, Siegel and Howell [39] and Penner [40]). The heat transfer community uses a large set of terms, each of which applies to a very specific variable or process. The astrophysics community has another set that agrees with some of the heat transfer nomenclature, but is not as large nor as dedicated. Laser diagnostics researchers often

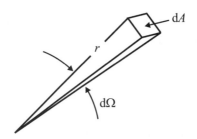

Figure 3.2: Definition of the solid angle.

use the astrophysical terms. As a result, people evaluating thermal radiation for combustor design may use different nomenclature to a laser diagnostician, even though the fundamental physics is the same. Here, we use the same nomenclature as other laser diagnostics sources [1, 2, 12, 40]. The classic heat transfer text by Siegel and Howell is a good source for the corresponding heat transfer nomenclature.

3.2.1   Geometric Terms

The image in Figure 3.1 is meant to convey the importance of the angular dependence of many optical phenomena. We begin, therefore, by defining the solid angle $d\Omega$ in order to emphasize the fact that collimated radiation is a limiting idealization. The solid angle is simply a three-dimensional version of the familiar two-dimensional radian-angle, which is defined as dl/r (in units of radians). In three dimensions, the solid angle [in units of steradians (sr)] is given by

$$d\Omega \equiv \frac{dA}{r^2} \quad (\text{sr}) \tag{3.1}$$

where the terms in the equation are defined in Figure 3.2. Consider a distributed emitter that can radiate into various angles with a radiation pattern that depends upon angle. Two cases can be found in the literature. In the first, radiation is treated as though it emanates from a surface, known as a "hemispherical emitter". This is the typical geometry used in heat transfer, for example, where a hot surface transfers heat to a cooler surface via radiation. It can also apply to astrophysics, where it is not possible to image radiation emitted from surfaces that are not visible from Earth. The surface area emitting towards Earth is the only emitter we can observe. Often, the same geometry is used by LIDAR researchers. The other viewpoint is to consider a three-dimensional emitter surface. This is the case used in LIF, for example, where the emitter volume radiates into all angles, and all of these angles are accessible.

The power carried by a projection of light is characterized numerous ways. Perhaps the most common quantity is the "irradiance" defined as the power per unit area at some location and time. Irradiance uses the variable "I" in W/m^2 in the SI system of units. A second useful quantity is the "radiance", which is defined as the power per unit area per unit solid angle at some location and time. Radiance uses the variable "J" in W/(m^2 sr) in the SI system of units. Note that this J is not the same as the "radiosity" from heat transfer, although they use the same variable. J is sometimes called "intensity", although that term is often applied to irradiance instead. It is also applied to the total power per unit sr. Because there is some confusion as to the specific meaning of "intensity", the word will not be used in this book.

Imagine a small hemispherical surface emitter with area dS, as shown in Figure 3.3. The power emitted by dS into $d\Omega$ at angles θ and ϕ is given by

$$J(\theta, \phi) \left[W/(m^2 sr) \right] \, dS \cos \theta \ (m^2) \, d\Omega \ (sr) \tag{3.2}$$

The $dS \cos \theta$ term represents a dot product between the surface normal $\hat{n}$ and $\hat{k}$. It is therefore the projection of area dS onto the direction $\hat{k}$. The projected area is a function of observation angle, falling to zero as θ approaches $\pi/2$. The total power per unit area (irradiance) emitted by dS is

$$I = \int_{\theta=0}^{\pi/2} \int_{\phi=0}^{2\pi} J(\theta, \phi) \cos \theta \, d\Omega(\theta, \phi) \tag{3.3}$$

Using the geometry in Figure 3.3, the differential solid angle is given by

$$d\Omega(\theta, \phi) = \frac{dA}{r^2} = \frac{r \sin \theta \, d\phi \, r \, d\theta}{r^2} = \sin \theta \, d\theta \, d\phi \tag{3.4}$$

and the irradiance becomes

$$I = \int_{\theta=0}^{\pi/2} \int_{\phi=0}^{2\pi} J(\theta, \phi) \cos \theta \sin \theta \, d\theta \, d\phi \tag{3.5}$$

Emission is often isotropic, and if so, the value of J does not depend upon angle. The integrals are then simply geometric. In that case, we

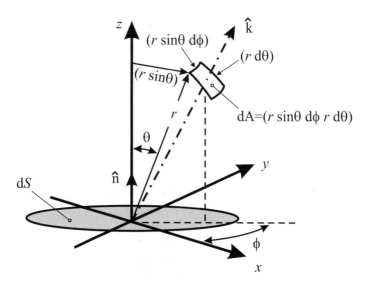

Figure 3.3: Hemispherical emitter surface.

get the simple result

$$I = \pi J \tag{3.6}$$

A full hemisphere is equivalent to π sr (if we must consider the effect of projected area).

As an alternative, we can consider spherical emission from a three-dimensional volume, which is usually the sample geometry in a flowfield measurement. The picture in Figure 3.3 would have to be modified to allow a full range of motion for $\hat{k}$ (i.e. $0 \le \theta \le \pi$). In addition, we no longer worry about the projection angle for the surface area; the projected area from a sphere is not angle dependent. In that case, equation (3.2) becomes

$$J(\theta, \phi) \left[\mathrm{W}/(\mathrm{m}^2\mathrm{sr}) \right] \; \mathrm{d}S \; (\mathrm{m}^2) \; \mathrm{d}\Omega \; (\mathrm{sr}) \tag{3.7}$$

The irradiance is

$$I = \int_{\theta=0}^{\pi} \int_{\phi=0}^{2\pi} J(\theta, \phi) \sin\theta \; \mathrm{d}\theta \; \mathrm{d}\phi \tag{3.8}$$

and for isotropic emission we find

$$I = 4\pi J \tag{3.9}$$

A full sphere is equivalent to 4π sr. This explains the term $\Omega/4\pi$ which one finds in expressions for the collected LIF signal. This fraction is simply the portion of the total isotropic spherical emission that has actually been collected. Ω is simple to calculate; it is the system-apertured, useful area on the first collection lens divided by the square of the distance between the lens and the sample volume.

It is probably obvious at this point that, if the hemispherical emitter had actually been a very small hemisphere (i.e. if we did not have to consider the effects of a projected flat area which falls to zero at $\theta = 0$), then the $\cos\theta$ term would be eliminated and irradiance would be equal to $2\pi J$ in the isotropic case.

3.2.2 Spectral Terms

For spectroscopy, we are usually concerned with the way in which light is distributed across the optical spectrum. Several different spectral terms are used interchangeably. The wavelength of the light is denoted by λ. The frequency of light is given by

$$\nu = \frac{v}{\lambda} \tag{3.10}$$

where v is the speed of light, and the optical frequency has units of (Hz), or cycles per second. This speed is dependent upon the index of refraction n (≥ 1) of the material through which the light passes via

$$v = \frac{c}{n} \tag{3.11}$$

where c is the speed of light in vacuum. Light will slow down in transmitting materials. For this reason, the frequency ν is a constant for the optical train, although λ will change with n. Spectroscopists, therefore, sometimes use the "vacuum wavelength" (denoted by $\lambda_\circ = c/\nu$) because it is a universal number.

It is sometimes convenient to use the angular frequency ω, given by

$$\omega \ (\text{rad/sec}) = 2\pi \ (\text{rad/cycle}) \ \nu \ (\text{cycles/s}) \tag{3.12}$$

when the Euler formalism $[\exp(i\omega t)]$ is used. The 2π ensures harmonicity. It is usually best, however, to continue to use ν (e.g. use $2\pi\nu$ in the Euler formalism) in order to avoid problems with spurious or missing values of 2π.

Spectroscopists often refer to the spectral location of light by the wavelength, usually in nm or μm. At other times wave numbers (cm^{-1}) are used (Europeans call the same unit a "Kayser"). Use of wave numbers seems most predominant among infrared spectroscopists. The energy of a photon is given by $h\nu = hc/\lambda_\circ$, where h is Planck's constant. If we normalize the photon energy by hc, we then have the wave number given by $1/\lambda_\circ$. There is no universal symbol used to denote wave numbers. Some authors use ν', but that can potentially cause confusion with the notation for the upper state vibrational quantum number v'. Chemists often use $\tilde{\nu}$ for this purpose, but this book indicates a complex number with $\tilde{}$ (consistent with most physics treatments). In this book, therefore, we will use $\breve{\nu}$ to denote wave numbers (ν because wave numbers are frequency divided by c, but with the " $\breve{}$ " to draw attention to the fact that this is not the same as ν itself).

Bandwidth is designated in the same sets of terms. If wavelength is used, then bandwidth is designated as $d\lambda$, $\delta\lambda$, or $\Delta\lambda$. If frequency is used, the bandwidth is designated by $d\nu$, $\delta\nu$, or $\Delta\nu$, and if wave number is used, then bandwidth is designated by $d\breve{\nu}$, $\delta\breve{\nu}$, or $\Delta\breve{\nu}$. One must be aware of the fact that ν and $\breve{\nu}$ have an inverse relationship to λ. In other words, for ν:

$$d\nu = d\left(\frac{c}{\lambda_\circ}\right) = -\left(\frac{c}{\lambda_\circ^2}\right) d\lambda_\circ \qquad (3.13)$$

The minus sign makes sense. As we stretch to higher wavelength, we are stretching to lower frequency. When using the magnitude of the bandwidth in calculations, the sign is irrelevant.

These spectral distinctions are no small matter. Sometimes authors are not clear which spectral location and/or bandwidth designation (frequency, angular frequency, wavelength or wave numbers) they are using. Worse yet, they sometimes use confusing notation. For example, ν is sometimes used to mean wave numbers, not frequency, and it is often necessary to check units in order to decide which it is, because authors will not call attention to it. Confusion or mixing of spectral systems can generate spurious or missing values of 2π, c, or $1/\lambda^2$. Such problems occur, for example, when the spectral radiance or irradiance (defined below) is used, especially in a differential way, or when a broadening function is not well defined. Careful attention to detail is required.

Spectral irradiance and spectral radiance are defined on a per unit spectral bandwidth (usually in Hz) basis. In that case we have

$$I_\nu \left[\frac{\text{power}}{(\text{unit area})(\text{unit bandwidth})} \right]$$

$$J_\nu \left[\frac{\text{power}}{(\text{unit area})(\text{sr})(\text{unit bandwidth})} \right]$$

The irradiance is equal to the integral over all frequencies of the spectral irradiance:

$$I = \int_{-\infty}^{+\infty} I_\nu \, d\nu \tag{3.14}$$

Similarly, the radiance is related to the spectral radiance:

$$J = \int_{-\infty}^{+\infty} J_\nu \, d\nu \tag{3.15}$$

The spectral irradiance and spectral radiance can be described in terms of the spectral energy density ρ_ν [J/(m³·Hz)] as follows:

$$I_\nu = \rho_\nu c \tag{3.16}$$

$$J_\nu = \frac{d\rho_\nu}{d\Omega} c \tag{3.17}$$

The dependence of J_ν on the solid angle is written as a differential because ρ_ν may be a continuous function of angle.

Finally, the spectral energy density can also be cast in terms of photon number density N_p:

$$N_p = \frac{\rho_\nu}{h\nu} \frac{[\text{J}/(\text{m}^3 \text{ Hz})]}{(\text{J/photon})} \, d\nu \ (\text{Hz}) \tag{3.18}$$

which assumes that the spectrum is flat within $d\nu$. Section 3.5 provides more detail on photons and photon statistics.

3.2.3 Relationship to Simple Laboratory Measurements

Occasionally, it is not obvious how one can relate these energy definitions to an actual laboratory measurement. In truth, many measurements are relative and the results are reported in "arbitrary units". In many other cases, the measurement is calibrated with a known technique. In such a case, absolute values are unimportant, as long as the system remains linear and simple expressions can be used to scale the numbers. The use of these definitions in an absolute sense really only matters for a truly absolute measurement. In this regard, it is somewhat easier to think in terms of a diffuse ("Lambertian") source (e.g. thermal radiation from a gas, Rayleigh scattering from small particles etc.). These sources radiate steadily and uniformly (or with a simple angular dependence) into 4π steradians. The expressions provided above can be used directly for such a source.

Laser beams, on the other hand, have specific spatial, spectral and temporal behavior. Because the source is strongly directional, can vary rapidly with time, and can be spectrally narrow, it becomes necessary to characterize the laser irradiance in a way that is adapted to the unique properties of the source. For this reason, the discussion will diverge for a few pages to present a very useful convention introduced by Partridge and Laurendeau [41] and modified somewhat by Settersten [42]. This formalism can be useful, for example, in cases where a Gaussian laser beam spatial distribution affects the measurement appreciably (e.g. when sampling from a portion of the nearly saturated LIF sample volume, in order to avoid linearity at the edges of the volume, or when evaluating the crossing region for crossed beam measurements) or in cases where the spectral width of the laser is approximately the same as the width of the spectral feature under study.

The spectral irradiance $I_\nu(x, y, z, \nu, t)$, which has units of W/(m^2 Hz), can be represented as the product of a normalized spectral irradiance $I_\nu^\circ(z)$, a dimensionless transverse profile $R(x, y)$, a dimensionless temporal distribution function $T(t)$, and a dimensionless spectral distribution function $L(\nu)$:

$$\boxed{I_\nu(x, y, z, \nu, t) = I_\nu^\circ(z)\ R(x, y)\ T(t)\ L(\nu)} \tag{3.19}$$

This approach isolates the spatial, temporal, and spectral dependences of the spectral irradiance. It is then possible to relate I_ν to quantities that can be measured or estimated.

The spatial dependence represented by $R(x,y)$ is the transverse profile of the laser beam, which is usually a Gaussian function of radius. The spectral profile represented by $L(\nu)$ contains the spectral distribution of energy emitted by the laser. Under many circumstances in spectroscopy, the laser spectrum is very narrow relative to an absorption linewidth and can be considered a delta function in frequency space. Often, however, this is not true, and it may be necessary to perform an overlap integration between the laser line shape and the absorption profile.

The time dependence $T(t)$ of the laser radiation can vary. Many lasers emit radiation continuously ("continuous wave", or "cw") while others are pulsed. Continuous wave beams do not require a temporal distribution function. In such a case, the time dependence in I_ν is eliminated, and $T(t)$ can be simply set to 1 in equation (3.19). Truly pulsed lasers are excited by pulsed sources such as flashlamps. They typically emit 10-50 ns pulses at repetition rates from tens of Hz to hundreds of Hz. Others are continuously pumped (and are sometimes referred to as cw lasers) but pulsed via a Q-switch (generating tens of ns pulses at up to 100 kHz) or they are mode-locked to produce very short pulses (50 fs-100 ps) at high repetition rates ($\sim$ 100 MHz). In many of these pulsed cases it becomes necessary to characterize $T(t)$.

It is useful to normalize $R(x, y)$, $L(\nu)$, and $T(t)$. The spatial profile is normalized so that its area integral is equal to the cross-sectional area A ($= \pi d_{\mathrm{TH}}^2$) of a "top-hat" function which is defined further below:

$$\int \int_{-\infty}^{+\infty} R(x, y) \, \mathrm{d}x \, \mathrm{d}y = A \tag{3.20}$$

Next, the temporal and spectral distribution functions are normalized as follows:

$$\int_{-\infty}^{+\infty} T(t) \, \mathrm{d}t = \Delta t_{\frac{1}{2}} \tag{3.21}$$

and

$$\int_{-\infty}^{+\infty} L(\nu) \, \mathrm{d}\nu = \Delta \nu_{\frac{1}{2}} \tag{3.22}$$

where $\Delta t_{\frac{1}{2}}$ is the temporal full width at half-maximum (FWHM) of the laser pulse, and $\Delta \nu_{\frac{1}{2}}$ is the spectral FWHM.

One can then relate I_ν° to measurable quantities. The total irradiance $I(x, y, z, t)$, which has units of $\mathrm{W/m^2}$, is equal to the integral over all frequencies of the spectral irradiance:

$$\begin{aligned} I(x, y, z, t) &= \int_{-\infty}^{+\infty} I_\nu(x, y, z, \nu, t) \, \mathrm{d}\nu \\ &= I_\nu^\circ(z) \, R(x, y) \, T(t) \, \Delta \nu_{\frac{1}{2}} \end{aligned} \tag{3.23}$$

The energy per unit area is given by the integral over all time of the total irradiance, and the energy $\mathcal{E}$ is given by the area integral of that result:

$$\begin{aligned} \mathcal{E} &= \int\int \left(\int I(x, y, z, t)\, \mathrm{d}t\right) \mathrm{d}x\, \mathrm{d}y \\ &= I_\nu^\circ(z)\, A\, \Delta\nu_{\frac{1}{2}}\, \Delta t_{\frac{1}{2}} \end{aligned} \tag{3.24}$$

Equation (3.24) relates the normalized spectral irradiance to parameters that can be experimentally determined:

$$\boxed{I_\nu^\circ(z) = \frac{\mathcal{E}}{A\, \Delta\nu_{\frac{1}{2}}\, \Delta t_{\frac{1}{2}}}} \tag{3.25}$$

This expression can then be used for $I_\nu^\circ(z)$ in equation (3.19), to produce a continuous function of position, frequency and time, for use during simulations for example. The pulse energy [or, for a cw system, energy per unit time (= power)] can be measured with a power meter. The area can be measured using standard beam characterization techniques, the spectrum can be observed with a spectrometer or interferometer, and the pulse can be measured with a fast photodetector or autocorrelator (depending upon the pulse width).

The spatial, spectral, and temporal distributions can also be described by simple models for laser performance. These properties of laser radiation are discussed broadly and in detail by Siegman [25] and Verdeyen [26], among others. Here we focus upon how these properties can be expressed within the ERT formalism.

With respect to spatial properties, it is common to consider a laser beam as a one-dimensional ray. This description suffices for many models, but in some cases it is necessary to investigate the effect of the transverse distribution of energy. Here we present the formalism that makes this possible. Both Gaussian and top-hat representations are developed.

The top-hat model assumes that the laser energy does not vary in the transverse direction, but instead is uniform over some cross-sectional area A:

$$R(x, y) = \begin{cases} 1 & : \quad x^2 + y^2 \leq d_{\mathrm{TH}}^2/4, \\ 0 & : \quad x^2 + y^2 > d_{\mathrm{TH}}^2/4 \end{cases} \tag{3.26}$$

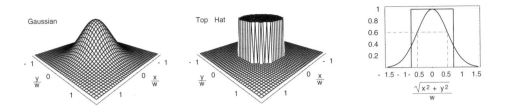

Figure 3.4: Gaussian and top-hat spatial profiles for a laser pulse.

This is the profile used to set the normalization condition given in equation (3.20). When a top-hat spatial profile is assumed, this definition transforms the problem such that it is only necessary to consider one spatial dimension, the beam propagation direction.

For more sophisticated modeling, a Gaussian profile is used. The lowest-order ($_{oo}$) solution to Maxwell's equations for a transverse electric and magnetic (TEM) wave is an azimuthally symmetric Gaussian function in cross-section. Many scientific lasers emit such a TEM$_{oo}$ beam [e.g. cw systems, diode lasers (which are astigmatic but Gaussian), some solid state and dye systems]. The profile function for a TEM$_{oo}$ laser beam with waist w is

$$R(x, y) = \frac{d_{\mathrm{TH}}^2}{2w^2} \exp\left[-\frac{2(x^2 + y^2)}{w^2}\right]$$
(3.27)

The normalization condition, equation (3.20), is satisfied by this function.

As it turns out, there is some level of arbitrariness in the definition of the top-hat diameter d_{TH}. Following the convention of Siegman [25], we set the Gaussian and top-hat functions equal on the beam axis. This assignment defines the top-hat diameter in terms of the Gaussian beam waist:

$$d_{\mathrm{TH}} = \sqrt{2}\, w$$
(3.28)

The two beam profiles are shown in Figure 3.4. Additionally, the two functions are plotted on a common axis to demonstrate the relationship between the Gaussian beam waist and top-hat diameter. Other lasers emit fairly complex beams (e.g. pulsed YAG lasers with pseudo-Gaussian output, and excimer lasers). These beams are typically modeled as either a top-hat or a Gaussian cross-section beam, depending upon which is closest to the actual laser output.

For short pulses, the spectral distribution $L(\nu)$ is determined by the temporal distribution, because the two are linked by the Fourier transform limit [25]. For laser pulses that can be described by hyperbolic secant squared temporal distributions, the temporal and spectral widths are related as follows:

$$
\begin{aligned}
\Delta t_{\frac{1}{2}} \Delta \nu_{\frac{1}{2}} &= \frac{4}{\pi^2} \left[\ln(\sqrt{2}+1) \right]^2 \\
&= 0.315
\end{aligned}
$$
(3.29)

Engineers familiar with Fourier transforms will recognize this notion. It presents a fundamental limit to the bandwidth (or pulse width) generated by a source. In order to achieve narrow bandwidth, it is thus necessary to have a very long pulse width (i.e. cw output). Conversely, in order to achieve short pulse width it is necessary to force broad bandwidth.

In the limit of narrow bandwidth, the laser should be a single longitudinal mode (single frequency) cw laser. The actual Fourier transform limited bandwidth associated with a cw laser (i.e. $\Delta \nu_{\frac{1}{2}} \approx 0$) is never reached in practice. In cw lasers, the spectral profile is determined by the finesse of the laser cavity, amplified spontaneous emission (ASE), and by a fluorescent background. The spectral bandwidth is a property of the device that is published by the manufacturer.

It is often reasonable to approximate a narrow spectral distribution function as a delta function. The normalized spectral irradiance is then modeled as

$$
I_\nu^\circ = \frac{I(z)}{\Delta \nu_{\frac{1}{2}}}
$$
(3.30)

and the spectral irradiance is then

$$
I_\nu(z, \nu) = I(z)\delta(\nu - \nu_\circ)
$$
(3.31)

where $\nu_\circ$ is the laser frequency.

In the intermediate bandwidth regime, it may be necessary to model the spectral profile of the laser somewhat more accurately. This can happen, for example, when a nanosecond pulse laser is used. In such cases, the temporal profile is sometimes ignored because the steady-state assumption can be applied. This assumption is usually a good one at atmospheric pressure where collision rates are much faster than the pulse. In this regime, however, the laser bandwidth can be of the same order of magnitude as the absorption linewidth. Such a spectral profile can be accurately described by a normalized Lorentzian function that has a width $\Delta\nu_{\frac{1}{2}}$ equal to the manufacturer's specification:

$$L(\nu) = \frac{1}{2\pi} \frac{1}{1 + \left[\frac{2(\nu-\nu_o)}{\Delta\nu_{\frac{1}{2}}}\right]^2} \tag{3.32}$$

where ν_o is the spectral location of the peak in the distribution. The form of equation (3.32) meets the normalization requirement of equation (3.22).

Under other conditions (e.g. at low pressure), the same Lorentzian function can be used for the spectral profile, but the temporal profile may prove important. To model the asymmetric profile of a nanosecond pulsed system is somewhat messy. The pulse formed by such a laser has a rapidly rising leading edge. This is due to the very large population inversion stored in the gain medium, well above threshold for the open cavity. The high gain level releases a large number of photons as soon as the cavity is opened (e.g. by a Q-switch). The very rapid rise in photon population sweeps out gain in the medium via stimulated emission. The photon number density then begins to slow its rate of rise as the laser gain approaches threshold, reaching the pulse peak just at threshold. The photon population then dies away exponentially inside the cavity, because gain has dropped below threshold. The pulse, therefore, has a very long trailing tail. This behavior can be modeled using rate equations (see, for example, Siegman [25]). It is important to point out that many pulsed laser systems operate with multiple longitudinal modes, each one at a slightly different frequency within the gain bandwidth of the laser. Each of these modes will Q-switch with slightly different gain levels, and hence at slightly different times. The output pulse is the coherent sum of these pulses, which can beat with each other to generate very rapid (hundreds of ps) and deep modulation, and this beating varies from pulse to pulse. This behavior can often be overlooked if a large number of averages are taken, but if mode beating is unacceptable one should use a single-longitudinal-mode laser. In summary, there is no simple, single expression one can use to describe the temporal profile of a nanosecond pulse. It might be best to measure it and use the measured profile to represent the average pulse waveform. Most pulse measurements will not have sufficient speed to detect mode beating, but such a measurement provides a good representation of an average pulse envelope.

The only reason to use pulses of even greater bandwidth is to generate short pulses (fs-ps). Short-pulse lasers typically produce nearly transform-limited pulses which are symmetric and follow a hyperbolic secant squared relationship [25]. Therefore, it is appropriate to use the transform limit given in equation (3.29) to relate the pulse width and bandwidth. The short-pulse spectral distribution function can be modeled as follows:

$$L(\nu) = \ln(\sqrt{2}+1) \operatorname{sech}^2 \left[2 \ln(\sqrt{2}+1) \frac{\nu - \nu_\circ}{\Delta\nu_{\frac{1}{2}}} \right] \qquad (3.33)$$

This expression is consistent with the normalization condition given in equation (3.22).

One can actually use two models for the temporal distribution function of a short pulse. The first model is a simple top-hat profile, for cases where the pulse is very short relative to other events:

$$T(t) = \begin{cases} 1 & : \quad \text{for } 0 < t \le \Delta t_{\frac{1}{2}}, \\ 0 & : \quad \text{for } t > \Delta t_{\frac{1}{2}} \end{cases} \qquad (3.34)$$

Obviously, this function satisfies the normalization condition, equation (3.21).

The second model would be used only in cases where the pulse shape is important, in density matrix based investigations for example (for reasons that will become clear in Chapter 14). It assumes, not surprisingly, the hyperbolic secant expression. The following distribution function satisfies the normalization condition given in equation (3.21):

$$T(t) = \ln(\sqrt{2}+1) \operatorname{sech}^2 \left[2 \ln(\sqrt{2}+1) \frac{t - t_\circ}{\Delta t_{\frac{1}{2}}} \right] \qquad (3.35)$$

3.3   Development of the ERT

The ERT is derived using a standard thermosciences approach. A control volume is drawn in space, and then the energy entering and leaving is cataloged, together with the associated energy changes within

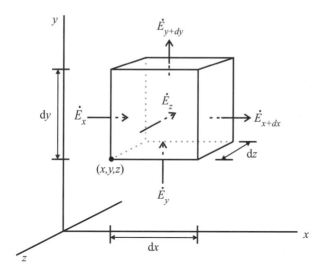

Figure 3.5: Three-dimensional control volume for development of the ERT.

the volume. Both a time derivative and a divergence term for the energy are found, together with source and sink terms. These sources and sinks use expressions for the resonant response of absorbers and emitters, and therein lies the difference with other energy equations. The following development follows somewhat the presentations in Measures [12], Penner [40], and to some extent Vincenti and Kruger [29]. The result is almost identical to the unsteady energy equation found in thermodynamics or heat transfer.

The elemental control volume to be used in this development is shown in Figure 3.5. In order to develop the ERT, it is necessary to track radiant energy at frequency ν within bandwidth $d\nu$, cataloging the fluxes $\dot{E}$ as well as creation and loss terms. Conservation of radiant energy within the control volume, E_{cv}, can then be written as a rate equation:

$$\frac{\partial E_{cv}}{\partial t} = \dot{E}_x + \dot{E}_y + \dot{E}_z - \dot{E}_{x+dx} - \dot{E}_{y+dy} - \dot{E}_{z+dz} + creation - loss \quad (3.36)$$

where "*creation*" describes the volumetric generation of radiant energy at frequency ν within bandwidth $d\nu$ (due to spontaneous emission for example) and "*loss*" describes the volumetric decay of radiant energy at frequency ν within bandwidth $d\nu$ (due to absorption for example). Constitutive relations for these two terms will be included after the basic ERT has been derived. Rearranging the equation produces

$$\frac{\partial E_{cv}}{\partial t} + (\dot{E}_{x+dx} - \dot{E}_x) + (\dot{E}_{y+dy} - \dot{E}_y) + (\dot{E}_{z+dz} - \dot{E}_z) = creation - loss$$

(3.37)

We begin with the left-hand side of equation (3.37). The term E_{cv} is the energy stored within the control volume, and it can be cast in terms of the spectral energy density by

$$E_{cv} = \rho_\nu \left[J/(m^3 Hz) \right] d\nu \; (Hz) \; dV \; (m^3)$$

(3.38)

where $dV = dx \; dy \; dz$. The time derivative in equation (3.37) is then

$$\frac{\partial E_{cv}}{\partial t} = \frac{\partial}{\partial t} (\rho_\nu \; d\nu \; dV) \, (J/s = W)$$

(3.39)

Now consider the left-hand face of the control volume in Figure 3.5, as shown in Figure 3.6. It is necessary to cast $\dot{E}_x$ in terms of spectral energy density as well. If one were to take the amount of energy entering the control volume within time Δt, it would be in the form of a small "brick" of energy (see Figure 3.7) which would be inserted into the volume. The depth of the brick would be $c_x \Delta t$. Note that the speed of light in vacuum is used here. Spectroscopic systems are typically modeled as individual absorber/emitter species (atoms or molecules) separated by free space (vacuum), hence the use of c (c_x here because we want the component in the x direction). The formalism then matches the definition of spectral energy density given in equation (3.16). The energy inserted into the control volume over time Δt would then be the energy density times the volume:

$$E_x = (\rho_\nu \; d\nu)(A c_x \Delta t)$$

(3.40)

Dividing both sides by Δt and taking the limit as $\Delta t \to 0$ will produce an expression for $\dot{E}_x(W)$, the energy convected across the boundary per unit time:

$$\dot{E}_x = \rho_\nu \left[J/(m^3 Hz) \right] \; d\nu \; (Hz) A \; (m^2) \; c_x \; (m/s)$$

(3.41)

which is quite similar to the mass convection term $\rho V A$ used in fluid mechanics. Now imagine that conditions at point (x, y, z) in Figures 3.5 and 3.6 are known. It would be fair to ask how an answer can

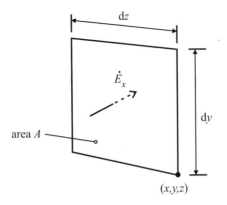

Figure 3.6: Left-hand face of cv, the input face for $\dot{E}_x$.

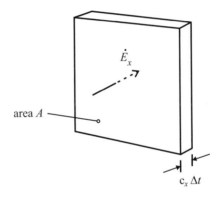

Figure 3.7: "Brick" of energy entering cv during Δt.

be assumed at the outset? The outcome of this exercise will be a differential equation for E as a function of time and position, requiring both boundary and initial conditions. The boundary conditions must be known at some location before the equation can be solved, hence it is legitimate for us to assume that the answer is known at a generic location (x, y, z). It is then necessary to calculate $\dot{E}_x$ based upon that knowledge. We calculate $\dot{E}_x$ at the center of the left-hand face, which is not the same as $\dot{E}_x$ at (x, y, z). Note that $\dot{E}_x = \rho_\nu \, d\nu c_x \, dA$, where $\rho_\nu \, d\nu$ is a scalar quantity, dA is a fixed constant, and c_x is a vector component.

At the center of the left-hand face (e.g. at $x, y + dy/2, z + dz/2$), we write $\dot{E}_x$ as a Taylor expansion about (y, z) (because x is fixed at that face):

$$\dot{E}_x \left(x, y + \frac{dy}{2}, z + \frac{dz}{2} \right) = \dot{E}_x(x, y, z) \tag{3.42}$$

$$+ \frac{\partial \dot{E}_x}{\partial y} \frac{dy}{2} + \text{terms of the order } y^2 \text{ and higher}$$

$$+ \frac{\partial \dot{E}_x}{\partial z} \frac{dz}{2} + \text{terms of the order } z^2 \text{ and higher}$$

The control volume is infinitesimally small, and the "terms of the order y^2 and higher", and "terms of the order z^2 and higher" are considered negligible. Equation (3.42) is thus linearized to

$$\dot{E}_x \left(x, y + \frac{dy}{2}, z + \frac{dz}{2} \right) = \dot{E}_x(x, y, z) + \frac{\partial \dot{E}_x}{\partial y} \frac{dy}{2} + \frac{\partial \dot{E}_x}{\partial z} \frac{dz}{2} \tag{3.43}$$

Using the same ideas, we cast $\dot{E}_{x+dx}$ (the value at the right-hand face, e.g. at $x + dx, y + dy/2, z + dz/2$):

$$\dot{E}_{x+dx} \left(x + dx, y + \frac{dy}{2}, z + \frac{dz}{2} \right)$$

$$= \dot{E}_x(x, y, z) + \frac{\partial \dot{E}_x}{\partial x} dx + \frac{\partial \dot{E}_x}{\partial y} \frac{dy}{2} + \frac{\partial \dot{E}_x}{\partial z} \frac{dz}{2} \tag{3.44}$$

This form assumes that terms of the order dx^2 and higher are also negligible.

Equation (3.37) takes the difference between $\dot{E}_{x+dx}$ and $\dot{E}_x$, which can be achieved by taking the difference between equations (3.44) and (3.43):

$$\dot{E}_{x+dx} - \dot{E}_x = \frac{\partial \dot{E}_x}{\partial x} dx$$

$$= \frac{\partial}{\partial x} (\rho_\nu \, d\nu c_x \, dA) \, dx$$

$$= \frac{\partial}{\partial x} \rho_\nu \, d\nu c_x \, dx \, dy \, dz \tag{3.45}$$

because $dA \, dx = dx \, dy \, dz$. Similarly

$$\dot{E}_{y+dy} - \dot{E}_y = \frac{\partial}{\partial y} \rho_\nu \, d\nu c_y \, dx \, dy \, dz \tag{3.46}$$

$$\dot{E}_{z+dz} - \dot{E}_z = \frac{\partial}{\partial z} \rho_\nu \, d\nu c_z \, dx \, dy \, dz \tag{3.47}$$

Now reassemble equation (3.37):

$$\frac{\partial}{\partial t}(\rho_\nu \, d\nu) \, dx \, dy \, dz + \frac{\partial}{\partial x}(\rho_\nu \, d\nu c_x) \, dx \, dy \, dz + \frac{\partial}{\partial y}(\rho_\nu \, d\nu c_y) \, dx \, dy \, dz$$

$$+\frac{\partial}{\partial z}(\rho_\nu \, d\nu c_z) \, dx \, dy \, dz = \; creation - loss \qquad (3.48)$$

It is then possible to divide by $dx \, dy \, dz$ to remove all dependence upon the size of the control volume, which puts the $creation - loss$ terms on a per-unit volume basis as well. Note also that c_x is associated with $\partial/\partial x$, c_y with $\partial/\partial y$, and c_z with $\partial/\partial z$. The sum of these three spatial terms can thus be written as $\vec{\nabla} \cdot (\rho_\nu \, d\nu \vec{c})$, and the ERT becomes

$$\frac{\partial}{\partial t}(\rho_\nu d\nu) + \vec{\nabla} \cdot (\rho_\nu d\nu \vec{c}) = \; creation - loss \qquad (3.49)$$

By using spectroscopic constitutive relations for the $creation - loss$ terms, the generalized ERT can be written as a spectroscopic ERT. For a two-level atom or molecule it is

$$\frac{\partial}{\partial t}\left(\rho_\nu \, d\nu\right) + \vec{\nabla} \cdot \left(\rho_\nu \vec{c} \, d\nu\right) \qquad\qquad (3.50)$$

$$= h\nu N_n A_{nm} Y_\nu \, d\nu + h\nu(N_n B_{nm} - N_m B_{mn})\rho_\nu Y_\nu \, d\nu \; \left(\text{W/m}^3\right)$$

where:

ρ_ν	=	spectral energy density $[(\text{Js})/\text{m}^3$ or $\text{J}/(\text{m}^3\text{Hz})]$
$d\nu$	=	bandwidth (1/s or Hz)
$\vec{c}$	=	velocity (vector term) of light (m/s)
ν	=	frequency, not necessarily ν_{nm} because we want to be able to scan across the line (1/s or Hz)
$h\nu$	=	photon energy (J)
n	=	designates the upper energy level
m	=	designates the lower energy level
N_n	=	total number density of atoms or molecules in the upper energy level $(1/\text{m}^3)$
N_m	=	total number density of atoms or molecules in the lower energy level $(1/\text{m}^3)$
A_{nm}	=	Einstein coefficient for spontaneous emission (1/s), the probability that state N_n will emit, i.e. the spontaneous emission rate coefficient

$$
\begin{aligned}
B_{nm} \;=\;\; & \text{Einstein coefficient for stimulated emission} \\
& [\mathrm{m^3/(Js^2)} \text{ or } \mathrm{(m^3Hz)/(Js)}], \\
& \text{driven by the optical field} \\
B_{mn} \;=\;\; & \text{Einstein coefficient for stimulated absorption} \\
& [\mathrm{m^3/(Js^2)} \text{ or } \mathrm{(m^3Hz)/(Js)}], \\
& \text{a coefficient for consumption of the optical field} \\
Y_{\nu} \;=\;\; & \text{lineshape function } (1/\mathrm{Hz} \text{ or } \mathrm{s})
\end{aligned}
$$

The left-hand side of equation (3.50) contains the time rate of change in energy density (within $d\nu$) in the volume plus the net convection of energy density (within $d\nu$) into or out of the volume. The right-hand side contains two source terms and one sink term. The right-hand side is a rate equation, based upon the densities of participating energy states, their rate coefficients (A and B terms), and the spectral energy density in the case of the stimulated (B) terms. These describe the response of an atom or molecule to resonant radiation. If the optical frequency is not at ν_{nm}, or not within the bandwidth of the spectral line, then there is no such response.

In the first term on the right-hand side, A_{nm} is the Einstein coefficient for spontaneous emission from excited-state atoms or molecules. We have not examined at this point how they got into the excited state. That would require additional, coupled rate expressions. As the N_n relax down to a lower level m, they are literally contributing to the optical field for which we are accounting, via spontaneous emission of radiation into the optical bandwidth $d\nu$. For a large collection of atoms or molecules, this emission is isotropic. The effect of ν and $d\nu$ is due to the lineshape function Y_ν, which describes the width of the lines in Figure 1.1. The function Y_ν is normalized in the following sense:

$$
\int_{-\infty}^{+\infty} Y_\nu \, d\nu = 1 \tag{3.51}
$$

Y_ν simply reduces the impact of A_{nm} as ν moves away from the line center frequency. The lineshape is typically described by a Gaussian or Lorentzian function, or the convolution of both (a Voigt profile, lineshapes are discussed in detail in Chapter 11). As such, these functions transition smoothly from zero in the wing to a maximum fractional value at the central wavelength and then back to zero in the opposite wing. Narrow line profiles generate stronger response at the central frequency than do wider line profiles, because the integrated response is always unity.

Next, the term B_{nm} in equation 3.50 is the Einstein coefficient for stimulated emission, a source term. The local optical field stimulates emission from atoms or molecules in the excited state, and the rate of emission is proportional both to the stimulating field strength, represented by ρ_ν, and to the relative population in the excited state, represented by N_n. In this case the emission is coherent with the field that induced it; it is at the same wavelength, has the same direction, and has a fixed phase relationship relative to the stimulating field. The stimulated response is modulated, again, by the lineshape function Y_ν. Here it may be necessary to consider the spectral profile of ρ_ν, as discussed in the previous subsection. Stimulated emission is responsible for optical gain within a laser resonator, and in fact one often hears that the beam "sweeps out the gain" via stimulated emission.

The term B_{mn} is the Einstein coefficient for stimulated absorption (or simply absorption), a sink term. Here, the rate of energy loss by absorption is also proportional to the field strength (modulated again by Y_ν) and to the relative population in the ground state, N_m.

The representations for A_{nm}, B_{nm}, and B_{mn} in equation (3.50) actually define the Einstein coefficients by describing their interaction with light and their contribution to the field energy. One should be aware that other authors can use somewhat different notation for the Einstein B coefficients. Some, for example, will define a quantity $b_{mn} \equiv B_{mn} Y_\nu$ or by $B_{mn}\, \rho_\nu Y_\nu$. More confusing still, some authors use a slightly different font and still capitalize the B. This goes back to a time when the illuminating light source was always broader than the absorption line, and so the observed absorption (it was usually absorption) was always a spectrally integrated signal. The original B coefficients were defined this way, and so the form b_{mn} was developed to call attention to the difference. Nowadays the difference is probably quite clear.

3.4 Implications of the ERT

It is important to note that equation (3.50) is an energy equation for just that portion of the radiation field that falls within $d\nu$ centered at some optical frequency ν. It is quite possible that other, coupled rate equations will be necessary to describe the entire system. As an example, the population of energy level n may not be controlled by equation (3.50) alone. There may be concurrent chemical formation reactions leading to chemiluminescence, or there may be collisional encounters that shift molecules from state n into other energy levels (collisional quenching or energy transfer, described by the rate Q). In the analysis of LIF, for example, coupled rate equations are used to describe populations, while a simplified version of equation (3.50) is used to generate (right-hand side) and then propagate (left-hand side) a signal to the detection system.

Rate Equations

Equation (3.50) is an energy conservation expression that couples the radiation field to spectroscopic terms. The terms on the right-hand side of the equation contain the rates of spectroscopic transitions in resonant atoms or molecules, and they are written in a form similar to chemical kinetics. We can therefore use the definitions for the Einstein coefficients to write expressions for population changes due to optical interactions. Collisional interactions can also be included. If we assume a two-level system (just one upper energy level and one lower energy level), and that the populations can be described by optical and collisional interactions alone, we can write the rates of change for the populations in levels n and m:

$$\frac{\mathrm{d}N_m(t)}{\mathrm{d}t} = -N_m(t)B_{mn}\int_\nu \rho_\nu Y_\nu\,\mathrm{d}\nu$$
$$+ \; N_n(t)\left[B_{nm}\int_\nu \rho_\nu Y_\nu\,\mathrm{d}\nu + A_{nm}\int_\nu Y_\nu\,\mathrm{d}\nu + Q_{nm}\right] \quad (3.52)$$

$$\frac{\mathrm{d}N_n(t)}{\mathrm{d}t} = N_m(t)B_{mn}\int_\nu \rho_\nu Y_\nu\,\mathrm{d}\nu$$
$$- \; N_n(t)\left[B_{nm}\int_\nu \rho_\nu Y_\nu\,\mathrm{d}\nu + A_{nm}\int_\nu Y_\nu\,\mathrm{d}\nu + Q_{nm}\right] \quad (3.53)$$

where Q_{nm} is the rate at which collisions (or other non-radiative phenomena in this simple model) remove molecules from N_n. Here, we have left the equation in terms of ρ_ν, Y_ν, and $\mathrm{d}\nu$ because we have not said anything about the spectral profile of the source or the absorption line. We have assumed that Q_{mn}, the collisional rate of excitation into level n, is negligible. If the energy gap between levels n and m is small and the medium is hot, it may be necessary to include Q_{mn}.

Equations (3.52) and (3.53) can often be found written in a different way. First, equation (3.16) can be rearranged to give $\rho_\nu = I_\nu/c$. Equation (3.52) can then be written as

$$
\begin{aligned}
\frac{dN_m(t)}{dt} \;=\; & -N_m(t)B_{mn}\int_\nu \frac{I_\nu}{c}Y_\nu\,d\nu \\
& +N_n(t)\left[B_{nm}\int_\nu \frac{I_\nu}{c}Y_\nu\,d\nu + A_{nm} + Q_{nm}\right] \quad (3.54)
\end{aligned}
$$

where we have assumed that the integral $\int_\nu Y_\nu\,d\nu$ following A_{nm} in equation (3.16) will be evaluated over all frequencies, and the normalization condition for Y_ν sets this integral to one. This equation is more commonly written as

$$
\frac{dN_m(t)}{dt} = -N_m(t)W_{mn} + N_n(t)[W_{nm} + A_{nm} + Q_{nm}] \quad (3.55)
$$

where

$$
W_{mn} \;\equiv\; B_{mn}\int_\nu \frac{I_\nu}{c}Y_\nu\,d\nu\ (1/\text{s}) \quad (3.56)
$$

$$
W_{nm} \;\equiv\; B_{nm}\int_\nu \frac{I_\nu}{c}Y_\nu\,d\nu\ (1/\text{s}) \quad (3.57)
$$

Similarly, equation (3.53) is written as

$$
\frac{dN_n(t)}{dt} = N_m(t)W_{mn} - N_n(t)[W_{nm} + A_{nm} + Q_{nm}] \quad (3.58)
$$

Equations (3.55) and (3.58) are in the form most commonly found in laser diagnostics texts.

For spontaneous emission in the absence of other population-changing phenomena (e.g. collisional quenching), for example, we can write equation (3.53) as

$$
\frac{dN_n}{dt} = -N_n A_{nm} \quad (3.59)
$$

Here we have considered all frequencies in the location of the resonance, and as before $\int_\nu Y_\nu\,d\nu = 1$. As with simple first-order chemical kinetics, we can calculate a lifetime from this rate expression by integrating over time:

$$N_n(t) = N_n(0)e^{-\frac{t}{\tau}}, \quad \text{where } \tau = \frac{1}{A_{nm}} \tag{3.60}$$

Hence, the Einstein coefficient for spontaneous emission is also the excited-state decay rate, or the inverse of the excited-state lifetime (sometimes called the fluorescence lifetime). If all the excited-state decay mechanisms are included (e.g. quenching), we obtain a full population decay time, often labeled T_1 (which is much shorter than $1/A_{nm}$ at atmospheric pressure).

Saturation

In gas-phase diagnostics, the stimulated emission term is often ignored when the energy gap $h\nu$ is large (e.g. the upper level is electronic, and it is then not highly populated at equilibrium), and/or ρ_ν is small (does not populate N_n), because N_n is then very small ($N_n \ll N_m$). As ρ_ν increases, even with a large energy gap, N_n grows. As N_n increases, stimulated emission can couple N_n back to N_m. Eventually, high levels of ρ_ν drive the system into nonlinear "saturation" where 50% of a two-level population is in state n and 50% is in state m. The rates of stimulated excitation and de-excitation are then matched. The irradiance at which saturation occurs can be estimated via the rate equations.

We begin with equations (3.52) and (3.53) (still assuming a two-level atom or molecule). For simplicity, assume that the bandwidth of the light source is much broader than the absorption profile, so that ρ_ν is approximately constant over the bandwidth of the absorption line. We can then integrate these expressions over ν from $-\infty$ to $+\infty$ and obtain

$$\frac{d}{dt}N_m(t) = -N_m(t)B_{mn}\rho_\nu + N_n(t)[B_{nm}\rho_\nu + A_{nm} + Q_{nm}] \tag{3.61}$$

$$\frac{d}{dt}N_n(t) = N_m(t)B_{mn}\rho_\nu - N_n(t)[B_{nm}\rho_\nu + A_{nm} + Q_{nm}] \tag{3.62}$$

In addition to these expressions, we can write mass conservation as

$$N_n(t) + N_m(t) = N_{\text{total}} \tag{3.63}$$

At saturation, the rates of excitation and de-excitation are balanced, at which point steady state is reached. We therefore assume steady state and set the time derivatives equal to zero. Equation (3.62) is then

$$N_m B_{mn} \rho_\nu = N_n [B_{nm} \rho_\nu + A_{nm} + Q_{nm}] \tag{3.64}$$

Using equation (3.63) to eliminate reference to N_m in equation (3.64), we obtain

$$\frac{N_n}{N_{\text{total}}} = \frac{B_{mn} \rho_\nu}{(B_{nm} + B_{mn}) \rho_\nu + A_{nm} + Q_{nm}} \tag{3.65}$$

Using the definition of ρ_ν given by equation (3.16), we rearrange to find

$$\frac{N_n}{N_{\text{total}}} = \frac{B_{mn}}{B_{nm} + B_{mn}} \frac{1}{1 + \frac{I_{\nu,\text{sat}}}{I_\nu}} \tag{3.66}$$

where:

$$\boxed{I_{\nu,\text{sat}} = \frac{(A_{nm} + Q_{nm})c}{B_{nm} + B_{mn}}} \tag{3.67}$$

Note in equation (3.66) that just when $I_\nu = I_{\nu,\text{sat}}$, then $N_n/N_{\text{total}} = 1/4$, in the two-level approximation where $B_{nm} = B_{mn}$. As $I_\nu \gg I_{\nu,\text{sat}}$, then $N_n/N_{\text{total}} = 1/2$. Equation (3.67) is simply a scaling law. It determines the irradiance that is necessary to make the stimulated terms match or dominate the spontaneous and non-radiative terms.

Absorption and emission

In order to reduce the ERT to common analytical expressions found in the literature, one often assumes steady state for the population levels (N_n and N_m). Conversely, in cases where steady state is an inappropriate assumption, a sequence of temporal differential equations can be solved numerically. As an example, energy transfer from a laser-pumped state into other states (vibrational and rotational energy transfer) can be relatively fast, and it is typically described numerically (see, for example, [43, 44]). Steady state is an appropriate assumption when the characteristic time for the optical signal is significantly longer than the response time of a molecule (excitation, energy transfer, collision and so forth [45]). It is not an appropriate assumption when, for example, the optical interaction is several hundred picoseconds or less (at atmospheric pressure). Settersten and Linne [46] have investigated the accuracy of the rate equation approach when applied to laser excitation by relatively short pulses. They found that the rate equations provide reasonable accuracy (a disagreement between the rate equations and the quantum mechanically correct density matrix equations of less than 10%) if the excited state population is less than 20%.

In what follows, standard analytical expressions are developed by assuming steady-state, at which point the $\partial/\partial t$ term in equation 3.50 is zeroed. Each term is written over the same frequency interval $d\nu$, and it can therefore be dropped. The coordinate system can then be arranged so that the light passes along z, and we assume that the radiation we are tracking passes through a relatively small solid angle Ω. The assumption of a small solid angle is almost always valid because collection optics are what determine Ω. Collection lenses typically come in sizes from one to four inches in diameter. Placed at a normal distance (typically two focal lengths) away from the collection volume, these subtend a relatively small solid angle (maximum of $f\# \sim 1.5$). That means the transverse gradients $\partial/\partial x$ and $\partial/\partial y$ in equation (3.50) are usually negligible. We now rewrite equation 3.50 in terms of $\Delta\rho$, the spectral energy density that is confined to Ω, and we apply the set of assumptions just discussed:

$$\frac{d}{dz}(c\rho_\nu) = h\nu N_n A_{nm} Y_\nu \frac{\Omega}{4\pi} + h\nu(N_n B_{nm} - N_m B_{mn})Y_\nu \rho_\nu \qquad (3.68)$$

Note that we have accounted for the portion of the spontaneous emission that falls within Ω only. Since spontaneous emission from a volume is isotropic, we can simply take the fraction of the total 4π steradians collected (i.e. $\Omega/4\pi$). We also know from the definition of ρ_ν that

$$J_\nu = c\frac{\rho_\nu}{\Omega} \qquad (3.69)$$

(assuming that ρ_ν is isotropic within Ω), so equation (3.68) becomes

$$\frac{dJ_\nu}{dz} = \frac{h\nu}{4\pi}N_n A_{nm} Y_\nu + \frac{h\nu}{c}(N_n B_{nm} - N_m B_{mn})Y_\nu J_\nu \qquad (3.70)$$

We then define a volume emission coefficient per steradian by

$$\boxed{\epsilon_\nu(z) \equiv \frac{h\nu}{4\pi}N_n A_{nm} Y_\nu \; \left[\mathrm{W/(m^3 srHz)}\right]} \qquad (3.71)$$

and a volume absorption coefficient by

$$\boxed{\kappa_\nu(z) \equiv \frac{h\nu}{c}(N_m B_{mn} - N_n B_{nm})Y_\nu \; (1/\mathrm{m})} \qquad (3.72)$$

These are sometimes written as mass absorption coefficients by dividing by the density of atoms or molecules involved. It is then necessary to multiply these mass-based coefficients by the density when they are used in ERT-based expressions as written here. Dimensional units can be used to determine whether this is necessary. Equation (3.70) can be written in terms of ϵ_ν and κ_ν as

$$\frac{dJ_\nu}{dz} = \epsilon_\nu(z) - \kappa_\nu(z)J_\nu \qquad (3.73)$$

Now we simplify equation (3.73) by defining a source function and an optical depth. The source function is

$$\boxed{S_\nu(z) \equiv \frac{\epsilon_\nu(z)}{\kappa_\nu(z)}} \qquad (3.74)$$

while the optical depth is given by

$$\boxed{\tau_\nu(z) \equiv \int_0^z \kappa_\nu(z')dz'} \qquad (3.75)$$

The total optical depth is then $\tau_\nu(\ell)$, where ℓ is the total interaction path length. Equation (3.73) then becomes

$$\frac{dJ_\nu}{d\tau_\nu} + J_\nu = S_\nu \qquad (3.76)$$

the solution to which is

$$\boxed{J_\nu(\ell) = J_\nu(0)e^{-\tau_\nu(\ell)} + e^{-\tau_\nu(\ell)} \int_{\tau_\nu(0)}^{\tau_\nu(\ell)} S_\nu(z)e^{\tau_\nu(z)} \, d\tau_\nu(z)} \qquad (3.77)$$

The medium is not necessarily homogeneous. Often in combustion systems it is not, but researchers may still perform line-of-sight measurements in order to obtain a path-averaged result. By assuming homogeneity, one can simplify equation (3.77). In this case, the source function and optical depth become

$$S_\nu(z) = S_\nu = \frac{\epsilon_\nu}{\kappa_\nu} \qquad (3.78)$$

$$\tau_\nu(\ell) = \kappa_\nu\ell \qquad (3.79)$$

and equation (3.77) becomes

$$J_\nu(\ell) = J_\nu(0)e^{-\kappa_\nu \ell} + S_\nu[1 - e^{-\kappa_\nu \ell}] \qquad (3.80)$$

We can make good use of this expression. In the case of a "cold" (i.e. not strongly emitting) gas, ϵ_ν and S_ν are both zero, and equation (3.80) becomes the classical Beer's law expression for absorption of radiation as a function of distance through the absorber:

$$J_\nu(\ell) = J_\nu(0)e^{-\kappa_\nu \ell} \qquad (3.81)$$

In other words, the entering radiance $J_\nu(0)$ will decrease exponentially with distance into an absorber. The strength of this energy absorption is controlled by the parameter κ_ν, which is assumed to be constant across the field of view. The absorption coefficient is a function of the energy per photon, of the Einstein rate coefficients for stimulated absorption and emission, of the populations of the lower energy level (given by Boltzmann statistics if we assume LTE) and the upper energy level (usually negligible in the linear regime), and of the wavelength (as described by the lineshape function).

If we assume that the solid angle Ω does not change with ℓ, then it can be eliminated from both sides to give

$$I_\nu(\ell) = I_\nu(0)e^{-\kappa_\nu \ell} \qquad (3.82)$$

If the medium is optically thin (i.e. $\kappa_\nu \ell$ is very small), we can express equation (3.81) as a series and truncate it to give

$$J_\nu(\ell) = J_\nu(0) [1 - \kappa_\nu \ell] \qquad (3.83)$$

If we can assume that $N_n = 0$ (which is often the case when the energy gap is large, e.g. for electronic transitions) then this expression becomes

$$J_\nu(\ell) = J_\nu(0) \left[1 - \frac{h\nu}{c} N_m B_{mn} Y_\nu \ell \right] \qquad (3.84)$$

Alternatively, if the energy gap is small (e.g. for ro/vibrational transitions in the infrared) it may become necessary to account for thermal population in the upper level. In such a case, Penner [40] uses equation

(3.96) (to be discussed below), together with statistical mechanics to generate a multiplier of

$$\left(1 - e^{\frac{-h\nu}{k_B T}}\right) \tag{3.85}$$

which is then used to modify N_m in equation (3.86):

$$J_\nu(\ell) = J_\nu(0) \left[1 - \frac{h\nu}{c} N_m B_{mn} Y_\nu \left(1 - e^{\frac{-h\nu}{k_B T}}\right) \ell\right] \tag{3.86}$$

If the medium is "hot" (i.e. emission is strong) and there is no external light source (i.e. we neglect line-of-sight absorption), then we can modify equation (3.80) to describe emission:

$$\boxed{J_\nu(\ell) = S_\nu \left[1 - e^{-\kappa_\nu \ell}\right]} \tag{3.87}$$

where we have folded the populations back into κ_ν. In the optically thin case we assume small optical depth ($\kappa_\nu \ell$) and linearize the expression:

$$\boxed{J_\nu(\ell) = S_\nu \left[1 - (1 - \kappa_\nu \ell)\right] = \epsilon_\nu \ell} \tag{3.88}$$

but, from equation (3.71) this becomes

$$J_\nu(\ell) = \frac{h\nu}{4\pi} N_n A_{nm} Y_\nu \ell \tag{3.89}$$

This $J_\nu(\ell)$ is what one collects in a diagnostic such as thermal emission, or LIF. In the absence of absorption, radiant emission from a specific location is thus controlled by the parameter ϵ_ν, which is assumed to be constant across the field of view in this expression. The emission coefficient is a function of the photon energy, the Einstein coefficient for spontaneous emission, the population of the upper energy level, and the wavelength (as described by the lineshape function). The population of the emitting level, N_n, is given by Boltzmann statistics if we are observing thermal emission, or by rate equations describing laser excitation in the case of LIF. Note that the radiance J_ν is on a per unit solid angle basis, thus the 4π in the denominator. To convert to irradiance I_ν, one simply calculates Ω for the experiment and multiplies J_ν by this Ω. That produces a recognizable $\Omega/4\pi$ term in the expression.

The optical depth is given by $\tau_\nu(\ell) = \kappa_\nu \ell$ in this case. Optical depth can affect both absorption through expression (3.81) and emission through expression (3.87). Note that these expressions do not require an optically thin limit. κ_ν is given by equation (3.72).

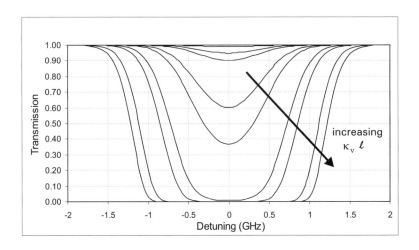

Figure 3.8: Effect of optical depth on absorption line shape.

Imagine we look over a very long path length, or we have a very large number density of absorbers (in the case of absorption) or emitters (in the case of emission). Path length and number density have the same effect on optical depth. At very large optical depth, absorption [via equation (3.81)] completely extinguishes the radiance. This happens first at the center of the lineshape function where Y_ν is largest, and then increases across the line as depth increases. Figure 3.8 contains several Lorentzian-broadened absorption profiles for increasing optical depth. One can readily observe that the curve reaches zero transmission and then begins to broaden as the wings of the line absorb more strongly.

Emission behaves in a similar way. Equation (3.87) shows that radiant emission increases as optical depth increases until the exponential term is so small that J_ν is effectively equal to S_ν. As optical depth increases, emission departs from linearity and reaches a limiting value of $J_\nu = S_\nu$. This phenomenon is depicted in Figure 3.9, where we have normalized equation (3.87) by dividing both sides by S_ν. In that case, the limiting value is 1.0 as shown.

This is the case for just one emission line. Thermal emission typically occurs in the IR, owing to ro/vibrational transitions in molecules. In a typical gas mixture, such as a flame, there are many molecules and they usually have a high density of states (i.e. complex spectral structure). This region is therefore very crowded with emission lines. As the system becomes optically thick, these lines can merge to form

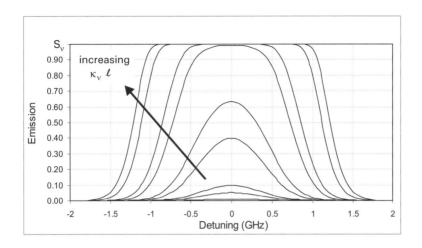

Figure 3.9: Effect of optical depth on emission line shape.

a continuum. One can easily imagine these dense but discrete lines converging into a smooth, very broad structure much like the Planck black-body function.

Black and gray body behavior

Several additional expressions can be developed using the rate equations. Using the definitions of S_ν, ϵ_ν and κ_ν [equations (3.74), (3.71) and (3.72)] we obtain

$$S_\nu = \frac{c}{4\pi} \frac{A_{nm}}{B_{nm}} \frac{1}{\frac{N_m B_{mn}}{N_n B_{nm}} - 1} \tag{3.90}$$

From statistical mechanics, the fraction N_m/N_n is given by

$$\frac{N_m}{N_n} = \frac{g_m}{g_n} \exp\left(\frac{h\nu_{nm}}{k_B T}\right) \tag{3.91}$$

Equation (3.90) can then be written as

$$S_\nu = \frac{c}{4\pi} \frac{A_{nm}}{B_{nm}} \frac{1}{\frac{B_{mn}}{B_{nm}} \frac{g_m}{g_n} \exp\left(\frac{h\nu_{nm}}{k_B T}\right) - 1} \tag{3.92}$$

The term

$$\frac{1}{\frac{B_{mn}}{B_{nm}} \frac{g_m}{g_n} \exp\left(\frac{h\nu_{nm}}{k_B T}\right) - 1} \tag{3.93}$$

begins to look like the Planck black-body distribution. At the transition frequency ν_{nm}, the Planck distribution is written with the term

$$\frac{1}{\exp\left(\frac{h\nu_{nm}}{k_B T}\right) - 1} \tag{3.94}$$

In order to make S_ν the same as the Planck distribution, the following would have to be true:

$$\boxed{\frac{A_{nm}}{B_{nm}} = \frac{8\pi h\nu^3}{c^3}} \tag{3.95}$$

and

$$\boxed{g_n B_{nm} = g_m B_{mn}} \tag{3.96}$$

Then equation (3.92) becomes

$$S_\nu = \frac{2h\nu^3}{c^2} \frac{1}{\exp\left(\frac{h\nu_{nm}}{k_B T}\right) - 1} \tag{3.97}$$

which is indeed the Planck black-body distribution at $\nu = \nu_{nm}$. Normally the Planck distribution is written as a continuous function of ν, not a function of discrete values ν_{nm}. There is a point, however, where the spectral behavior denoted by the subscript nm becomes a continuous function. This will be discussed below.

Equation (3.87) (for a hot, emitting gas in the absence of absorption) is worth revisiting first. It was written as

$$J_\nu(\ell) = S_\nu \left[1 - e^{-\kappa_\nu \ell}\right]$$

which means that the term:

$$\left[1 - e^{-\kappa_\nu \ell}\right]$$

would have to be the emissivity of the gas. In infrared spectroscopy, in fact, it is quite common to write expressions like (3.87) in terms of the Planck distribution function and an emissivity that is based upon the spectrum of the emitting gas. For an uncrowded but structured spectrum, this means the emissivity goes to zero between each emission line. An expression combining the Planck distribution and the gas emissivity will give the power emitted per unit area per unit solid angle, which can then be used with a model for the optical collection system to simulate a signal level at the detector.

Note from equations (3.71), (3.72) and (3.73) that both κ_ν and S_ν have built into them the populations of the two energy levels linked by the absorption/emission process. Relative populations given by Boltzmann statistics are thus embedded in the Planck distribution.

Combine this emissivity notion with the image of Figure 3.9 and the discussion surrounding it. The emitted radiance described by equation (3.87) increases as the optical depth increases, finally reaching a limiting value of S_ν, which we now know is the Planck black-body function. As a crowded spectrum becomes optically thick, the structure fills in, the emissivity ceases to have discrete lines, and the spectrum becomes a continuum. The emission spectrum of the gas becomes the smooth, unstructured Planck distribution one often finds in heat transfer textbooks. In fact, when measured emissivity is shown as a function of wavelength in such texts, it has very weak spectral structure.

3.5 Photon Statistics

In the last section, several relations that tied the ERT to the Planck distribution [equations (3.95) and (3.96)] were passed over quickly. We simply said that they must be true for the ERT to produce the Planck distribution. A proof that the ERT can generate the black-body distribution requires statistical mechanics.

Following the treatment by Vincenti and Kruger [29], we begin by defining $N_\mathrm{p}(r, t)$ as the number of photons per unit volume at some location r and time t. Unlike gas molecules, photons all have the same speed ($v = c/n$, where this n is the refractive index), so there is no kinetic energy distribution. There is, however, a distribution of energy, $h\nu$, momentum, $h\nu/c$, and direction, $\hat{k}$ (as defined in Figure 3.3). It is possible to define a photon distribution function f by

$$\frac{\#\text{photons within } d\Omega \; d\nu}{\text{unit volume}} = N_\mathrm{p} f \; d\Omega \; d\nu \qquad (3.98)$$

The distribution f has units of $1/(\text{sr Hz})$, and it is a function of position r, time t, direction $\hat{k}$, and frequency ν. It contains the dependence on solid angle and frequency that is embedded in J_ν, and it is normalized via

$$\int_0^\infty \int_0^{4\pi} f \; d\Omega \; d\nu = 1 \qquad (3.99)$$

The spectral radiance J_ν can be related to the photon flux, $cN_p f$, and the energy per photon $h\nu$. In effect, optical energy is transported by the movement of discrete photons. The photon flux times the energy per photon is then the energy flux:

$$J_\nu = cN_p f \, h\nu \; \left[\text{W}/(\text{m}^2 \text{ sr Hz}) \right] \tag{3.100}$$

This result allows the photon number density to be related to the spectral radiance. One can use the expression for f from equation (3.100) in equation (3.99) and multiply by N_p to obtain

$$\boxed{N_p = \frac{1}{c} \int_0^\infty \int_0^{4\pi} \frac{J_\nu}{h\nu} \, d\Omega \, d\nu} \tag{3.101}$$

By combining equations (3.16) and (3.100), one can find an expression relating the spectral energy density to the photon energy density:

$$\rho_\nu = \frac{1}{c} \int_0^{4\pi} J_\nu \, d\Omega = \int_0^{4\pi} h\nu \, N_p f \, d\Omega \tag{3.102}$$

An equation that simply conserves photons can be written as:

$$\frac{\partial}{\partial t} \left[N_p f \, d\Omega \, d\nu \, dV \right] + \vec{\nabla} \cdot \left[\hat{k} N_p c f \, d\Omega \, d\nu \, dV \right]$$
$$= creation - loss \tag{3.103}$$

where dV is the differential volume element. This is the ERT in a different form, where we have neglected to provide details on the right-hand side for a moment. If we integrate this expression over 4π steradians, we will find the same ERT we had before [equation (3.50)]. That makes sense; in the development of equation (3.50), we used irradiance because we were assuming that all directions would be included. If we assume that dV is constant, it cancels and we can write the following:

$$\frac{\partial}{\partial t} \left[\frac{1}{h\nu} \int_0^{4\pi} h\nu \, N_p f \, d\Omega \, d\nu \right] + \vec{\nabla} \cdot \left[\frac{c\hat{k}}{h\nu} \int_0^{4\pi} h\nu \, N_p f \, d\Omega \, d\nu \right]$$
$$= creation - loss \tag{3.104}$$

Using equation (3.102), this simplifies to

$$\frac{\partial}{\partial t} \left[\rho_\nu \, d\nu \right] + \vec{\nabla} \cdot \left[c\hat{k} \rho_\nu \, d\nu \right] = creation - loss \tag{3.105}$$

This is identical to equation (3.49), because $\vec{c} = c\hat{k}$. One could easily copy over the right-hand side from equation (3.50) to complete this expression as the ERT.

Conversely, to reproduce equation (3.103) one needs to recast the form of equation (3.50) into a per unit steradian basis. Consider the spontaneous emission term, for example. The photon distribution generated by spontaneous emission, for use in the right-hand side of equation (3.103), could be given by

$$N_{\mathrm{p}}f = \frac{\mathrm{d}}{\mathrm{d}\Omega}(N_n A_{nm} Y_\nu) \qquad (3.106)$$

There is really no need to do this, equation (3.50) serves quite well. More important is the incorporation of statistical mechanics using the photon representation.

The Planck distribution

Statistical mechanics describes the state of a system at LTE. In this case, the equilibrium expression must include both molecules and photons. If radiation is at equilibrium, the photon energy distribution is given by Bose-Einstein statistics [29]. The equilibrium number density of photons within an energy interval $\mathrm{d}\nu$ about the energy $h\nu$ is [29]

$$N_{\mathrm{p}}^* = 8\pi \, \frac{\nu^2}{c^3} \, \mathrm{d}\nu \, \frac{1}{\mathrm{e}^{\frac{h\nu}{k_{\mathrm{B}}T}} - 1} \qquad (3.107)$$

where the asterisk (*) indicates an equilibrium quantity that is not on a per unit solid angle per unit bandwidth basis. At equilibrium, and for a statistical quantity, the distribution should be isotropic.

Equation (3.98) defines a photon distribution function per unit solid angle and unit bandwidth. We can therefore define a new distribution function that is independent of solid angle, by assuming that f is isotropic:

$$f^* \equiv \int_0^{4\pi} f \, d\Omega = 4\pi f \qquad (3.108)$$

At equilibrium, the total number density within spectral bandwidth $\mathrm{d}\nu$ is written in terms of f^*:

$$N_{\mathrm{p}}^* = N_{\mathrm{p}} f^* \, \mathrm{d}\nu = N_{\mathrm{p}} 4\pi f \, \mathrm{d}\nu \qquad (3.109)$$

Equation (3.100) relates photon number density to the spectral radiance. Using it, equation (3.109) becomes

$$N_{\mathrm{p}}^* = \frac{4\pi}{ch\nu} \, J_\nu \, \mathrm{d}\nu \tag{3.110}$$

Combining equations (3.107) and (3.110) produces

$$\boxed{J_\nu = \frac{2h\nu^3}{c^2} \frac{1}{e^{h\nu/k_{\mathrm{B}}T} - 1}} \tag{3.111}$$

Note that equation (3.111) is identical to equation (3.97), except that equation 3.97 is written in terms of S_ν while 3.111 is written in terms of J_ν. Equation (3.111) is simply the distribution function for photons in radiative equilibrium at some temperature. Radiative equilibrium occurs within a black body, where emissivity is 1. In the black-body limiting case, $J_\nu = S_\nu$, so the two equations are consistent. The ERT treatment does generate the Planck distribution if we assume that light can be quantized and that these photons obey statistical mechanics.

An outcome of this finding is the fact that equations (3.95) and (3.96) must also be true. They form, in fact, a very useful system for relating the various Einstein coefficients to each other.

3.6   Conclusions

Example applications of the ERT can be found in almost every article on laser diagnostics in combustion (see, for example, the review by Kohse-Höinghaus [9]). As a diagnostic-specific example, Daily [38] develops the expressions for LIF using the ERT, and he does it for several cases including the dynamics of multiple levels. His notation is different from ours. Unfortunately, there are several sets of notation used to describe the same physics, using the same equations. Here we have used one very common set.

As mentioned before, the ERT is quite useful for describing linear phenomena such as absorption, emission, particulate scattering, LIF and so forth. It is simply an extension of classical thermosciences. To this point, however, we have not said anything about how the critically important Einstein coefficients are generated. The way in which they are related to the absorption coefficient κ_ν and how that enters into

Beer's law is quite clear, but there is much more to resonance response. It is not really possible to say more, however, unless electromagnetism is invoked. This is true because the most basic interaction between an atom or molecule and light is an interaction between an electric dipole and an oscillating electric field.

Chapter 4

OPTICAL ELECTROMAGNETICS

4.1 Introduction

The notion that light is a propagating electromagnetic wave was given a solid foundation by J. C. Maxwell when he unified the theories of electricity and magnetism into the four equations that bear his name. Nowadays we accept the electromagnetic wave nature of light and rely upon it for many applications (e.g. interferometry, holography and others), but such acceptance was a long time developing. An interesting history of this development can be found in the introductions to the books on optics by Born and Wolf [47] and by Hecht [48].

Here, we discuss Maxwell's equations and the outcomes that are useful for diagnostics. There are two purposes for this discussion. First, the more fundamental approaches to linear spectroscopy, and any description of nonlinear spectroscopy, will necessarily make use of electromagnetic theory. An introduction is therefore in order. Second, there is potential for confusion about exactly what values one must use in diagnostics expressions, when the time comes actually to insert numbers into equations. Much of that confusion can arise because there are several possible sets of units used in electromagnetism, and they can change the appearance of the equations (including expressions for A_{mn}, B_{mn}, B_{nm}, and Y_{ν}) . The only way to overcome this difficulty is to develop a full understanding of the differences between systems of units in electromagnetism. For this reason, Appendix A discusses

75

systems of units in detail following the pattern of development in this chapter. Throughout this text, expressions are developed in the MKSA system (a rationalized SI system of units).

Several very good, complete treatments of electromagnetism can be found in common textbooks. Optical diagnostics rely upon the dynamics of electromagnetism, but most texts put that off for hundreds of pages, focusing initially upon electro- and magneto-statics. Among such texts, both Jackson [49] and Wangsness [50] provide very readable treatments in the MKSA unit system, and they include a chapter on electromagnetic units. For those who would prefer to move directly to optical interactions, the optics text by Klein and Furtak [51] provides simply those portions of electromagnetism that apply to optical phenomena, in MKSA units.

This chapter begins with a discussion of Maxwell's equations as they apply to the propagation of electromagnetic waves through vacuum. Following the vacuum treatment, the same series of equations as they apply to propagation through a transmitting material are presented. That development lays the foundation for the treatment of an electromagnetic (optical) field as it interacts with the classical electron oscillator, or Lorentz atom in Chapter 5. Finally, we treat the Lorentz atom using Hamiltonian mechanics in Chapter 6, as a preparatory step towards quantum mechanics.

4.2   Maxwell's Equations in Vacuum

There was in fact a controversy over whether it was possible to propagate electromagnetic waves through vacuum, or whether outer space was actually filled with a "luminiferous æther" that supported electromagnetic waves, much as a gas or liquid supports acoustic waves. It is now accepted that electromagnetic waves do propagate through vacuum, thanks to the finite value of the electric permittivity $\epsilon_\circ$ and magnetic permeability $\mu_\circ$ in free space, or vacuum (the subscript $\circ$ on ϵ and μ denotes free space in this context). A somewhat reduced set of Maxwell's equations can then be used to describe this propagation.

Maxwell's equations consist simply of two field conservation expressions (Gauss' electric and magnetic laws) and two expressions for induction (Faraday's law for magnetic induction of current and Ampere's law for electric induction of a magnetic field). They have their foundation in basic empirical observations, such as Coulomb's law for charge interaction and the Biot-Savart law for magnetic interaction.

Gauss' electric law

Gauss' electric law is simply Coulomb's law written in field terms. Coulomb's law describes the force acting between two charged particles, and it is based upon empirical observation. Coulomb's law simply states that the force between two charges acts along the line connecting the two, and that force is proportional to the strength of the two charges q_1 and q_2 and the square of the inverse of the distance between the two:

$$\vec{F} = \frac{1}{4\pi\epsilon_o}\frac{q_1 q_2}{R^2}\hat{R} \tag{4.1}$$

where $\hat{R} = \vec{R}/R$. Note that the constant of proportionality in MKSA units is $(4\pi\epsilon_o)^{-1}$. The same constant is simply 1 in the Gaussian system of units. In MKSA units, $\vec{F}$ is in newtons (N) q is in coulombs (C) and R is in meters (m). The fundamental constant ϵ_o is called the electric permittivity of free space, and it is equal to 8.85×10^{-12} F/m in MKSA units (it is equal to 1 in Gaussian units). As an aside, this constant is sometimes reported with the 4π multiplied by the value just supplied, and it is important to watch for that. A farad (F) is equal to 1C/V. The base units for a farad are $A^2 s^4/kgm^2$. An ampere (A) is the base unit for current, equal to 1 C/s. If these definitions are applied to equation (4.1), the units will work out.

To write a field expression based upon this result, we close a control volume about a region in space that can contain point charges (sources or sinks, see Figure 4.1). Conservation of the electric field is written in terms of the surface integral of over the volume:

$$\int_{Area} \vec{E} \cdot d\vec{A} \tag{4.2}$$

The definition of an electric field is

$$\vec{E} \equiv \frac{\vec{F}}{q} \tag{4.3}$$

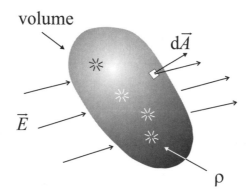

Figure 4.1: Control volume closed about a region in space containing point charges.

(where q is a single "sense charge", in C, that moves about in space). In the MKSA system, the units for electric field are V/m, and the base units for a volt (V) are kg m^2/A s^3.

The surface integral in equation (4.2) can be expressed in terms of charge using Coulomb's law and equation (4.3). Constants come out of the integration, and the surface integral then reduces to an integration of the solid angle $(dA/R^2 = d\Omega)$ over 4π steradians. This 4π then cancels the 4π in equation (4.1). This is a hallmark of the rationalized MKSA system - when 4π is found in an expression, it often cancels a 4π steradians. Allowing for distributed charge, written in terms of a charge density ϱ (in C/m^3), Gauss' law in integral form is

$$\int_{\text{Area}} \vec{E} \cdot d\vec{A} = \frac{1}{\epsilon_\circ} \int_{\text{Vol}} \varrho \, dV \qquad (4.4)$$

One can then convert to a differential form using Stoke's law and the divergence theorem [52] (just as they can be used to express the integral form of the Navier-Stokes equations in differential form):

$$\vec{\nabla} \cdot \vec{E} = \frac{\varrho}{\epsilon_\circ} \qquad (4.5)$$

This conservation expression for the electric field is missing any time derivatives. It may appear to be incorrect for this reason, but it is not a charge conservation expression. Conservation of charge is written in a more recognizable form as

$$\boxed{\frac{\partial \varrho}{\partial t} + \vec{\nabla} \cdot \vec{J} = 0} \qquad (4.6)$$

where J is the current density (A/m^2), not radiance as used in Chapter 3. Conflicting usage of variables is unfortunately quite common. We will revisit the issue of charge conservation after all four equations are developed. The outcome will be a new "displacement current".

Gauss' magnetic law

Gauss' magnetic law is quite similar, except that there are no known magnetic monopoles. Magnetism does not have individual point "charges", so the right-hand side of the associated conservation expression is now zero:

$$\vec{\nabla} \cdot \vec{B} = 0 \qquad (4.7)$$

The magnetic induction $\vec{B}$ has the MKSA units of Wb/m^2 (or T, for Tesla). The base units for a weber (Wb) are $kg\ m^2/A\ s^2$.

Faraday's law of induction

Magnetic flux is defined by

$$\Phi = \int_{\text{Area}} \vec{B} \cdot d\vec{A} \qquad (4.8)$$

If the magnetic flux passing through a closed conducting loop (see Figure 4.2) is constant, then $d\Phi/dt = 0$, and no current is observed in the loop. If the flux varies with time $(d\Phi/dt \neq 0)$, current is produced. Current in the loop depends upon the resistance of the material, which makes it difficult to describe in a general way. A more useful expression, therefore, can be written in terms of electromotive force (EMF). The EMF is defined as the work done on a charge per unit charge (in either J/C or V):

$$\text{EMF} \equiv \frac{W}{q} = \frac{1}{q} \oint_c \vec{F} \cdot d\vec{\ell} = \oint_c \vec{E} \cdot d\vec{\ell} \qquad (4.9)$$

where we have made use of equation (4.3). A proportionality between $-d\Phi/dt$ and EMF has been experimentally observed, and this forms

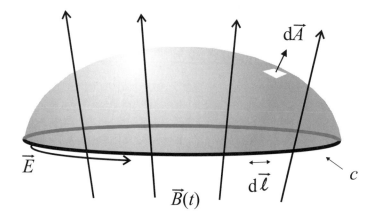

Figure 4.2: Magnetic flux passing through a loop and inducing a current.

the basis for Faraday's law. In MKSA units, we let the constant of proportionality be 1. In integral form the law is then written as

$$\oint_c \vec{E} \cdot \mathrm{d}\vec{\ell} = -\frac{\partial}{\partial t} \int_{\text{Area}} \vec{B} \cdot \mathrm{d}\vec{A} \tag{4.10}$$

One nagging question is how we can "let" a constant of proportionality be simply "1". There are four equations in this set, and they describe the behavior of propagating electromagnetic fields. Our requirement is that Maxwell's equations simply reproduce observation. There are actually several ways to do this. If we fix the constants of proportionality as described for Gauss' laws, then we are free to let the constant be 1 here, as long as we ensure that the units work out. This topic is discussed in detail in Appendix A.

Stoke's law and the divergence theorem can be applied again to give the differential form of Faraday's law:

$$\vec{\nabla} \times \vec{E} = -\frac{\partial \vec{B}}{\partial t} \tag{4.11}$$

Ampere's law of induction

Finally, Ampere's law of induction and the Biot-Savart law are both generalizations based, again, upon empirical observation. Current passing through a conductor can induce a magnetic field, and the two are related by a constant of proportionality. According to the Biot-Savart law, the contribution to the magnetic induction $\vec{B}$ due to a small length of conductor $d\vec{\ell}$ carrying current I (in A) can be expressed as

$$d\vec{B} = \frac{\mu_o}{4\pi} I \frac{d\vec{\ell} \times \vec{R}}{|\vec{R}|^3} = \frac{\mu_o}{4\pi} I \frac{d\vec{\ell} \times \hat{R}}{|\vec{R}|^2} \qquad (4.12)$$

where $\vec{R}$ is the vector from the element to the location at which the field is sensed. The fundamental constant μ_o is called the magnetic permeability of free-space, and it is equal to $4\pi \times 10^{-7}$ (H/m) (or N s^2/C^2) in the MKSA system (it is equal to 1 in the Gaussian system). One henry (H) has base units of kg m^2/s^{2}A^2.

This is a good place to make a note about units. Based upon equation (4.12), a long straight wire will induce a magnetic field strength given by

$$B = \frac{\mu_o}{2\pi} \frac{I}{L} \qquad (4.13)$$

where L is the distance away from the wire. A small, current-conducting element that is held in such a magnetic field will experience a force due to that field, or

$$d\vec{F} = I \, d\vec{\ell} \times \vec{B}(\vec{r}) \qquad (4.14)$$

This is simply the Lorentz force, induced by the magnetic field. If a second current-carrying wire (number 2) is placed in the magnetic field of the first wire (number 1), it will therefore experience a force given by:

$$F_{12} = \frac{\mu_o}{2\pi} \frac{I_1 I_2}{L} l \qquad (4.15)$$

where l is the length of wire. This expression, in fact, is used to define 1 A of current. The constant of proportionality linking the magnetic field to current in expression (4.12) contains a fundamental physical constant (μ_o) which was originally observed to be $4\pi \times 10^{-7}$ H/m. More careful measurements placed it close to this number but not exactly equal to it. Apparently, the value $4\pi \times 10^{-7}$ was simply too tidy to do without. Despite the measurements, μ_o has been defined as this value. In order to accommodate this notion, equation (4.15) is used to define 1 A of current. When two long wires (to minimize end effects) are placed 1 m apart, and the currents passing through both are equal and set to produce a force of 2×10^{-7} N/m, then that current is defined to be 1

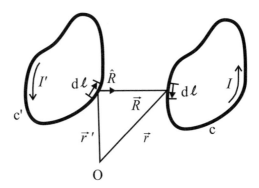

Figure 4.3: Current passing through two different loops. Current I' passing around loop c' induces a magnetic field which then induces current I in loop c (adapted with permission from Wangsness [50]).

A. This allows us to define the magnetic permeability $\mu_\circ$ as described. Since current is also dq/dt, this further defines 1 C. Note that a single electron has a charge of 1.6×10^{-19} C. There is an imbalance here; $\epsilon_\circ$ remains a decidedly messy number while $\mu_\circ$ is adjusted for tidiness. That, however, is what is done.

The pattern used previously to develop Gauss' laws and Faraday's laws can be repeated for the development of Ampere's law if we consider current loops as magnetic field sources (see Wangsness [50] for this approach). The loop labeled by c' in Figure 4.3 generates a magnetic field as a function of position given by

$$\vec{B}(\vec{r}) = \frac{\mu_\circ}{4\pi} \oint_{c'} \frac{I' \, d\vec{\ell}\,' \times \hat{R}}{|\vec{R}|^2} \tag{4.16}$$

where we have simply summed the contributions of small elements around the loop, consistent with the Biot-Savart law [equation (4.12)].

To write a more field-oriented expression for Ampere's law, we integrate the magnetic induction around the unprimed loop (c), using equation (4.16) for $\vec{B}$:

$$\oint_c \vec{B} \cdot d\vec{\ell} = \frac{\mu_\circ}{4\pi} \oint_c \oint_{c'} \frac{I' \, d\vec{\ell} \cdot (d\vec{\ell}\,' \times \hat{R})}{|\vec{R}|^2} \tag{4.17}$$

Another way to say this is that the magnetic field around loop c is somehow related to current in the vicinity of loop c (in the form of I' in Figure 4.3).

Since $\vec{A} \cdot (\vec{B} \times \vec{C}) = (\vec{A} \times \vec{B}) \cdot \vec{C}$, this expression can be rearranged to give

$$\oint_c \vec{B} \cdot d\vec{\ell} = -\frac{\mu_o}{4\pi} I' \oint_c \oint_{c'} \frac{\left((-d\vec{\ell} \times d\vec{\ell'}) \cdot \hat{R}\right)}{|\vec{R}|^2} \qquad (4.18)$$

The inner line integral, around c', can be reduced to the solid angle Ω using geometric arguments, the outcome being the following:

$$\oint_c \vec{B} \cdot d\vec{\ell} = -\frac{\mu_o}{4\pi} I' \Omega \qquad (4.19)$$

The necessary geometric arguments are not reproduced here because they consume a significant amount of text and they can be found in Wangsness [50]. In short, however, if the primed loop remains outside the unprimed loop, then Ω is simply zero, and there is no net contribution to induced magnetic field around the unprimed loop. If, however, the current I' passes through the unprimed loop, then $\Omega = -4\pi$ and

$$\oint_c \vec{B} \cdot d\vec{\ell} = \mu_o I' \qquad (4.20)$$

Finally, if there are multiple current sources, we simply add their individual contributions as a current density $\vec{J}$ (A/m^2) passing across area element dA (see Figure 4.4):

$$\oint_c \vec{B} \cdot d\vec{\ell} = \mu_o \int_{\text{Area}} \vec{J} \cdot d\vec{A} \qquad (4.21)$$

In short, a current passing through an area will induce a magnetic field in a loop that encloses the area. The loop need not be an electrical conductor. The field follows the right-hand rule with respect the direction of current. Some texts treat this statement and equation (4.21) as the most basic expression of Ampere's law, and thus avoid the previous treatment that invoked the Biot-Savart law. We have chosen to include the previous treatment (which has been shortened) because electromagnetic units conventions (discussed in Appendix A) invoke the Biot-Savart law.

Stokes law and the divergence theorem provide a differential version of Ampere's law:

$$\vec{\nabla} \times \vec{B} = \mu_o \vec{J} \qquad (4.22)$$

One should be forewarned that this equation is correct only at steady state. The next section will explain why this is so and produce a correct, generalized expression.

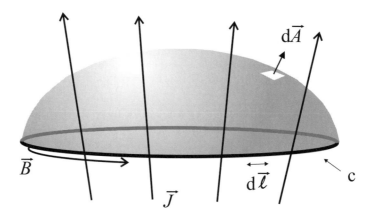

Figure 4.4: Electric current passing through a loop and inducing a magnetic field.

Maxwell's equations

Maxwell realized that Gauss' two laws, Faraday's law, and Ampere's law could unify a field theory that would explain many physical phenomena. He also realized that there was a problem with charge conservation. For example, the divergence of equation (4.22) is

$$\vec{\nabla} \cdot (\vec{\nabla} \times \vec{B}) = \mu_\circ \vec{\nabla} \cdot \vec{J} \qquad (4.23)$$

Since $\vec{\nabla} \cdot (\vec{\nabla} \times \vec{A}) = 0$, then $\vec{\nabla} \cdot \vec{J} = 0$. This is true only for steady state situations. The full expression for charge conservation is given by equation (4.6). There is a problem, therefore, with a missing $\partial \varrho / \partial t$; the unsteady case. Charge density ϱ can be found in Gauss' electric law. Differentiating equation (4.5) with respect to time produces

$$\frac{\partial \varrho}{\partial t} = \epsilon_\circ \vec{\nabla} \cdot \frac{\partial \vec{E}}{\partial t} \qquad (4.24)$$

This indicated to Maxwell that if the current density $\vec{J}$ was replaced by

$$\vec{J} \Longrightarrow \vec{J}_f + \epsilon_\circ \frac{\partial \vec{E}}{\partial t} \qquad (4.25)$$

then charge conservation in Ampere's equation could be assured even in unsteady situations. Maxwell thus introduced the notion of a "displacement current" into Ampere's law. This notion is consistent with physical observation. In a capacitor, for example, when the electric field varies as a function of time, even though there is no current, a magnetic field will be produced. The displacement current is defined by equation (4.24) as

$$\vec{J}_D = \epsilon_o \frac{\partial \vec{E}}{\partial t} \tag{4.26}$$

and so Ampere's law is given by

$$\vec{\nabla} \times \vec{B} = \mu_o \vec{J}_f + \mu_o \epsilon_o \frac{\partial \vec{E}}{\partial t} \tag{4.27}$$

In equation (4.27), we have changed the former $\vec{J}$ to $\vec{J}_f$. The subscript f denotes free current in order to clarify the difference between $\vec{J}_f$ and $\vec{J}_D$. Maxwell's equations in MKSA units are then:

$$\boxed{\vec{\nabla} \cdot \vec{E} = \frac{\varrho_f}{\epsilon_o}} \tag{4.28}$$

$$\boxed{\vec{\nabla} \cdot \vec{B} = 0} \tag{4.29}$$

$$\boxed{\vec{\nabla} \times \vec{E} = -\frac{\partial \vec{B}}{\partial t}} \tag{4.30}$$

$$\boxed{\vec{\nabla} \times \vec{B} = \mu_o \vec{J}_f + \mu_o \epsilon_o \frac{\partial \vec{E}}{\partial t}} \tag{4.31}$$

Other published versions of Maxwell's equations may appear to disagree, which can seem confusing to the uninitiated. This may be due to units (see Appendix A), or one set may be for propagation through vacuum while another is for propagation through transmitting material (see below), or simplifying assumptions may have been used to produce an altered equation set. In one such case, the spatial and time dependence are separated by assuming the temporally harmonic form (the Euler formalism):

$$
\begin{aligned}
\vec{J}_f(\vec{r}, t) &= \vec{J}_f(\vec{r})e^{i\omega t} \\
\vec{E}(\vec{r}, t) &= \vec{E}(\vec{r})e^{i\omega t} \\
\vec{B}(\vec{r}, t) &= \vec{B}(\vec{r})e^{i\omega t}
\end{aligned}
$$

where $i \equiv \sqrt{-1}$. Often this assumption is combined with the notion that a wave propagates through a space where the free-charge density is zero ($\varrho_f = 0$, this is a common occurrence in optics). When the above expressions for $\vec{J}_f$, $\vec{E}$, and $\vec{B}$ are inserted into equations (4.28) to (4.31), while setting $\varrho_f = 0$, we obtain

$$\vec{\nabla} \cdot \vec{E}(\vec{r}) = 0$$
$$\vec{\nabla} \cdot \vec{B}(\vec{r}) = 0$$
$$\vec{\nabla} \times \vec{E}(\vec{r}) = -i\omega\vec{B}(\vec{r})$$
$$\vec{\nabla} \times \vec{B}(\vec{r}) = \mu_\circ\vec{J}_f + i\omega\mu_\circ$$

These expressions look different but in fact are consistent as long as the assumptions are known.

4.3   Basic Conclusions from Maxwell's Equations

The wave equation

Before moving on to material interactions, several useful conclusions can be drawn regarding Maxwell's equations. For optical interactions, we assume that the free charge density in Gauss' electric law, ϱ_f and that the free current in Ampere's law, $\vec{J}_f$ are both zero. Taking the curl of equation (4.30) produces

$$\vec{\nabla} \times \vec{\nabla} \times \vec{E} = -\frac{\partial}{\partial t}(\vec{\nabla} \times \vec{B}) \tag{4.32}$$

Then substitute for $(\vec{\nabla} \times \vec{B})$ from equation (4.31) (recalling that we have set $\vec{J}_f = 0$):

$$\vec{\nabla} \times \vec{\nabla} \times \vec{E} = -\frac{\partial}{\partial t}\left(\mu_\circ\epsilon_\circ\frac{\partial\vec{E}}{\partial t}\right) \tag{4.33}$$

From vector calculus

$$\vec{\nabla} \times \vec{\nabla} \times \vec{E} = \vec{\nabla}(\vec{\nabla} \cdot \vec{E}) - \vec{\nabla}^2\vec{E} \tag{4.34}$$

but $\vec{\nabla} \cdot \vec{E} = 0$ from equation (4.28) (with $\varrho_f = 0$), and hence

$$\boxed{\nabla^2\vec{E} = \mu_\circ\epsilon_\circ\frac{\partial^2\vec{E}}{\partial^2 t}} \tag{4.35}$$

This is the wave equation for an electric field. We can find the wave equation for a magnetic field by performing the same operations on equations (4.29) - (4.31). That will produce

$$\nabla^2 \vec{B} = \mu_\circ \epsilon_\circ \frac{\partial^2 \vec{B}}{\partial^2 t} \qquad (4.36)$$

Note the presence of $\mu_\circ \epsilon_\circ$ in a location normally occupied by $1/(\text{velocity})^2$ in both wave equations. That means that the speed of propagation of an electromagnetic wave through free space ought to be $1/(\mu_\circ \epsilon_\circ)^{1/2}$. Maxwell combined independent measurements of $\mu_\circ$ and $\epsilon_\circ$ and found that this speed matched a measurement of the speed of light conducted by Fizeau [48] at about the same time. This did not immediately convince skeptics, but nowadays the relationship

$$c = \frac{1}{\sqrt{\mu_\circ \epsilon_\circ}} \qquad (4.37)$$

is an accepted expression for the speed of light in vacuum. This is in fact an observation that sets the value of the product $\mu_\circ \epsilon_\circ$.

Plane waves

Simple sinusoidal functions are solutions to the wave equation, and superposition allows one to construct more complex solutions to the same wave equation, using combinations of these basis functions. Two expressions of great utility can be found via the Euler formalism $[\exp(i\theta) = \cos\theta + i\sin\theta, \exp(-i\theta) = \cos\theta - i\sin\theta]$ to produce generalized harmonic solutions:

$$\vec{E} = \text{Re}\left(\vec{E}_\circ e^{i(\vec{k}\cdot\vec{r} - \omega t)} \right) \qquad (4.38)$$

$$\vec{B} = \text{Re}\left(\vec{B}_\circ e^{i(\vec{k}\cdot\vec{r} - \omega t)} \right) \qquad (4.39)$$

where $\vec{E}_\circ$ and $\vec{B}_\circ$ are amplitudes. As an alternative, we can write

$$\vec{E} = \left(\vec{E}_\circ e^{i(\vec{k}\cdot\vec{r} - \omega t)} + c.c. \right) \qquad (4.40)$$

$$\vec{B} = \left(\vec{B}_\circ e^{i(\vec{k}\cdot\vec{r} - \omega t)} + c.c. \right) \qquad (4.41)$$

where *c.c.* denotes the complex conjugate. The goal in equations (4.38) and (4.39) or in equations (4.40) and (4.41) is to recover a real value in order to use the formalism to describe realistic, physical phenomena. Both forms do this, but one should be aware that the outcome will include amplitudes of $\vec{E}_\circ$ and $\vec{B}_\circ$ in the first two equations, but $2\vec{E}_\circ$ and $2\vec{B}_\circ$ will result from the second two.

Sometimes equations (4.38) - (4.41) are written with $(\omega t - \vec{k} \cdot \vec{r})$ in the exponential. The difference between the two is inconsequential. In equations (4.38) and (4.39), the real part never changes when the sign is flipped, because $\cos \theta = \cos(-\theta)$. In equations (4.40) and (4.41), by switching the ωt and $\vec{k} \cdot \vec{r}$, we are simply taking the complex conjugate. Equations (4.40) and (4.41) always add an expression with its own complex conjugate, so the sum does not change when ωt and $\vec{k} \cdot \vec{r}$ are switched. Throughout optics and electromagnetism, the two forms are freely interchanged to make certain calculations more straightforward.

In equations (4.38) - (4.41), time dependence is given by ωt and spatial dependence is provided by $\vec{k} \cdot \vec{r}$. The terms ω, ν, and λ are defined as before. The "wave vector" $\vec{k}$ has magnitude $2\pi/\lambda$ (in order to make the wave spatially harmonic with wavelength λ), and it points normal to the planar wavefronts.

The term $\vec{k} \cdot \vec{r}$ in equations (4.38) - (4.41) defines the spatial properties of a plane wave, for the following reason. Imagine that $\vec{r}_\circ$ defines a fixed vector from the origin and ending on a plane, as depicted in Figure 4.5. A second, variable vector $\vec{r}$ will always end in the same plane if $\vec{k} \cdot (\vec{r} - \vec{r}_\circ) = 0$. Rearranging, this simply says that the expression $\vec{k} \cdot \vec{r} =$ constant defines a plane normal to $\vec{k}$. This constant can change in value as we step along the optical ray that is conjugate with each instantaneous direction vector $\vec{k}$. In fact, if we let $\sin(\vec{k} \cdot \vec{r})$ oscillate with a harmonic wavelength λ, then the field amplitude associated with each plane can change in a harmonic way, as given in equations (4.38) and (4.39). This happens when $|\vec{k}| = 2\pi/\lambda$.

Another way to depict this is to imagine a series of planes propagating along z (then $\vec{k}$ points along z) that extend to infinity in the x and y directions (Figure 4.6a). The field amplitude changes sinusoidally along z as shown in Figure 4.6b. Typically, the amplitude terms in equations (4.38)-(4.41), $\vec{E}_\circ$ and $\vec{B}_\circ$, are dependent upon position (x, y) and time t. Positional dependence allows energy to decrease with distance away

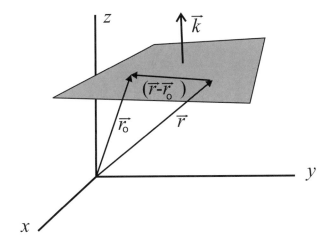

Figure 4.5: Plane wave formalism.

from the axis, in order to conserve energy. The time dependence of the amplitude terms is usually much slower than the time dependence of the sinusoidal terms.

This plane wave formalism is especially useful when we use it as a basis function to describe real problems. As a very simple example, the wavefronts emitted by a laser do approximate plane waves, and we use an amplitude term $[\vec{E}_o(r)$ in this case] to describe Gaussian decay with distance r away from z. This result is an outcome of diffraction theory (see, for example, reference [51]). As a second example, various plane wave solutions can be combined coherently to describe more complex diffraction problems (see, for example, the text by Goodman [53]). In coherent addition, the waves are in a fixed phase relationship, so that constructive and destructive interference are allowed during the super-position of the waves. This description forms the basis for the field of Fourier optics, in which the individual $\vec{k}$ vectors are the Fourier spatial frequencies of the composite field. The rays one uses in geometric optics to describe the propagation of light are normal to the wavefronts, so these rays are conjugate with the $\vec{k}$ vectors at all points in space. Planar wavefronts then have parallel rays (at least locally). The wavefront picture in Figure 4.6a is then another way to describe collimated light.

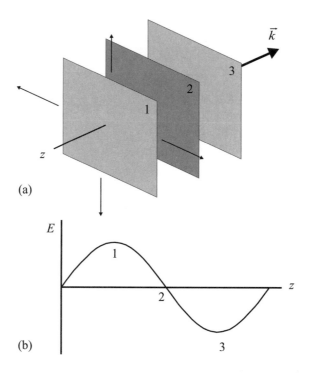

Figure 4.6: Harmonic amplitudes of plane waves.

Transverse electromagnetic waves

To continue the discussion on outcomes of Maxwell's equations (still holding ϱ_f and $\vec{J}_f$ zero) we can insert equation (4.38) into (4.28), and equation (4.39) into (4.29), to obtain

$$\vec{\nabla} \cdot \vec{E} = -i\vec{k} \cdot \vec{E} = 0 \qquad (4.42)$$

$$\vec{\nabla} \cdot \vec{B} = -i\vec{k} \cdot \vec{B} = 0 \qquad (4.43)$$

$\vec{E}$ and $\vec{B}$ are both normal to $\vec{k}$, and $\vec{k}$ points in the direction of plane wave propagation; $\vec{E}$ and $\vec{B}$ are therefore transverse oscillations. Now, inserting equations (4.38) and (4.39) into (4.30) yields

$$\vec{B} = \frac{\vec{k} \times \vec{E}}{c|\vec{k}|} \qquad (4.44)$$

Since $\vec{E}$ is normal to $\vec{k}$, $\vec{E}$ and $\vec{B}$ are normal to each other. Finally

$$|\vec{B}| = \frac{1}{c}|\vec{E}| \qquad (4.45)$$

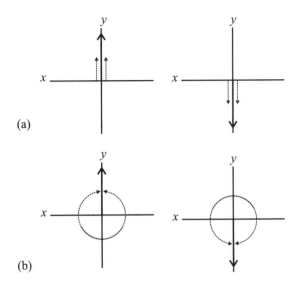

Figure 4.7: Linear polarization (solid vectors) decomposed into two polarization states (dotted vectors).

The conclusion is that $|\vec{E}|$ and $|\vec{B}|$ are proportional to each other and in phase.

The three vectors $\vec{E}$, $\vec{B}$, and $\vec{k}$ form a right-handed coordinate system. The $\vec{E}$ field induces (or propagates) the $\vec{B}$ field and vice versa. Vectors $\vec{E}$ and $\vec{B}$ are polarized. In the discussion of equations (4.38)–(4.41), mention of the fact that $\vec{E}_o$ and $\vec{B}_o$ are actually vectors was not made. Their vector nature is the polarization state of the field. The image to keep in mind, therefore, is a propagating wave with sinusoidally varying electric and magnetic field amplitudes that are normal to each other and to the direction of propagation. It is not enough to describe the field amplitude as a plane, as shown in Figure 4.6. Each of the planes shown has a polarization direction vector that resides in the plane (normal to $\vec{k}$).

The picture we have just described is for a single polarization state, for a single propagating plane wave. Any more complex polarization state of a field can be described by the superposition of two basis polarization states, since the waves are transverse and perpendicular to $\vec{k}$. Linear polarization, for example, can be described by two linear states (Figure 4.7a) or two circular states (Figure 4.7b). Alternatively, circular or elliptic polarization can be described by two circular states

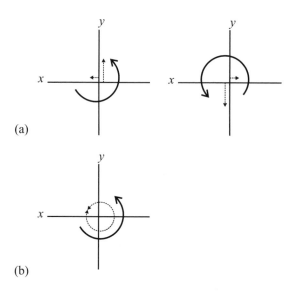

(a)

(b)

Figure 4.8: Circular polarization (solid vectors) decomposed into two polarization states (dotted vectors).

or two linear states (see Figures 4.8a and 4.8b). As an example, if the two linear vectors in Figure 4.8a have the same amplitude and are out of phase with each other by $\pi/2$, they will add coherently to give circular polarization.

Before leaving this section, we develop one more useful expression. If we assume again that the electric field can be described as a harmonic wave, we can describe it using equation (4.38):

$$\vec{E}(\vec{r}, t) = \vec{E}(\vec{r})e^{i(\vec{k}\cdot\vec{r} - \omega t)}$$

If we then insert this into wave equation (4.35), we can cancel the time-dependent exponential terms that fall out and obtain

$$\boxed{\nabla^2 \vec{E} = -k^2 \vec{E}} \tag{4.46}$$

Equation (4.46) is sometimes called the spatial wave equation, or the Helmholtz equation. It describes the spatial dependence of a wave without the time dependence. We will refer back to this result when we justify the form used for the time-independent Schrödinger equation in quantum mechanics.

4.4 Material Interactions

Now we turn our attention to propagation of an electromagnetic wave through transmissive material. The fields in the wave will then interact with the material. There are three basic phenomena that arise. First, conduction relies upon the movement of free electrons in the electric field, and we can express the resultant current by the constitutive relationship $J = \sigma E$, where σ is the electrical conductivity in the medium. Usually, J_f is equated with σE, and for optical interactions, J_f is usually zero.

Next, the material can have an electric polarization state. This forms the basis for spectroscopic interactions. The material polarization is defined as the volume-averaged dipole moment of the individual atomic dipoles (operating within the electric dipole approximation). Electric quadrupoles and magnetic dipoles can exist in a medium but their interactions are much weaker. The material polarization is defined by

$$\vec{P} \equiv \lim_{\Delta V \to 0} \frac{1}{\Delta V} \sum_{i=1}^{N\Delta V} \vec{\mu}_i \qquad (4.47)$$

where ΔV is a volume element; N is the atom or molecule number density, and $\vec{\mu}_i$ is the dipole moment of the individual atom or molecule. It is unfortunate, but $\vec{\mu}_i$ is used for the atomic or molecular dipole moment, and μ is used to represent the magnetic permeability. Sometimes $\vec{p}_i$ is used to represent the dipole moment, to alleviate this problem, but more often one will encounter $\vec{\mu}_i$. Major spectroscopic interactions are electronic, so references to the magnetic permeability will cease as the discussion progresses.

Electric dipoles can exist for several reasons. In a nonresonant interaction, the field can simply displace electrons in an atom or molecule, creating a dipole. Alternatively, the atom or molecule may have an intrinsic dipole and the electric field simply aligns it. Finally, the electric field can temporarily separate ions (NaCl, for example). In a resonant case, the dipole is associated with a spectroscopic transition in the material. There are also nonlinear phenomena (the response of the dipole is not linear with change in electric field amplitude) that can occur if the field strength is large enough.

Whatever phenomena may cause material polarization in the presence of the electric field, they are all represented by an electric susceptibility, χ_e:

$$\boxed{\vec{P} = \epsilon_0 \chi_e \vec{E}}$$
(4.48)

This polarization is then analogous to the displacement current source term in Ampere's law, and it must be located in the same place:

$$\vec{\nabla} \times \vec{B} = \mu_0 \left(\vec{J}_f + \epsilon_0 \frac{\partial \vec{E}}{\partial t} + \epsilon_0 \chi_e \frac{\partial \vec{E}}{\partial t} \right)$$
(4.49)

To be complete, we should consider material magnetization as well. This can arise owing to electron spin states, for example. We write a magnetic displacement current as

$$\vec{J} = \vec{\nabla} \times \vec{M}$$
(4.50)

where magnetization $\vec{M}$ is given by the constitutive relation

$$\vec{M} = \frac{\chi_m}{(1 + \chi_m)\mu_0} \vec{B}$$
(4.51)

Magnetization is usually negligible in optics.

These terms obviously complicate Maxwell's equations. There are two ways to deal with this. First, we can define new terms by

$$\vec{D} = \epsilon_0 \vec{E} + \vec{P}$$
(4.52)

$$\vec{H} = \frac{\vec{B}}{\mu_0} - \vec{M}$$
(4.53)

$\vec{D}$ is called the electric displacement, and $\vec{H}$ is called the magnetic field. These definitions produce the following version of Maxwell's equations:

$$\vec{\nabla} \cdot \vec{D} = \varrho_f$$
(4.54)

$$\vec{\nabla} \cdot \vec{B} = 0$$
(4.55)

$$\vec{\nabla} \times \vec{E} = -\frac{\partial \vec{B}}{\partial t}$$
(4.56)

$$\vec{\nabla} \times \vec{H} = \sigma \vec{E} + \frac{\partial \vec{D}}{\partial t}$$
(4.57)

These look very much like Maxwell's equations in vacuum, except that they now contain $\vec{D}$ and $\vec{H}$.

Another approach is simply to define a new permittivity and permeability for the material by

$$\epsilon = \epsilon_{\circ}(1 + \chi_{\text{e}}) \tag{4.58}$$

$$\mu = \mu_{\circ}(1 + \chi_{\text{m}}) \tag{4.59}$$

These also define the dielectric constant κ_{e} and relative permeability κ_{m} as

$$\kappa_{\text{e}} = (1 + \chi_{\text{e}}) \tag{4.60}$$

and

$$\kappa_{\text{m}} = (1 + \chi_{\text{m}}) \tag{4.61}$$

For optical interactions, we can assume that $\varrho_{\text{f}} = \sigma = M = 0$. We can then rewrite Maxwell's equations for transmitting materials:

$$\vec{\nabla} \cdot \vec{E} = \frac{\varrho_{\text{f}}}{\epsilon} = 0 \tag{4.62}$$

$$\vec{\nabla} \cdot \vec{B} = 0 \tag{4.63}$$

$$\vec{\nabla} \times \vec{E} = -\frac{\partial \vec{B}}{\partial t} \tag{4.64}$$

$$\vec{\nabla} \times \vec{B} = \mu\sigma\vec{E} + \mu\epsilon\frac{\partial \vec{E}}{\partial t} = \mu\epsilon\frac{\partial \vec{E}}{\partial t} \tag{4.65}$$

and by inspection

$$\nabla^2 \vec{E} = \mu\epsilon\frac{\partial^2 \vec{E}}{\partial t^2} \tag{4.66}$$

We now have the wave equation with a new speed in the medium given by $v = (\mu\epsilon)^{1/2}$. The ratio of the speed in vacuum to the speed in the material is then

$$n = \frac{c}{v} = \sqrt{\frac{\mu\epsilon}{\mu_{\circ}\epsilon_{\circ}}} \tag{4.67}$$

where n is the index of refraction. $\vec{M}$ is usually zero for optical interactions (i.e. $\kappa_m = 1$). The index can than be related to other material properties by

$$n^2 = (1 + \chi_e) = \kappa_e \qquad (4.68)$$

At this point it is fairly common to drop the subscript e in χ_e. Indeed, one often finds reference simply to "the susceptibility", without specifying the fact that it is the electric susceptibility under discussion. Note that we can reorganize equation (4.66) to give

$$\nabla^2 \vec{E} - \mu_\circ \epsilon_\circ \frac{\partial^2 \vec{E}}{\partial t^2} = \mu_\circ \frac{\partial^2 \vec{P}}{\partial t^2} \qquad (4.69)$$

On the left-hand side of equation (4.69) we have the wave equation describing the propagation of an electromagnetic wave in vacuum. On the right-hand side we have a source term given by the material polarization. In short, as a non-resonant optical field propagates through a material and interacts with material dipoles, the electric field sets up a material polarization via the susceptibility. This oscillating dipole becomes an emitter, which then propagates additively into the field. A large portion of the field simply bypasses the dipoles, however, because in a transmitting medium they are relatively dispersed.

4.5   Brief Mention of Nonlinear Effects

At this point, it is appropriate to make some mention of nonlinear effects. If the amplitude of the electric field is sufficiently strong, it can induce a nonlinear polarization response. This occurs through the susceptibility:

$$\vec{P} = \epsilon_\circ \left[\chi^{(1)} \vec{E} + \chi^{(2)} \vec{E}^2 + \chi^{(3)} \vec{E}^3 ... \right] \qquad (4.70)$$

Another commonly encountered version of the induced polarization is

$$\vec{P} = \epsilon_\circ \chi^{(1)} \vec{E} + \chi^{(2)} \vec{E}^2 + \chi^{(3)} \vec{E}^3 ... \qquad (4.71)$$

where the $\chi^{(i)}$ for $i = 2, 3...$ are defined differently in the two formalisms. The differences are discussed in more detail in Appendix A.

Each of the terms in equation (4.70) or (4.71) are due to different physical phenomena. These nonlinear susceptibilities are small, and consequently they went unnoticed until lasers with sufficiently large energy were invented. Some of these phenomena can be significantly enhanced by a material resonance and some are not. Second harmonic generation (frequency doubling) and third harmonic generation (tripling) are nonresonant $\chi^{(2)}$ and $\chi^{(3)}$ processes respectively. Degenerate four-wave mixing, on the other hand, is a resonant $\chi^{(3)}$ process.

Nonlinear interactions behave according to a slightly modified wave equation, in which the nonlinear polarization is the source term for a new electromagnetic wave:

$$\nabla^2 \vec{E} - \mu_\circ \epsilon_\circ \frac{\partial^2 \vec{E}}{\partial t^2} = \mu_\circ \frac{\partial^2 \vec{P}_{\mathrm{NL}}}{\partial t^2} \tag{4.72}$$

In frequency doubling of Nd:YAG laser output, for example, the 1.06 μm wavelength radiation from the laser will induce a $\vec{P}_{\mathrm{NL}}$ in the doubling crystal (usually KDP or KTP, depending upon the style of laser). This volumetric $\vec{P}_{\mathrm{NL}}$ acts as a source term for a new electromagnetic wave at 532 nm, according to equation (4.72). Exactly how a $\vec{P}_{\mathrm{NL}}$ is set up to be a source term for the new wave has to do with the physics of the particular interaction in question, of which there are many. A major goal is the arrangement of the nonlinear source dipoles in such a way as to allow their signals to add coherently, thus producing a signal beam of maximum intensity. This is done through phase matching.

Our intention here is to provide a background that can help in understanding nonlinear optics, but a full treatment of nonlinear optics is well outside the scope of this text. Here we deal with the linear response of atoms and molecules to an optical field. Students of nonlinear optics are encouraged to consult one of the available texts on the subject (e.g. Boyd [22]).

4.6 Irradiance

Optical sensors do not respond to the electric field amplitude, they respond to the irradiance of light. If an electromagnetic field model describes the propagation of light, there must be a way to relate this formalism to the optical irradiance to which sensors actually respond.

The energy flux of a propagating electromagnetic wave is given by Poynting's theorem (holding $\vec{M} = 0$ again):

$$\vec{S} = \vec{E} \times \vec{H}^* = \frac{1}{\mu_\circ}\text{Re}[\vec{E} \times \vec{B}^*] \tag{4.73}$$

Here, the asterisk indicates a complex conjugate. If $\vec{E}$ and $\vec{H}$ (or $\vec{B}$) are complex, then, in order to find the magnitude of the cross-product correctly, it is necessary to use the complex conjugate as shown. Using equation (4.44), and taking the real component of $\vec{E}$ and $\vec{B}$ from equations (4.38) and (4.39), the cross-product can be given by

$$\vec{S} = \frac{1}{\mu_\circ}\vec{E}_\circ \times \frac{\vec{k} \times \vec{E}_\circ}{c|\vec{k}|}\cos^2(\vec{k} \cdot \vec{r} - \omega t) \tag{4.74}$$

We never actually measure oscillations at the frequency of light, and so any measurement of energy will necessarily average over rapid oscillations. Such an average produces

$$\overline{\cos^2(\vec{k} \cdot \vec{r} - \omega t)} = 1/2 \tag{4.75}$$

which leads to

$$\overline{\vec{S}} = \frac{\epsilon_\circ c}{2}|\vec{E}_\circ|^2 \hat{s} \tag{4.76}$$

where $\hat{s}$ is the unit Poynting vector. The units of $\vec{S}$ are W/unit area, so $\vec{S}$ is a power vector.

Equations (4.38) and (4.39) [the formalism using $\text{Re}(\vec{E}_\circ e^{i(\vec{k}\cdot\vec{r}-\omega t)})$ for an electric field] are not the only way to represent an electric or magnetic field. Equations (4.40) and (4.41) [the formalism using $(\vec{E}_\circ e^{i(\vec{k}\cdot\vec{r}-\omega t)} + c.c.)$ for an electric field] do so as well. The discussion surrounding those equations stated that "one should be aware that the outcome will include amplitudes of $\vec{E}_\circ$ and $\vec{B}_\circ$ in the first two equations, but $2\vec{E}_\circ$ and $2\vec{B}_\circ$ will result from the second two". This is true if the same $\vec{E}_\circ$ is used in both formalisms. Sometimes an author means $\vec{E}_\circ/2$ in the complex conjugate formalism. In a case where the same $\vec{E}_\circ$ is used in both formalisms, equation (4.76) becomes

$$\overline{\vec{S}} = 2\epsilon_\circ c|\vec{E}_\circ|^2 \hat{s} \tag{4.77}$$

The magnitude of $\vec{S}$ represents the irradiance, and the letter "S" is usually replaced by the letter "I". For the sake of completeness, when equations (4.38) and (4.39) represent the fields, irradiance is given by

$$I = \frac{\epsilon_\circ c}{2}|\vec{E}_\circ|^2 \qquad (4.78)$$

and, when equations (4.40) and (4.41) represent the fields, irradiance is given by

$$I = 2\epsilon_\circ c|\vec{E}_\circ|^2 \qquad (4.79)$$

These expressions hold when $E_\circ$ is defined as shown in equations (4.38) - (4.41). This may seem confusing, but it is incumbent upon author and reader alike to understand which representation for the field is being used, and to remain consistent about it within a specific piece of work. When the time comes to insert specific values for the field strength, it should then be obvious what value to use.

Equations (4.78) and (4.79) are written for vacuum propagation. When waves propagate through dispersive media, these equations are multiplied by n (see Klein and Furtak [51]). As mentioned in Chapter 3, however, spectroscopic interactions are modeled as free space propagation followed by absorption or emission at an atom or molecule (neglecting dispersive effects). For this reason, one will always find vacuum expressions like equations (4.78) and (4.79) in the spectroscopic literature. In addition, conservation of energy dictates that, if a wave passes from vacuum into dispersive media, the value of I specified in equations (4.78) and (4.79) must be preserved at the interface. Unless the index of the dispersive medium equals 1, however, the medium will reflect. Not all of the vacuum energy can enter a truly dispersive medium, but a self-consistent treatment of the two cases will indeed conserve energy. One can either track a vacuum wave and find I as shown, or track a dispersive wave and find I as described in Klein and Furtak. In spectroscopy, again, the vacuum expressions are used because dispersive effects are not of immediate interest.

Note in both equations (4.78) and (4.79) that the irradiance is proportional to the wave amplitude squared $|\vec{E}_\circ|^2$. These expressions connect the parallel formalisms that describe optical interactions—one being the electromagnetic wave equation and the other being the ERT. When they are used to describe the same phenomenon, the two are linked through this expression and they must agree with each other.

An example link between electromagnetism and the ERT

As one demonstration of this last point, it is possible to develop electromagnetic versions of the formalism introduced originally by Partridge and Laurendeau [41], presented in Section 3.2.3 of this book. Settersten and Linne [46] have developed the link between electromagnetic expressions and the ERT formalism for pulsed lasers, as presented here. We begin with the field representation in equations (4.40) and (4.41). The time dependence of the electric field at some fixed location can be represented by a modified version of equation (4.40):

$$E(t) = \left(\tilde{E}_\circ(t) e^{(-i2\pi\nu_\circ t)} + c.c. \right) \tag{4.80}$$

where the ~ over $E_\circ$ indicates a complex quantity for generality, and $\nu_\circ$ is the central frequency of the light. Equations (3.19) and (3.25) can be rearranged to provide the spectral irradiance as a function of frequency and time:

$$I_\nu(\nu, t) = \frac{\mathcal{E}}{A} \frac{T(t) L(\nu)}{\Delta t_{\frac{1}{2}} \Delta \nu_{\frac{1}{2}}} \tag{4.81}$$

where the spatial profile $R(x, y)$ has been divided out (hence the presence of $1/A$) because we are specifically interested in time and frequency response in this example. The connection between the spatial profile of the electric field and the irradiance has been discussed elsewhere [25, 26]. Other variables (e.g. $\Delta \nu_{\frac{1}{2}}$) have been defined in the discussion contained in section 3.2.3.

The temporal irradiance distribution function can be developed using equation (4.79):

$$T(t) = \frac{2\epsilon_\circ c \mid \tilde{E}_\circ(t) \mid^2 \Delta t_{\frac{1}{2}}}{2\epsilon_\circ c \int_{-\infty}^{+\infty} \mid \tilde{E}_\circ(t) \mid^2 dt} = \frac{\mid \tilde{E}_\circ(t) \mid^2 \Delta t_{\frac{1}{2}}}{\int_{-\infty}^{+\infty} \mid \tilde{E}_\circ(t) \mid^2 dt} \tag{4.82}$$

which satisfies the normalization criterion in equation (3.21). The spectral distribution function can be written in an analogous fashion:

$$L(\nu) = \frac{\mid \tilde{E}_\circ(\nu - \nu_\circ) \mid^2 \Delta \nu_{\frac{1}{2}}}{\int_{-\infty}^{+\infty} \mid \tilde{E}_\circ(\nu - \nu_\circ) \mid^2 d\nu} \tag{4.83}$$

which satisfies the normalization criterion in equation (3.22). If the pulse is transform limited, the spectral distribution $\tilde{E}_o(\nu - \nu_o)$ is produced by taking the Fourier transform of $\tilde{E}_o(t)$:

$$\tilde{E}_o(\nu) = \int_{-\infty}^{+\infty} \tilde{E}_o(t) \, e^{i2\pi\nu t} \, \mathrm{d}t \tag{4.84}$$

The pulse energy per unit area is also provided by equation (4.79):

$$\frac{\mathcal{E}}{A} = 2\epsilon_o c \int_{-\infty}^{+\infty} |\, \tilde{E}_o(t) \,|^2 \, \mathrm{d}t \tag{4.85}$$

By inserting these expressions into equation (4.81), the spectral irradiance becomes

$$I_\nu(\nu, t) = 2\epsilon_o c \frac{|\, \tilde{E}_o(t) \,|^2 |\, \tilde{E}_o(\nu - \nu_o) \,|^2}{\int_{-\infty}^{+\infty} |\, \tilde{E}_o(\nu - \nu_o) \,|^2 \, \mathrm{d}\nu} \tag{4.86}$$

Rayleigh's theorem (see reference [46]) states that the integral over all frequency of $|\, \tilde{E}_o(\nu - \nu_o) \,|^2$ is equal to the integral over all time of $|\, \tilde{E}_o(t) \,|^2$. Inserting this into equation (4.86) provides a second expression for spectral irradiance:

$$I_\nu(\nu, t) = 2\epsilon_o c \frac{|\, \tilde{E}_o(t) \,|^2 |\, \tilde{E}_o(\nu - \nu_o) \,|^2}{\int_{-\infty}^{+\infty} |\, \tilde{E}_o(t) \,|^2 \, \mathrm{d}t} \tag{4.87}$$

Without an analytical expression for $\tilde{E}_o(t)$, however, equations (4.86) and (4.87) are not easy to apply. As discussed in Section 3.2.3, a Q-switched laser pulse cannot be described by a simple analytical function. Other pulses (e.g. mode-locked laser pulses) can be described, however, by Gaussian or hyperbolic secant functions. The Gaussian expression is sometimes used as an approximation to a Q-switched laser pulse.

The Gaussian function is given by

$$E_o(t) = E_o \exp\left[-4 \ln 2 \left(\frac{t - t_o}{\Delta t_{\frac{1}{2}}}\right)^2\right] \tag{4.88}$$

where E_o is the peak in the electric field. The energy per unit area for this pulse is then:

$$\frac{\mathcal{E}}{A} = 2\epsilon_o c \, E_o^2 \, \Delta t_{\frac{1}{2}} \sqrt{\frac{\pi}{4 \ln 2}} \tag{4.89}$$

After proper normalization, the distribution functions for the Gaussian pulse are:

$$T(t) = \sqrt{\frac{4 \ln 2}{\pi}} \exp\left[-4 \ln 2 \left(\frac{t - t_o}{\Delta t_{\frac{1}{2}}}\right)^2\right] \tag{4.90}$$

$$L(\nu) = \sqrt{\frac{4 \ln 2}{\pi}} \exp\left[-4 \ln 2 \left(\frac{\nu - \nu_o}{\Delta \nu_{\frac{1}{2}}}\right)^2\right] \tag{4.91}$$

which generates a time-bandwidth product of

$$\Delta t_{\frac{1}{2}} \Delta \nu_{\frac{1}{2}} = \frac{2 \ln 2}{\pi} \tag{4.92}$$

Alternatively, mode-locked laser pulses (50 fs - 100 ps) are usually described with a hyperbolic secant function:

$$E_o(t) = E_o \operatorname{sech}\left[2 \ln(\sqrt{2}+1) \left(\frac{t - t_o}{\Delta t_{\frac{1}{2}}}\right)\right] \tag{4.93}$$

and the energy per unit area for this pulse is

$$\frac{\mathcal{E}}{A} = 2\epsilon_o c\, E_o^2\, \Delta t_{\frac{1}{2}} \frac{1}{\ln(\sqrt{2}+1)} \tag{4.94}$$

After proper normalization, the distribution functions for the hyperbolic secant pulse are

$$T(t) = \ln(\sqrt{2}+1) \operatorname{sech}^2\left[2 \ln(\sqrt{2}+1)\frac{t - t_o}{\Delta t_{\frac{1}{2}}}\right] \tag{4.95}$$

$$L(\nu) = \ln(\sqrt{2}+1) \operatorname{sech}^2\left[2 \ln(\sqrt{2}+1)\frac{\nu - \nu_o}{\Delta \nu_{\frac{1}{2}}}\right] \tag{4.96}$$

which generates a time-bandwidth product of

$$\Delta t_{\frac{1}{2}} \Delta \nu_{\frac{1}{2}} = \frac{4}{\pi^2}\left[\ln(\sqrt{2}+1)\right]^2 \tag{4.97}$$

These results are identical to those presented in Section 3.2.3.

4.7 Conclusions

This chapter has presented an abbreviated, classical electromagnetic formalism related to optical interactions. As just mentioned, it is a somewhat more fundamental treatment of light transmission, but it is a parallel formalism to the ERT. Electromagnetism and the wave equation can explain phenomena that cannot be described by the simpler but more intuitive ERT. Our development of electromagnetism continues in the next section, where we discuss the classical dipole oscillator. The topic is then picked up in the chapter on density matrix equations (Chapter 14), where we connect a quantum dipole with an electromagnetic wave.

Chapter 5

THE LORENTZ ATOM

5.1 Classical Dipole Oscillator

The classical dipole oscillator, or Lorentz atom, is an extremely simple model. Despite this fact, it establishes concepts that will carry throughout this book. Indeed, the notion of an oscillator strength rests on a desire to "tweak" the classical oscillator solution into agreement with measurement, or with quantum calculations.

We begin our treatment of the Lorentz atom using an engineering approach. Newtonian mechanics are applied to a simple "spring-mass-dashpot" problem. We will also assume a dilute, nonpolar and non-magnetic gas. For a single molecule (denoted as molecule j), we modify equation (4.48) as follows:

$$\vec{\mu}_j = \alpha_j \vec{E}_{\mathrm{m}}(\vec{r}_j) \tag{5.1}$$

where $\vec{\mu}_j$ and α_j are the dipole moment and the polarizability, respectively, of individual atom or molecule j; α_j is the microscopic version of the macroscopic term $\epsilon_\circ \chi$. We define $\vec{E}_{\mathrm{m}}(\vec{r}_j)$ as the localized, microscopic electric field; $\vec{E}_{\mathrm{m}}$ can be different from $\vec{E}$ owing to localized phenomena. We assume, however, that $\vec{E}_{\mathrm{m}}$ does not vary around an individual atom. This field then interacts with the simplified atom depicted in Figure 5.1. The atom has a neutral overall charge, and the electron is connected to the large nucleus by both a spring (representing a potential field that allows electron motion but holds the electron within the orbit of the nucleus) and a "dashpot" (an energy loss mechanism). We discuss the need for this loss mechanism below,

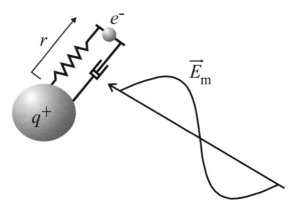

Figure 5.1: Classical electron oscillator.

but for now we will simply say that it replaces the excited-state decay
of a real atom. It is this decay that produces natural broadening of
spectral lines in real atoms and molecules. We then define the equilibrium separation between the electron and nucleus by $r = 0$. This would
correspond to the separation at the zero-point of energy in the actual
potential well. The electric field then provides a forcing function to
which the electron responds (assuming the nucleus is so massive that
only the electron moves). When Newtonian mechanics are applied to
the problem, the equation of motion is

$$m_e \frac{\mathrm{d}^2 \vec{r}}{\mathrm{d}t^2} = -k_s \vec{r} - \frac{m_e}{\tau} \frac{\mathrm{d}\vec{r}}{\mathrm{d}t} + \vec{F} \qquad (5.2)$$

where k_s is the spring constant, τ is the characteristic damping (decay)
time, and m_e is the mass of the electron. In simple dynamics problems,
it is most common to define a damping force by $F_{\text{damping}} = -b \, \mathrm{d}r/\mathrm{d}t$,
and the decay time becomes $\tau = 2m_e/b$. Here, following spectroscopic
conventions, we use $\tau = m_e/b$. The damping factor b has not been
rigorously defined, so the distinction is not critical. Using the definition
of an electric field [equation (4.3)], the force acting on the electron is
$\vec{F} = q_e \vec{E}_m$. We denote the charge by q_e, recognizing that it could be
different from the elementary charge e. The oscillating $\vec{E}$ field literally
drives the dipole into oscillation, with an appropriate amplitude and
phase lag.

Engineers have solved this same equation of motion numerous times.
When there is no damping, the system has a natural oscillation fre-

quency given by

$$\omega_o = \sqrt{\frac{k_s}{m_e}} \tag{5.3}$$

We assume harmonic behavior. Therefore, at the electron, $\vec{E}_m = \vec{E}_{mo}e^{i\omega t}$ and $\vec{r} = \vec{r}_o e^{i\omega t}$. The solution to the equation of motion (5.2) is then

$$\vec{r} = \frac{q_e/m_e}{(\omega_o^2 - \omega^2 + i\omega/\tau)}\vec{E}_m \tag{5.4}$$

The atomic polarization is given by

$$\vec{\mu}_j = q_e\vec{r}_j = \alpha_j\vec{E}_m \tag{5.5}$$

and the atomic polarizability is then

$$\alpha_j = \frac{q_e\vec{r}_j}{\vec{E}_m} \tag{5.6}$$

The volume-averaged dipole moment is written $\vec{P}(\vec{r}) = N\overline{\vec{\mu}_j(\vec{r}_j)}$ (the overbar denotes an average). Using equations (5.4) and (5.5), this macroscopic polarization can be written as

$$\vec{P}(\vec{r}) = \frac{Nq_e^2/m_e}{(\omega_o^2 - \omega^2 + i\omega/\tau)}\overline{\vec{E}_m} \tag{5.7}$$

where the microscopic field around the collection of atoms or molecules involved is averaged. If we assume a very dilute gas, $\vec{E}_m$ becomes $\vec{E}$ (the macroscopic field we have discussed previously), and

$$\tilde{\alpha} = \frac{q_e^2/m_e}{(\omega_o^2 - \omega^2 + i\omega/\tau)} \tag{5.8}$$

where the subscript j has been removed to reduce confusion in what follows. The polarizability is thus a complex number (denoted by a tilde). In order to separate this expression into real and imaginary parts, one can multiply the numerator and denominator by the complex conjugate of the denominator in equation (5.8) (e.g. by $\omega_o^2 - \omega^2 - i\omega/\tau$), to obtain

$$\tilde{\alpha} = \alpha^R + i\alpha^I \tag{5.9}$$

$$\alpha^{\mathrm{R}} = \left(\frac{q_e^2}{m_e}\right)\frac{(\omega_o^2 - \omega^2)}{(\omega_o^2 - \omega^2)^2 + \omega^2/\tau^2} \qquad (5.10)$$

$$\alpha^{\mathrm{I}} = -\left(\frac{q_e^2}{m_e}\right)\frac{\omega/\tau}{(\omega_o^2 - \omega^2)^2 + \omega^2/\tau^2} \qquad (5.11)$$

Moreover, since the dielectric constant is

$$\kappa_e = 1 + \chi_e = 1 + \frac{N\tilde{\alpha}}{\epsilon_o} \qquad (5.12)$$

we find a complex dielectric constant via

$$\tilde{\kappa}_e = \kappa^{\mathrm{R}} + i\kappa^{\mathrm{I}} \qquad (5.13)$$

$$\kappa^{\mathrm{R}} - 1 = \left(\frac{Nq_e^2}{m_e\epsilon_o}\right)\frac{(\omega_o^2 - \omega^2)}{(\omega_o^2 - \omega^2)^2 + \omega^2/\tau^2} \qquad (5.14)$$

$$\kappa^{\mathrm{I}} = -\left(\frac{Nq_e^2}{m_e\epsilon_o}\right)\frac{\omega/\tau}{(\omega_o^2 - \omega^2)^2 + \omega^2/\tau^2} \qquad (5.15)$$

Clearly, we can also write expressions for a complex susceptibility. The real and imaginary parts of the complex susceptibility are traditionally denoted as χ' and χ'' respectively:

$$\tilde{\chi}_e = \chi' + i\chi'' \qquad (5.16)$$

and then

$$\chi' = \left(\frac{Nq_e^2}{m_e\epsilon_o}\right)\frac{(\omega_o^2 - \omega^2)}{(\omega_o^2 - \omega^2)^2 + \omega^2/\tau^2} \qquad (5.17)$$

$$\chi'' = -\left(\frac{Nq_e^2}{m_e\epsilon_o}\right)\frac{\omega/\tau}{(\omega_o^2 - \omega^2)^2 + \omega^2/\tau^2} \qquad (5.18)$$

Finally, since the index of refraction is $n = \sqrt{\kappa_e}$, the index is also complex. We then write $\tilde{n} = n^{\mathrm{R}} + in^{\mathrm{I}}$. Using the approximation $\sqrt{1 + \delta} \cong (1 + \frac{1}{2}\delta)$ for small δ:

$$n^{\mathrm{R}} - 1 = \left(\frac{\omega_p^2}{2}\right)\frac{(\omega_o^2 - \omega^2)}{(\omega_o^2 - \omega^2)^2 + \omega^2/\tau^2} \qquad (5.19)$$

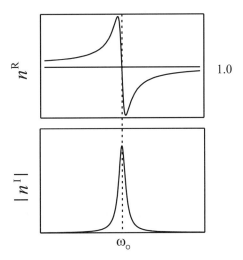

Figure 5.2: Real and imaginary index as a function of detuning from resonance.

$$n^{\mathrm{I}} = -\left(\frac{\omega_{\mathrm{p}}^2}{2}\right) \frac{\omega/\tau}{(\omega_{\mathrm{o}}^2 - \omega^2)^2 + \omega^2/\tau^2} \qquad (5.20)$$

where ω_{p} is called the plasma frequency given by $\omega_{\mathrm{p}} = \sqrt{Nq_{\mathrm{e}}^2/m_{\mathrm{e}}\epsilon_{\mathrm{o}}}$. The name "plasma frequency" can be misleading. In this context, it does not indicate the oscillation of an ionized gas. Indeed, most of the flowfields of interest to readers will not be plasmas at all, but this name is generally applied to $\sqrt{Nq_{\mathrm{e}}^2/m_{\mathrm{e}}\epsilon_{\mathrm{o}}}$.

The real and imaginary components of the index are plotted in Figure 5.2. It is important to point out that equations (5.19) and (5.20) rely upon the linearization $\sqrt{1+\delta} \cong (1 + \frac{1}{2}\delta)$. This does not apply to strong transitions. That will not pose a problem, however, because the simple Lorentz formalism is not applied to spectroscopy without first relating it to measurements or quantum calculations.

One can learn a good deal about atomic response to resonant optical fields from this simple model. It is necessary first, however, to point out that the ω_{o} appearing in the equations for the index is a mechanical system resonant frequency. As such, there is only one for each Lorentz atom, and the functions for the index vary continuously with frequency. This is not really the case for atoms and molecules, which have many discrete resonances corresponding to transitions between quantum levels.

5.2 Wave Propagation Through Transmitting Media

Having made this disclaimer, we now discuss the implications of the result shown in Figure 5.2. To begin, it is necessary to link this index to the propagation of an electromagnetic field. The wave equation for propagation through transparent (although potentially absorbing) media is given in equation (4.66), repeated here:

$$\vec{\nabla}^2 \vec{E} = \mu\epsilon \frac{\partial^2 \vec{E}}{\partial t^2}$$

A general solution is the plane wave expression given by $\vec{E} = \vec{E}_o e^{i(\omega t - \vec{k}\cdot\vec{r})}$. Plugging this into the wave equation, we find that

$$k = \frac{\tilde{n}\omega}{c} \tag{5.21}$$

The propagation constant k is also complex. The plane wave can now be described by

$$\vec{E} = \vec{E}_o e^{n^{\mathrm{I}}\frac{\omega}{c}\hat{k}\cdot\vec{r}} e^{i(\omega t - n^{\mathrm{R}}\frac{\omega}{c}\hat{k}\cdot\vec{r})} \tag{5.22}$$

There are now two portions of this propagation expression. The first exponential is real. Since the imaginary index is negative [as shown in equation (5.20)], this describes a plane wave with an amplitude that decays with propagation distance. If we assume that the wave propagates along z, we can rewrite the propagation expression as

$$\vec{E} = \vec{E}_o e^{n^{\mathrm{I}}\frac{\omega}{c}z} e^{i(\omega t - n^{\mathrm{R}}\frac{\omega}{c}z)} \tag{5.23}$$

We apply the definition of irradiance given above [equation (4.78)] to find

$$I = \frac{\epsilon_o c}{2}|\vec{E}|^2 = \frac{\epsilon_o c}{2}|\vec{E}_o|^2 e^{(2n^{\mathrm{I}}\frac{\omega}{c})z} \tag{5.24}$$

This is Beer's law for absorption again (where $I_o \equiv (1/2)\epsilon_o c\, |\vec{E}_o|^2$), and it agrees with our previous finding [equation (3.82)]. The absorption coefficient for the Lorentz atom is then

$$\kappa \equiv 2\,|n^{\mathrm{I}}|\,\frac{\omega}{c} \tag{5.25}$$

Before simply plugging n^{I} into this expression, we can simplify n^{I}. By doing this, we can also show why there are several variations of the Lorentz atom result in the literature. Starting with the equation $\tilde{n} = \sqrt{\tilde{\kappa}_e}$ and recalling that $\sqrt{1+\delta} \cong (1+\frac{1}{2}\delta)$ for small δ, we can write:

$$\tilde{n} = 1 + \frac{Nq_e^2}{2m_e\epsilon_o} \left[\frac{1}{(\omega_o^2 - \omega^2) + i\omega/\tau} \right] \tag{5.26}$$

Multiply and divide the term in braces by the complex conjugate of the denominator:

$$\frac{(\omega_o^2 - \omega^2) - i\omega/\tau}{(\omega_o^2 - \omega^2) - i\omega/\tau} \tag{5.27}$$

to extract the real and imaginary parts:

$$\tilde{n} = 1 + \frac{Nq_e^2}{2m_e\epsilon_o} \left[\frac{(\omega_o^2 - \omega^2) - i\omega/\tau}{(\omega_o^2 - \omega^2)^2 + \omega^2/\tau^2} \right] \tag{5.28}$$

This can be simplified by separating the squared terms in the square braces:

$$\tilde{n} = 1 + \frac{Nq_e^2}{2m_e\epsilon_o} \left[\frac{(\omega_o - \omega)(\omega_o + \omega) - i\omega/\tau}{[(\omega_o - \omega)(\omega_o + \omega)]^2 + \omega^2/\tau^2} \right] \tag{5.29}$$

Since ω will be in the vicinity of ω_o, we can assume that $\omega + \omega_o \cong 2\omega$. This allows a simplification to

$$\tilde{n} = 1 + \frac{Nq_e^2}{2m_e\epsilon_o\omega} \left[\frac{2(\omega_o - \omega) - i/\tau}{4(\omega_o - \omega)^2 + 1/\tau^2} \right] \tag{5.30}$$

From this we can extract a simplified real and imaginary index:

$$\boxed{n^R - 1 = \frac{Nq_e^2}{4m_e\epsilon_o\omega} \left[\frac{\omega_o - \omega}{(\omega_o - \omega)^2 + 1/(2\tau)^2} \right]} \tag{5.31}$$

$$\boxed{n^I = -\frac{Nq_e^2}{4m_e\epsilon_o\omega} \left[\frac{1/2\tau}{(\omega_o - \omega)^2 + 1/(2\tau)^2} \right]} \tag{5.32}$$

Hence we find

$$\boxed{\kappa \equiv 2|n^I|\frac{\omega}{c} = \frac{Nq_e^2}{4m_e\epsilon_o c} \left[\frac{1/\tau}{(\omega_o - \omega)^2 + 1/(2\tau)^2} \right]} \tag{5.33}$$

Compare this with equation (3.72). There is a clear relationship between the collected terms $(Nq_e^2)/(4m_e\epsilon_o c)$ in equation (5.33) and the

Einstein coefficients, as given in equation (3.72). The term inside the brackets is the lineshape function Y_ν, which is expressed as a Lorentzian here.

Note that the full width at half-maximum (FWHM) of a Lorentzian line is $\Delta\omega = 1/\tau$. Most of the line broadening phenomena one encounters in spectroscopy are associated with some characteristic time, and the line profiles follow the form of equation (5.33). These are homogeneous broadening processes; every atom or molecule experiences the same broadening process, and therefore every atom or molecule responds in the same way at each value of ω. This broadening function is commonly called a Lorentzian profile.

While this expression for the absorption coefficient is not quantitatively correct, it has the right basic form. For this reason, we define an oscillator strength in terms of the Lorentzian, or "classical", absorption coefficient. The oscillator strength is simply a correction to the classical solution to bring it into alignment with measurement or with quantum calculations. Roughly speaking, the oscillator strength adjusts equation (5.33) by

$$\kappa_{\text{actual}} = f_\circ \kappa_{\text{Lorentz}} \tag{5.34}$$

This is discussed in detail in Chapter 10.

The imaginary portion of the index leads to a modification of the wave equation:

$$\vec{E} = \vec{E}_\circ e^{i(\omega t - n^{\text{R}} \frac{\omega}{c} \hat{k}\cdot\vec{r})} \tag{5.35}$$

This is simply a path-length dependent phase shift. The dispersive part of the index is thus doing what is expected in geometric optics—it slows the speed of the wave. If a portion of a plane wave passes through a higher index than another portion of the wave, then the first will be slowed relative to the second. Since the ray is normal to the local surface, the rays of the wave are no longer parallel, and the rays are either diverging or converging in some way. Figure 5.3 illustrates this notion for a simple biconvex lens.

The last paragraph is almost complete. Note that the real part of the index changes rapidly, going below 1 for $\omega > \omega_\circ$, just for a short distance in frequency space. This phenomenon is called anomalous dispersion, because the rate of change in the real index with wavelength is negative in that small region centered about $\omega_\circ$. Away from any

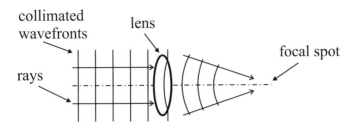

Figure 5.3: Wave propagation through a thin lens.

resonance, the real part of the index increases slowly with wavelength, and so the trend in the vicinity of ω_o is termed anomalous.

This phenomenon does not violate the maxim that wave speeds cannot exceed the speed of light in vacuum, despite the fact that the real portion of the index dips below 1 in Figure 5.2. The reason is presented in several variations in almost every text on this subject. The discussion by Born and Wolf [47] is quite good and it is simply quoted here:

> Now according to the theory of relativity, signals can never travel faster than c. This implies that the phase velocity cannot correspond to a velocity with which the signal is propagated. It is, in fact, **easy** to see that the phase velocity can not be determined experimentally and must therefore be considered to be void of any direct physical significance. For in order to measure this velocity, it would be necessary to affix a mark to the infinitely extended smooth wave and to measure the velocity of the mark. This would, however, mean the replacement of the infinite harmonic wave train by another function of space and time.

The phase velocity of which Born and Wolf speak is the velocity of individual phase fronts. The point they make bears upon the fact that no optical wave is really a plane wave as described by equations (4.38) and (4.39). As mentioned above, such plane waves are an artificial construct, but combinations of these waves can describe real situations. There is always bandwidth to light, even in the most narrow band case. The wave is therefore a mixture of plane waves at various frequencies. The mixture of frequencies then interacts with the dispersion curve shown in Figure 5.2. The dispersion curve can then modify the speed of this group of plane waves (called "group velocity"). It can in fact modify just a portion of the group, but relativity is not violated in an overall sense.

5.3   Dipole Emission

5.3.1   Dipole Emission Formalism

Not only can an electromagnetic wave be absorbed by a classical dipole, an oscillating dipole can emit radiation. This phenomenon is closely related to the Einstein A_{nm} coefficient, which is the excited-state decay rate for an atom or molecule (in the absence of other decay phenomena). Just as we have developed a classical absorption coefficient (related to the Einstein stimulated absorption coefficient B_{mn}), we can develop a classical expression for the spontaneous emission coefficient.

Unfortunately, in order to do this we will need to consider new electromagnetic expressions that are written in terms of potential fields. This development will occupy several paragraphs. Readers who would prefer to skip forward to the results can find them in equations (5.88) and (5.90).

Potential field theory renders some expressions in a form that is simpler to solve in certain geometries. This topic is discussed in detail by Wangsness [50], Jackson [49], and Born and Wolf [47]. Here, we review just the material that relates to the problem at hand. The following should not be taken as a general development of potential field theory, because we make simplifying assumptions as early as we can. We take a few short cuts as well.

Consider a very small volume in space, ΔV, containing a dipole emitter. This volume has charge inside it, given in a general way by $q_e = \varrho \Delta V$. Outside this volume there is no charge, externally imposed current, and so forth. One can apply vacuum electromagnetics to that region. This is an appropriate picture for analysis of an oscillating dipole that emits into free space.

To start, we develop scalar and vector potential field expressions, assuming that the electric and magnetic fields are both space and time dependent. Gauss' magnetic law is given by equations (4.5) and (4.29):

$$\vec{\nabla} \cdot \vec{B}(\vec{r}, t) = 0 \tag{5.36}$$

but, from vector mathematics

$$\vec{\nabla} \cdot (\vec{\nabla} \times \vec{A}) = 0 \tag{5.37}$$

One can therefore define a vector potential field $\vec{A}$ by:

$$\vec{B}(\vec{r}, t) = \vec{\nabla} \times \vec{A}(\vec{r}, t) \tag{5.38}$$

Now consider Faraday's law, equation (4.30):

$$\vec{\nabla} \times \vec{E}(\vec{r}, t) = -\frac{\partial \vec{B}(\vec{r}, t)}{\partial t}$$

Insert equation (5.38):

$$
\begin{aligned}
\vec{\nabla} \times \vec{E}(\vec{r}, t) &= -\frac{\partial}{\partial t}(\vec{\nabla} \times \vec{A}(\vec{r}, t)) \\
&= -\vec{\nabla} \times \frac{\partial \vec{A}(\vec{r}, t)}{\partial t}
\end{aligned}
\tag{5.39}
$$

This can be rearranged to give

$$\vec{\nabla} \times \left[\vec{E}(\vec{r}, t) + \frac{\partial \vec{A}(\vec{r}, t)}{\partial t} \right] = 0 \tag{5.40}$$

From vector calculus, $\vec{\nabla} \times \vec{\nabla}\varphi = 0$, where φ is a scalar field. The term in square braces in equation (5.40) is then equal to the gradient of some scalar field. Equation (5.40) can then be rewritten:

$$\vec{E}(\vec{r}, t) = -\vec{\nabla}\phi(\vec{r}, t) - \frac{\partial \vec{A}(\vec{r}, t)}{\partial t} \tag{5.41}$$

where equation (5.41) is actually a definition of the scalar field $\phi(\vec{r}, t)$. Thus far we have used two of Maxwell's equations (Gauss' magnetic law and Faraday's law) to define one vector potential field and one scalar potential field. Gauss' electric law [equation (4.28)] for a region with charge density is

$$\vec{\nabla} \cdot \vec{E}(\vec{r}, t) = \frac{\varrho}{\epsilon} \tag{5.42}$$

We are purposely leaving the electromagnetic expressions in a general form for now because we will solve for results within and without ΔV. Using equation (5.41) for $\vec{E}$ gives

$$\vec{\nabla} \cdot \left[-\vec{\nabla}\phi(\vec{r}, t) - \frac{\partial \vec{A}(\vec{r}, t)}{\partial t} \right] = \frac{\varrho}{\epsilon} \tag{5.43}$$

and upon rearrangement

$$\nabla^2\phi(\vec{r}, t) + \vec{\nabla} \cdot \frac{\partial \vec{A}(\vec{r}, t)}{\partial t} = -\frac{\varrho}{\epsilon} \tag{5.44}$$

Now we introduce Ampere's law [equation (4.31)]:

$$\vec{\nabla} \times \vec{B}(\vec{r}, t) = \mu \vec{J_f} + \mu\epsilon \frac{\partial \vec{E}(\vec{r}, t)}{\partial t} \tag{5.45}$$

Insert equation (5.38) for $\vec{B}$ and equation (5.41) for $\vec{E}$:

$$\vec{\nabla} \times \vec{\nabla} \times \vec{A}(\vec{r}, t) = \mu \vec{J_f} + \mu\epsilon \frac{\partial}{\partial t} \left[-\vec{\nabla}\phi(\vec{r}, t) - \frac{\partial \vec{A}(\vec{r}, t)}{\partial t} \right] \tag{5.46}$$

From vector calculus, $\vec{\nabla} \times \vec{\nabla} \times \vec{A} = \vec{\nabla}(\vec{\nabla} \cdot \vec{A}) - \nabla^2 \vec{A}$. Introducing this identity and multiplying by -1 produces

$$-\vec{\nabla}(\vec{\nabla}\cdot\vec{A}(\vec{r}, t)) + \nabla^2 \vec{A}(\vec{r}, t) = -\mu\vec{J_f} + \mu\epsilon\vec{\nabla}\frac{\partial\phi(\vec{r}, t)}{\partial t} + \mu\epsilon\frac{\partial^2 \vec{A}(\vec{r}, t)}{\partial t^2} \tag{5.47}$$

which can be rearranged to give

$$\nabla^2 \vec{A}(\vec{r}, t) - \mu\epsilon\frac{\partial^2 \vec{A}(\vec{r}, t)}{\partial t^2} - \vec{\nabla}\left[\vec{\nabla} \cdot \vec{A}(\vec{r}, t) + \mu\epsilon\frac{\partial\phi(\vec{r}, t)}{\partial t}\right] = -\mu\vec{J_f} \tag{5.48}$$

It is useful to rearrange equation (5.44) to look like equation (5.48), by adding and subtracting $\mu\epsilon\,(\partial^2\phi/\partial t^2)$:

$$\nabla^2 \phi(\vec{r}, t) - \mu\epsilon\frac{\partial^2\phi(\vec{r}, t)}{\partial t^2} + \frac{\partial}{\partial t}\left[\vec{\nabla} \cdot \vec{A}(\vec{r}, t) + \mu\epsilon\frac{\partial\phi(\vec{r}, t)}{\partial t}\right] = -\frac{\varrho}{\epsilon} \tag{5.49}$$

These two equations contain all four of Maxwell's equations. Unfortunately, they are coupled through $\vec{A}$ and ϕ in complicated ways. We can separate them using the Helmholtz theorem, which states that $\vec{A}$ is not completely specified until both $\vec{\nabla} \times \vec{A}$ and $\vec{\nabla} \cdot \vec{A}$ have been specified. Thus far we have only specified $\vec{\nabla} \times \vec{A}$, and we are free to specify $\vec{\nabla} \cdot \vec{A}$. A convenient condition would be

$$\vec{\nabla} \cdot \vec{A}(\vec{r}, t) + \mu\epsilon\frac{\partial\phi(\vec{r}, t)}{\partial t} = 0 \tag{5.50}$$

which is termed the Lorentz condition. If we apply the Lorentz condition, we can convert equations (5.48) and (5.49) to three simpler equations:

$$\vec{\nabla} \cdot \vec{A}(\vec{r}, t) + \mu\epsilon\frac{\partial\phi(\vec{r}, t)}{\partial t} = 0 \tag{5.51}$$

$$\nabla^2 \vec{A}(\vec{r}, t) - \mu\epsilon \frac{\partial^2 \vec{A}(\vec{r}, t)}{\partial t^2} = -\mu \vec{J}_{\text{f}} \tag{5.52}$$

$$\nabla^2 \phi(\vec{r}, t) - \mu\epsilon \frac{\partial^2 \phi(\vec{r}, t)}{\partial t^2} = -\frac{\varrho}{\epsilon} \tag{5.53}$$

These equations [combined with equations (5.38) and (5.41)] are equivalent to Maxwell's equations, but they are easier to solve in certain cases.

To solve for the scalar potential field around ΔV, begin with equation (5.53). Outside of ΔV, there is no charge and equation (5.53) becomes

$$\nabla^2 \phi(\vec{r}, t) - \mu\epsilon \frac{\partial^2 \phi(\vec{r}, t)}{\partial t^2} = 0 \tag{5.54}$$

We assume that the scalar potential is symmetric about ΔV since it represents a point source. We thus eliminate the angular dependence in ∇^2 to obtain

$$\frac{1}{r^2} \frac{\partial}{\partial r} \left(r^2 \frac{\partial \phi(r, t)}{\partial r} \right) - \mu\epsilon \frac{\partial^2 \phi(r, t)}{\partial t^2} = 0 \tag{5.55}$$

Now perform a transformation given by

$$\phi(r, t) = \frac{\beta(r, t)}{r} \tag{5.56}$$

to obtain

$$\frac{\partial^2 \beta(r, t)}{\partial r^2} - \mu\epsilon \frac{\partial^2 \beta(r, t)}{\partial t^2} = 0 \tag{5.57}$$

This is a wave equation for β versus r. A general solution would be two counter-propagating waves of the form

$$\beta(r, t) = f \left(t - \frac{r}{c} \right) + g \left(t + \frac{r}{c} \right) \tag{5.58}$$

or

$$\phi(r, t) = \frac{f \left(t - \frac{r}{c} \right)}{r} + \frac{g \left(t + \frac{r}{c} \right)}{r} \tag{5.59}$$

These functions represent spherical waves traveling at speed c. We wish to describe a wave traveling away from the charge, so we simplify the expression to

$$\phi(r, t) = \frac{f \left(t - \frac{r}{c} \right)}{r} \tag{5.60}$$

We now introduce a boundary condition. Equation (5.53) will be solved in the space surrounding ΔV, where there are no sources of charge or current. The solution in this region [expression (5.60)], therefore has to match the result at the surface of ΔV, as r shrinks to zero. We assume a very heavy, positively charged nucleus, and nuclear motion is therefore neglected. A very light, negatively charged electron is bound to this nucleus, but it can oscillate within the potential well that holds it. As a result, the dipole behaves like an accelerating charge, which is the electron in an oscillatory motion. As r shrinks to zero at a specific point in time, therefore, we must find a scalar potential that matches the field due to the single charge at the same point in time. We will still denote the charge by q_e, recognizing that it could be different from the elementary charge e. From equation (4.1), therefore, we can write

$$\vec{E} = \frac{q_e}{4\pi\epsilon_o}\frac{\hat{r}}{r^2} \tag{5.61}$$

We have used a small r to indicate the specific Lorentz atom electronic position, replacing the former and more generic large R. It is simple to show that

$$\left(\frac{\hat{r}}{r^2}\right) = -\nabla\left(\frac{1}{r}\right) \tag{5.62}$$

so the electric field becomes

$$\vec{E} = -\nabla\left(\frac{q_e}{4\pi\epsilon_o r}\right) \tag{5.63}$$

If in this case

$$\phi = \frac{q_e}{4\pi\epsilon_o r} \tag{5.64}$$

then $\vec{E} = -\nabla\phi$. Compare this with equation (5.41). It is clear that we have developed the stationary case for the more general equation (5.41). We therefore need a solution of the form in expression (5.60) that generates expression (5.64) at small r, at some fixed time:

$$\phi(r,t) = \frac{1}{4\pi\epsilon_o r}q_e\left(t - \frac{r}{c}\right) = \frac{\varrho\left(r, t - \frac{r}{c}\right)}{4\pi\epsilon_o r}\Delta V \tag{5.65}$$

The parentheses in this equation represent functional dependence. In words, the potential ϕ at time t depends on the charge at an earlier time $t' = (t - r/c)$, which is called the retarded time. We can find $\phi(r, t)$ by evaluating the integral

$$\phi(r, t) = \frac{1}{4\pi\epsilon_\circ} \int_{\text{Volume}} \frac{\varrho\left(r, t - \frac{r}{c}\right)}{r} dV \tag{5.66}$$

A similar expression can be derived for $\vec{A}$ by starting with equation (5.52):

$$\vec{A}(r, t) = \frac{\mu_\circ}{4\pi} \int_{\text{Volume}} \frac{\vec{J}\left(r, t - \frac{r}{c}\right)}{r} dV \tag{5.67}$$

Now assume that ϱ and $\vec{J}_\text{f}$ vary harmonically with time:

$$\varrho(r, t) = \varrho_\circ(r)e^{-i\omega t}$$
$$\vec{J}_\text{f}(r, t) = \vec{J}_\circ(r)e^{-i\omega t} \tag{5.68}$$

Using the retarded time (represented by the variable t in what follows) and the former definition of the wave vector $k = \omega/c = 2\pi/\lambda$, then

$$\phi(r, t) = \frac{e^{-i\omega t}}{4\pi\epsilon_\circ} \int_{\text{Volume}} \frac{e^{ikr} \varrho_\circ(r)}{r} dV$$

$$\vec{A}(r, t) = \frac{\mu_\circ e^{-i\omega t}}{4\pi} \int_{\text{Volume}} \frac{e^{ikr} \vec{J}_\circ(r)}{r} dV \tag{5.69}$$

It is now necessary to extract $\vec{E}$ and $\vec{B}$ from ϕ and $\vec{A}$. We can then find a solution for the radiation field around ΔV from expressions (5.69), and then find $\vec{E}$ and $\vec{B}$. The Poynting vector will then give power radiated.

To find $\vec{E}$ and $\vec{B}$, start by recognizing that any function that behaves as shown in equation (5.68) (e.g. $\varphi = \varphi_\circ e^{-i\omega t}$) has time derivatives given by

$$\frac{\partial \varphi}{\partial t} = -i\omega\varphi \tag{5.70}$$

This is a generally useful mathematical trick; to take the time derivative of φ, simply multiply by $-i\omega$. The Lorentz condition [equation (5.51)] then becomes

$$\vec{\nabla} \cdot \vec{A} - \mu\epsilon i\omega\phi = 0 \tag{5.71}$$

so

$$\phi = -\frac{i}{\mu\epsilon\omega}\vec{\nabla} \cdot \vec{A} \tag{5.72}$$

This expression can be inserted into equation (5.41), and, by using expression (5.38) with some vector mathematics, we can find $\vec{E}$ as follows:

$$
\begin{aligned}
\vec{E} &= -\vec{\nabla}\phi - \frac{\partial \vec{A}}{\partial t} \\
&= -\vec{\nabla}\phi + i\omega\vec{A} \\
&= \frac{i}{\epsilon\mu\omega}\left[\vec{\nabla}(\vec{\nabla}\cdot\vec{A}) + \omega^2\mu\epsilon\vec{A}\right]
\end{aligned}
\tag{5.73}
$$

We wish to find the field outside ΔV, in vacuum. Therefore, $\epsilon\mu = \epsilon_0\mu_0 = 1/c^2$, and $J_f = 0$. Equation (5.52) is then

$$
\nabla^2\vec{A} - \frac{1}{c^2}\frac{\partial^2\vec{A}}{\partial t^2} = 0
\tag{5.74}
$$

Using the mathematical trick:

$$
\nabla^2\vec{A} + \frac{\omega^2}{c^2}\vec{A} = 0
\tag{5.75}
$$

As before, $\vec{\nabla}\times(\vec{\nabla}\times\vec{A}) = \vec{\nabla}(\vec{\nabla}\cdot\vec{A}) - \nabla^2\vec{A}$, and using equation (5.38) yields

$$
\vec{\nabla}\times\vec{B} = \vec{\nabla}(\vec{\nabla}\cdot\vec{A}) - \nabla^2\vec{A}
\tag{5.76}
$$

leading to

$$
\vec{\nabla}(\vec{\nabla}\cdot\vec{A}) = \vec{\nabla}\times\vec{B} + \nabla^2\vec{A} = \vec{\nabla}\times\vec{B} - \frac{\omega^2}{c^2}\vec{A}
\tag{5.77}
$$

Using these result in equation (5.73), and recalling that $k = \tilde{n}\omega/c$ [equation (5.21)] with $\tilde{n} = 1$ in vacuum, gives

$$
\vec{E} = \frac{ic^2}{\omega}\left[\vec{\nabla}\times\vec{B} - \frac{\omega^2}{c^2}\vec{A} + \frac{\omega^2}{c^2}\vec{A}\right] = i\frac{c}{k}\vec{\nabla}\times\vec{B}
\tag{5.78}
$$

We now have what is required. The vector potential can be found using equation (5.69), following which the magnetic induction is found using equation (5.38), and the electric field is then found using expression (5.78). Unfortunately, the integrals in equation (5.69) are not straightforward. In order to evaluate them, the integrands are expanded in r, paying special attention to the differences between the functional dependences (e.g. $\exp(r)$ versus $1/r$). This decomposition can be found in Wangsness [50]. The outcome is a division of the vector potential into two components, $\vec{A} = \vec{A}_I + \vec{A}_{II}$. As it turns out, the term $\vec{A}_I$ contains the electric dipole term, while $\vec{A}_{II}$ contains magnetic dipole and electric quadrupole terms. In the dipole approximation, it is appropriate to assume that the magnetic dipoles and electric quadropoles can be neglected, so that $\vec{A} = \vec{A}_I$. The $\vec{A}_I$ term is

$$\vec{A}_I = \frac{\mu_\circ e^{i(kr - \omega t)}}{4\pi r} \int_{\text{Volume}} \vec{J}_\circ(r) \mathrm{d}V \tag{5.79}$$

The integral can be evaluated by rearrangement and the use of known expressions. Over the volume ΔV, charge conservation gives

$$\begin{aligned}
\vec{\nabla} \cdot \vec{J} + \frac{\partial \varrho}{\partial t} &= 0 \\
\vec{\nabla} \cdot \vec{J} - i\omega \varrho_\circ &= 0 \\
\vec{\nabla} \cdot \vec{J} &= i\omega \varrho_\circ
\end{aligned} \tag{5.80}$$

Now some mathematics is required. The integral

$$\oint_S \vec{r}(\vec{J}_\circ \cdot d\vec{A}) \tag{5.81}$$

is zero because there is no current in the free space just outside $\Delta V \rightarrow \mathrm{d}V$. From vector mathematics we have

$$\oint_S \vec{r}(\vec{J}_\circ \cdot d\vec{A}) = \int_{\text{Volume}} [(\vec{J}_\circ \cdot \vec{\nabla})\vec{r} + \vec{r}(\vec{\nabla} \cdot \vec{J}_\circ)] \, \mathrm{d}V \tag{5.82}$$

When $(\vec{J}_\circ \cdot \vec{\nabla})\vec{r}$ is broken into its Cartesian components and evaluated, it becomes clear that

$$(\vec{J}_\circ \cdot \vec{\nabla})\vec{r} = \vec{J}_\circ \tag{5.83}$$

which leads to

$$\int_{\text{Volume}} [\vec{J}_\circ + \vec{r}(\vec{\nabla} \cdot \vec{J}_\circ)] \, \mathrm{d}V = 0 \tag{5.84}$$

We then use equation (5.80) to give

$$\int_{\text{Volume}} [\vec{J}_\circ + \vec{r}(i\omega \varrho_\circ)] \, \mathrm{d}V = 0 \tag{5.85}$$

and then

$$\int_{\text{Volume}} \vec{J}_\circ \, dV = -i\omega \int_{\text{Volume}} \varrho_\circ \vec{r} \, dV = -i\omega \vec{\mu}_{\text{ed}} \qquad (5.86)$$

where $\vec{\mu}_{\text{ed}}$ is the electric dipole moment. Finally, equation (5.79) becomes

$$\vec{A} = \vec{A}_I = \frac{-i\mu_\circ \omega \vec{\mu}_{\text{ed}}}{4\pi r} e^{i(kr-\omega t)} \qquad (5.87)$$

This expression is used in equation (5.38) to find $\vec{B}$, and that result is used in equation (5.78) to find $\vec{E}$. In the far-field (which is where we observe emission) we assume that $kr \gg 1$. In the limit of large kr, the magnitude of the Poynting vector is:

$$\boxed{ \mid \overline{\vec{S}} \mid = I(\theta) = \frac{1}{2\mu_\circ} \text{Re}[\vec{E} \times \vec{B}^*] = \frac{\omega^4 |\vec{\mu}_{\text{ed}}|^2}{32\pi^2 \epsilon_\circ c^3} \frac{\sin^2 \theta}{r^2} } \qquad (5.88)$$

This result gives emissive power as a function of angle and radial position. Interestingly, the solution for the vector potential was symmetric about the origin, but $\overline{\vec{S}}$ is not. This result perhaps best illustrates the utility of using vector potentials in a case such as this. The result given by equation (5.88) is in fact a toroidal shape (see Figure 5.4). The hole in the center of the toroid is centered at the axis of the dipole (the axis of polarization). This makes sense: a detector pointing down the axis of a pure oscillating dipole (with collection solid angle $\Omega \sim 0$) will not detect emission. There is no electronic acceleration along the horizontal plane in Figure 5.4. This is a well-known result for Rayleigh scattering. Here, we have modeled the radiation from an oscillating dipole, which can result from resonant or nonresonant induced polarization. The result works for both, as long as we can describe the dipole moment that is induced.

From equation (5.88) we can also find the radiance J (not current) defined implicitly by:

$$I = J d\Omega = J \sin \theta \, d\theta \, d\phi = \frac{\omega^4 |\vec{\mu}_{\text{ed}}|^2}{32\pi^2 \epsilon_\circ c^3} \frac{\sin \theta}{r^2} \qquad (5.89)$$

The total emissive power from a classical oscillating dipole can be found by integrating $\overline{\vec{S}}$ over a closed surface:

$$\boxed{ \int_{\text{Area}} \overline{\vec{S}} \cdot d\vec{A} = P = \frac{\omega^4 |\vec{\mu}_{\text{ed}}|^2}{12\pi \epsilon_\circ c^3} } \qquad (5.90)$$

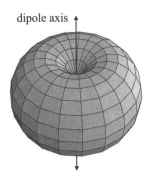

dipole axis

Figure 5.4: Radiation pattern from a simple oscillating dipole.

This solution will reappear when we apply classical Hamiltonian dynamics to the Lorentz atom. Using the rate equations, we can also say that $P = h\nu_{nm}A_{nm}$, which then leads to

$$A_{nm} = \frac{P}{h\nu_{nm}} = \frac{\omega^3|\vec{\mu}_{ed}|^2}{6h\epsilon_\circ c^3} \qquad (5.91)$$

Note that this expression is related to equation (3.71), an expression for the emission coefficient based upon the ERT. This result is also connected to the absorption coefficient via equations (3.72), (3.95) and (3.96). A full comparison of various expressions for resonance response will be presented in Chapter 10.

5.3.2 Dipole Radiation Patterns

The image presented in Figure 5.4 is based purely upon a theory for dipole emission. In general, these results have been confirmed by measurement of time and space averaged emission patterns. Rayleigh scattering from a polarized laser beam, for example, behaves as an ensemble average from single emitters that follow equation 5.88. One could reasonably ask, however, whether individual dipoles actually emit the pattern in Figure 5.4. Recent work by Dickson and co-workers [54, 55] has shown that the pattern in Figure 5.4 is correct for individual dipole emitters (see Figure 5.5).

Figure 5.5 contains images of emission patterns from several dipoles that are oriented in random ways. These dipoles were generated by dye molecules suspended in polymethyl-methacrylate and excited by laser

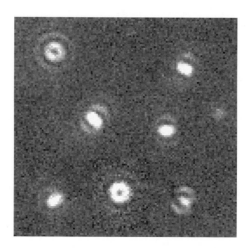

Figure 5.5: Radiation patterns from randomly oriented, individual dye molecules in suspension (reproduced with the permission of Professor Robert Dickson at the Georgia Institute of Technology, see description in text).

radiation with mixed polarization. The images were captured by a microscope objective, as described in references [54] and [55]. They are by nature of the process out of focus, and the authors describe techniques for analyzing the third dimension based upon the degree of defocus. Despite the defocusing in this two-dimensional image, one can clearly identify dipole axes oriented normal to the page; the camera viewed them along the dipole axis. These patterns are visible because the microscope objective had a sufficiently large solid angle Ω that it captured radiation at angles off of the dipole axis. Other images in Figure 5.5 clearly have the dipole axis within the plane of the page.

These images demonstrate conclusively that the classical model presented here successfully describes the spatial dependence of the emitted light. The amplitude terms are then adjusted agree with quantum calculations. As discussed in Chapter 7, the light emitted by a quantum dipole is also a quantized unit, a photon. The notion of a continuously variable emission pattern and single-photon emission may seem contradictory, but in the quantum description the patterns depicted in Figures 5.4 and 5.5 represent an angular distribution of probability density for photon emission. The images in Figure 5.5 thus represent many photons (as they must, in order to generate sufficient signal levels to acquire an image).

5.4 Conclusions

This chapter has completed the classical development of light/matter interaction. Expressions for both an absorption and an emission coefficient can be derived using the electromagnetic formalism. While they are not fully accurate, they are quite useful for describing observed phenomena. Many expressions for the resonance interaction are based upon these results. In fact, one approach to finding the quantum mechanical Einstein A coefficient is to find a quantum mechanical dipole moment (the "transition moment") and insert that into equation (5.91). That result (presented in Chapter 10) agrees with other quantum approaches, one example of which can be found in Chapter 14.

Chapter 6

CLASSICAL HAMILTONIAN DYNAMICS

6.1 Introduction

To this point, the material in this text has been consistent with the educational background of engineers. We have used concepts from, and drawn analogies to, elementary physics, dynamics, transform methods, fluid mechanics, radiation heat transfer, and so on. In order to describe atomic or molecular response to resonant optical fields accurately, however, it is necessary to invoke quantum mechanics. At this point, it is common simply to introduce the Hamiltonian operator as an expression of energy in Scrödinger's equation and move on to elegant solutions. This is typically done as though the Hamiltonian operator and Schrödinger's equation are a perfectly natural extension of classical mechanics. The connection, however, is not entirely obvious. It is the author's view that one can not use the results of quantum mechanics with confidence if the formalism is not fully accepted. This book, therefore, includes sections like this chapter that attempt to justify the quantum approach.

It turns out that the Hamiltonian is as natural as anything can be in quantum mechanics, if the reader has studied classical particle dynamics as taught in a physics course sequence. To cross this boundary between classical Newtonian dynamics and quantum mechanics requires a significant leap, which is even more disjointed if one has no experience with Hamiltonian dynamics. In order to follow more closely the trail of a physicist, therefore, we will briefly describe classical Hamiltonian dynamics and apply it to the Lorentz atom in this chapter. The following treatment is not comprehensive; it is meant simply to introduce the topic. Readers who prefer to forego the development of Hamiltonian dynamics can find the outcome in Section 6.3.

6.2   Overview of Hamiltonian Dynamics

The Hamiltonian formalism is a modification of Lagrangian dynamics. Marion and Thornton [56] describe both in their text on classical dynamics. Neither Lagrangian nor Hamiltonian dynamics introduces new laws of physics. They are an extension of Newtonian mechanics designed to make certain problems more tractable. The Hamiltonian formalism is useful for the description of atoms and molecules because it accounts for the energy of interaction within system constraints (including potential fields). To describe the vector forces acting upon individual electrons and a nucleus, within an ensemble of atoms or molecules, in a Newtonian sense can be very difficult. To describe atomic or molecular scalar energy states within a potential field, however, can be quite straightforward. A statement of Hamilton's principle (excerpted from Marion and Thornton) is:

> Of all the possible paths along which a dynamical system may move from one point to another, within a specified time interval (consistent with any constraints) the actual path followed is that which minimizes the time integral of the difference between the kinetic and potential energies (minimizes the Lagrangian function of the system).

This may sound quite different from the postulates of Newtonian mechanics, but it will always produce results that agree with Newtonian mechanics.

The problem of minimization of some function within a system of constraints is best approached using Lagrange's method. In the terminology of the calculus of variations, Hamilton's principle can be written as

$$\delta \left[\int_{t_1}^{t_2} (T - V) \, \mathrm{d}t \right] = 0 \qquad (6.1)$$

where the term δ [fn] $= 0$ denotes a Lagrangian minimization of "fn", T is kinetic energy, and V is potential energy (an unfortunate choice of variables, but it is the convention).

Here we consider only conservative forces. Nonconservative forces are those associated with a loss term that cannot be recovered, friction for example. Gravitational force, as an example, is conservative. Potential fields can not be defined in terms of nonconservative forces, but they can be easily defined in terms of conservative forces. While it is not entirely obvious from this discussion, Lagrangian mechanics do not apply to situations involving non-conservative forces.

We define particle position in a Cartesian coordinate system by $x_i(t)$ for $i = 1, 2, 3$. Velocity is then defined as $\dot{x}_i = dx_i(t)/dt$. The Lagrangian operator is defined by

$$L \equiv T - V = L(x_i, \dot{x}_i) \tag{6.2}$$

We then seek to minimize the integral

$$\int_{t_1}^{t_2} L(x_i, \dot{x}_i) \, dt \tag{6.3}$$

This is minimized by satisfying the following

$$\frac{\partial L}{\partial x_i} - \frac{d}{dt} \frac{\partial L}{\partial \dot{x}_i} = 0, \qquad i = 1, 2, 3 \tag{6.4}$$

Simple methods in calculus are used to generate this result (see Marion and Thornton [56]). Equation (6.4) is normally called Lagrange's equation, written here for one particle in Euclidean space.

One of the useful aspects of this scalar-based approach is that we can define various coordinate systems without worrying about vector representation. The number of degrees of freedom for a collection of n particles moving in three dimensions is

$$s = 3n - m \tag{6.5}$$

where m is the number of constraint relationships a system may have. We can define s generalized coordinates, denoted by w_j for $j = 1, 2, \ldots, s$ (the variable q is normally used for this coordinate, but that would cause confusion with charge, so we use w in this text). We can then define s velocities by $\dot{w}_j$ for $j = 1, 2, \ldots, s$. These coordinates do not need to be spatial. They could be length squared, energy, or even dimensionless numbers. There are, of course, transformation equations that link the various x_i to the values of w_j. If we say there are $i = 1, 2, 3$ spatial dimensions (as before) and $\alpha = 1, 2, 3 \ldots n$ point particles in the system, then

$$
\begin{aligned}
x_{\alpha,i} &= x_{\alpha,i}(w_1, w_2, w_3 \ldots w_s, t) \quad \text{or} \\
x_{\alpha,i} &= x_{\alpha,i}(w_j, t) \\
\dot{x}_{\alpha,i} &= \dot{x}_{\alpha,i}(w_j, \dot{w}_j, t)
\end{aligned}
$$

The Lagrange equation for a collection of particles in generalized coordinates is then

$$
\frac{\partial L}{\partial w_j} - \frac{\mathrm{d}}{\mathrm{d}t} \frac{\partial L}{\partial \dot{w}_j} = 0, \qquad j = 1, 2, 3 \ldots, s \tag{6.6}
$$

There are s such Lagrange equations. Together with the m constraints, and initial conditions, they completely specify the problem.

The expression for the Hamiltonian $(H = T + V)$ we often encounter in quantum mechanics is an outcome of classical energy conservation. We begin with a discussion of kinetic energy. The total system kinetic energy written in Cartesian coordinates is

$$
T = \frac{1}{2} \sum_{\alpha=1}^{n} \sum_{i=1}^{3} m_\alpha \dot{x}_{\alpha,i}^2 \tag{6.7}
$$

but

$$
x_{\alpha,i} = x_{\alpha,i}(w_j, t)
$$

and

$$
\dot{x}_{\alpha,i} = \sum_{j=1}^{s} \frac{\partial x_{\alpha,i}}{\partial w_j} \dot{w}_j + \frac{\partial x_{\alpha,i}}{\partial t}
$$

If the Cartesian frame itself is not time dependent (as one would expect), then $\partial(x_{\alpha,i})/\partial t = 0$, giving

$$
T = \frac{1}{2} \sum_{\alpha} \sum_{i} \sum_{j,k} m_\alpha \frac{\partial x_{\alpha,i}}{\partial w_j} \frac{\partial x_{\alpha,i}}{\partial w_k} \dot{w}_k \dot{w}_j \tag{6.8}
$$

We can define a variable $a_{j,k}$ by incorporating the summations over α and i to give

$$T = \sum_{j,k} a_{j,k} \dot{w}_k \dot{w}_j \tag{6.9}$$

At this point we can generate an expression that will be useful later, so the discussion will diverge for a moment. Differentiate T with respect to $\dot{w}_\ell$:

$$\frac{\partial T}{\partial \dot{w}_\ell} = \sum_k a_{\ell,k} \dot{w}_k + \sum_j a_{j,\ell} \dot{w}_j \tag{6.10}$$

Multiply by $\dot{w}_\ell$ and sum over all values of ℓ:

$$\sum_\ell \dot{w}_\ell \frac{\partial T}{\partial \dot{w}_\ell} = \sum_{k,\ell} a_{\ell,k} \dot{w}_k \dot{w}_\ell + \sum_{j,\ell} a_{j,\ell} \dot{w}_j \dot{w}_\ell = 2T \tag{6.11}$$

With equation (6.11) in hand, we can address full energy conservation. The total time derivative of L is

$$\frac{dL}{dt} = \sum_j \frac{\partial L}{\partial w_j} \dot{w}_j + \sum_j \frac{\partial L}{\partial \dot{w}_j} \ddot{w}_j \tag{6.12}$$

We know that

$$\frac{\partial L}{\partial w_j} = \frac{d}{dt} \frac{\partial L}{\partial \dot{w}_j} \tag{6.13}$$

from Lagrange's equation. Equation (6.12) is then

$$\frac{dL}{dt} = \sum_j \frac{d}{dt} \left(\dot{w}_j \frac{\partial L}{\partial \dot{w}_j} \right) \tag{6.14}$$

Collecting everything on the left-hand side:

$$\frac{d}{dt} \left(L - \sum_j \dot{w}_j \frac{\partial L}{\partial \dot{w}_j} \right) = 0 \tag{6.15}$$

We can therefore state that the term in parentheses is constant in time. We call it "$-H$". If the potential energy is velocity independent, then $\partial V / \partial \dot{w} = 0$, which leads to the following sequence [making use of equation (6.11) and remembering that $L = T - V$]:

$$\frac{\partial L}{\partial \dot{w}_j} = \frac{\partial T}{\partial \dot{w}_j}, \text{ then}$$

$$\left(L - \sum_j \dot{w}_j \frac{\partial L}{\partial \dot{w}_j} \right) = T - V - \sum_j \dot{w}_j \frac{\partial T}{\partial \dot{w}_j}$$

$$= T - V - 2T = -T - V = -H$$

thus

$$\boxed{H = T + V} \tag{6.16}$$

The total energy of the system is constant if coordinate transformations are independent of time, and potential energy is independent of velocity. Both requirements are met in simple models for atomic or molecular systems.

The classical Hamiltonian formalism is a simple transformation of equations, converting the single, second-order Lagrange equation into two first-order Hamiltonian equations. First, since V is not a function of velocity, the particle momentum p_j is given by

$$p_j = \frac{\partial L}{\partial \dot{w}_j} \tag{6.17}$$

Lagrange's equation is written as

$$\frac{\mathrm{d}}{\mathrm{d}t} \left(\frac{\partial L}{\partial \dot{w}_j} \right) = \frac{\partial L}{\partial w_j}, \quad \text{so } \dot{p}_j = \frac{\partial L}{\partial w_j} \tag{6.18}$$

We can write the differential of H two different ways. First, because we can write $H = H(w_k, p_k, t)$, the differential of H is

$$\mathrm{d}H = \sum_k \left(\frac{\partial H}{\partial w_k} \, \mathrm{d}w_k + \frac{\partial H}{\partial p_k} \, \mathrm{d}p_k \right) + \frac{\partial H}{\partial t} \, \mathrm{d}t \tag{6.19}$$

Second, from equations (6.16) and (6.17) we know that

$$H(w_k, p_k, t) = \sum_j p_j \dot{w}_j - L(w_k, \dot{w}_k, t) \tag{6.20}$$

which, in differential form, is

$$\mathrm{d}H = \sum_k \left(\dot{w}_k \, \mathrm{d}p_k + p_k \, \mathrm{d}\dot{w}_k - \frac{\partial L}{\partial w_k} \, \mathrm{d}w_k - \frac{\partial L}{\partial \dot{w}_k} \, \mathrm{d}\dot{w}_k \right) - \frac{\partial L}{\partial t} \, \mathrm{d}t$$

$$= \sum_k (\dot{w}_k \, \mathrm{d}p_k - \dot{p}_k \, \mathrm{d}w_k) - \frac{\partial L}{\partial t} \, \mathrm{d}t \tag{6.21}$$

Comparing coefficients of dw_k, dp_k, and dt in equations (6.19) and (6.21), we find

$$\boxed{-\dot{p}_k = \frac{\partial H}{\partial w_k}} \tag{6.22}$$

$$\boxed{\dot{w}_k = \frac{\partial H}{\partial p_k}} \tag{6.23}$$

which are Hamilton's equations of motion, and they can replace Lagrange's equation. To solve these equations, one constructs an expression for H in terms of w_k and p_k (e.g. position and momentum). This can be done directly, or one can solve for L to obtain p_k from $\partial L/\partial w_k$. Then Hamilton's equations (6.22) and (6.23) are solved. Often it is easier to simply use Lagrange's equation, but one advantage of Hamiltonian mechanics is that w_k and p_k are independent quantities. Lagrangian dynamics use w_k and $\dot{w}_k$, which are coupled.

There are cases where it is actually easier to use Lagrangian or Hamiltonian rather than Newtonian formalisms. It works well for atomic representation because it uses scalar potential fields.

Before applying Hamiltonian dynamics to an example problem, we use equations (6.22) and (6.23) to develop a classical expression that will parallel a development in quantum mechanics. If we have some generalized function of momentum, position and time given by $f(p, w, t)$ then we can write

$$df = \frac{\partial f}{\partial p}\, dp + \frac{\partial f}{\partial w}\, dw + \frac{\partial f}{\partial t}\, dt \tag{6.24}$$

Taking a differential with respect to time:

$$\frac{df}{dt} = \frac{\partial f}{\partial p}\frac{dp}{dt} + \frac{\partial f}{\partial w}\frac{dw}{dt} + \frac{\partial f}{\partial t} \tag{6.25}$$

Then use equations (6.22) and (6.23) for $\dot{p}$ and $\dot{w}$ to produce

$$\frac{df}{dt} = -\frac{\partial f}{\partial p}\frac{dH}{dw} + \frac{\partial f}{\partial w}\frac{dH}{dp} + \frac{\partial f}{\partial t} \tag{6.26}$$

The first two terms are written as

$$\boxed{\{H, f\} \equiv \frac{dH}{dp}\frac{\partial f}{\partial w} - \frac{\partial f}{\partial p}\frac{dH}{dw}} \tag{6.27}$$

This is termed a "Poisson bracket" and it is the classical analog to the quantum-mechanical commutator. Equation (6.26) is then written:

$$\boxed{\frac{\mathrm{d}f}{\mathrm{d}t} = \{H, f\} + \frac{\partial f}{\partial t}} \tag{6.28}$$

We will encounter a strikingly similar expression in quantum mechanics.

6.3   Hamiltonian Dynamics and the Lorentz Atom

Consider again the Lorentz atom as depicted in Figure 5.1. We can apply Hamiltonian dynamics to this problem if we remove the friction term. Hamiltonian dynamics apply only to conservative forces, and the Lorentz atom is a mechanical model. It may seem at this point that Hamiltonian dynamics have just lost their utility, but one can find a way to describe the energy transfer and this example will demonstrate that. First, we write the Hamiltonian for a modified Lorentz atom which has no damping (remove the dashpot in Figure 5.1). The Hamiltonian is then

$$H = T + V = \left(\frac{p^2}{2m_\mathrm{e}}\right) + \left(\frac{k_s r^2}{2} - q_\mathrm{e}\vec{r}\cdot\vec{E}(t,\vec{r})\right) \tag{6.29}$$

because the potential energy deposited into a dipole by an electric field is $-q_\mathrm{e}\vec{r}\cdot\vec{E}(t,\vec{r})$. Furthermore, $w = r$ and $\dot{w} = \dot{r}$. Using equations (6.23) and (6.29), we obtain

$$\dot{r} = \frac{\partial H}{\partial p} = \frac{p}{m_\mathrm{e}} \tag{6.30}$$

or

$$p = m_\mathrm{e}\dot{r} \tag{6.31}$$

Momentum is thus mass times velocity, which is consistent with Newtonian mechanics. In addition we have

$$\dot{p} = m_\mathrm{e}\ddot{r} \tag{6.32}$$

which is again consistent with Newtonian mechanics. Now, using equation (6.22) gives

$$-\dot{p} = -m_\mathrm{e}\ddot{r} = \frac{\partial H}{\partial r} = k_s r - q_\mathrm{e}E \tag{6.33}$$

which can be rearranged to give

$$\ddot{r} + \omega_o^2 r = \frac{q_e}{m_e} E \tag{6.34}$$

where $\omega_o = \sqrt{k_s/m_e}$ again. This expression is the same as equation (5.2) when damping is ignored. The process occurs along a linear path parallel to the electric field polarization, and so the vectors in equation (5.2) have become scalars here.

There is a problem in that a dipole, once set into oscillation, will emit electromagnetic radiation and hence lose the energy that was deposited. The Lorentz model as described in equation (6.34) is therefore not physically correct. We can estimate the rate of energy loss (after Allen and Eberly, [20]) and use that to modify the equation of motion. This energy loss term is then fully characterized even though it is non-conservative. We do this by closing a small spherical control volume around the dipole and accounting for the energy flow:

$$\int_{Area} \vec{S} \cdot \vec{n} \, dA + \frac{\partial W_{em}}{\partial t} + \frac{\partial W_{osc}}{\partial t} = 0 \tag{6.35}$$

where $\vec{S}$ is the Poynting vector, W_{em} is the energy of the electromagnetic field, and W_{osc} is the energy of the oscillator. The rate at which energy is lost by the dipole is assumed to be slow relative to the rate of dipole oscillation ω_o. Then the amount of electromagnetic energy in the control volume can be assumed steady relative to the dipole lifetime, and $\partial W_{em}/\partial t \sim 0$. Since energy is lost slowly relative to dipole oscillations, we can also assume that the oscillations are harmonic:

$$W_{osc}(t) \cong m_e \omega_o^2 \, \overline{r^2(t)} \tag{6.36}$$

where $\overline{r^2(t)}$ has been averaged over a large number of oscillations at $2\omega_o$. Equation (5.90) gives the rate of energy loss by an oscillating electric dipole via radiation. Here, we insert it into a modified equation (6.35) to obtain

$$\int_{Area} \vec{S} \cdot \vec{n} \, dA = -\frac{\partial W_{osc}}{\partial t} = \frac{\omega_o^4 q_e^2}{12\pi\epsilon_o c^3} \, \overline{r^2(t)} \tag{6.37}$$

Taking $\overline{r^2(t)}$ from equation (6.36) produces

$$\int_{Area} \vec{S} \cdot \vec{n} \, dA = -\frac{\partial W_{osc}}{\partial t} = \frac{\omega_o^2 q_e^2}{12\pi m_e \epsilon_o c^3} W_{osc}(t) \tag{6.38}$$

In words, equation (6.38) states that the rate of energy loss from the oscillating dipole is proportional to the energy stored in the dipole. We can thus simplify this conclusion by defining a decay time $\tau_\circ$ via

$$\frac{\partial W_{\text{osc}}}{\partial t} = -\frac{2}{\tau_\circ} W_{\text{osc}}(t) \tag{6.39}$$

which produces

$$W_{\text{osc}}(t) = W_{\text{osc}}(0)e^{-2t/\tau_\circ} \tag{6.40}$$

The characteristic energy decay rate is given by combining equations (6.38) and (6.39):

$$\frac{2}{\tau_\circ} = \frac{\omega_\circ^2 q_e^2}{12\pi m_e \epsilon_\circ c^3} \tag{6.41}$$

Using this expression, it is easy to show that $1/\tau_\circ \ll \omega_\circ$, which was the starting assumption. The characteristic decay rate in equation (6.41) can be inserted into the Hamiltonian-based Lorentz expression [equation (6.34)] as a classical damping term:

$$\ddot{r} + \frac{2}{\tau_\circ}\dot{r} + \omega_\circ^2 r = \frac{q_e}{m} E \tag{6.42}$$

This expression is identical to equation (5.2) if we insert $\mathbf{F} = q_e E$ into (5.2).

6.4   Conclusions

This development has demonstrated several points. First, the Hamiltonian formalism is entirely consistent with the earlier Newtonian treatment. Second, the decay time introduced in the Newtonian approach has been linked again to electromagnetic emission from the dipole. We have also demonstrated that energy loss at the atomic or molecular level can be accounted for in a Hamiltonian approach. Finally, and most importantly, both energy and momentum have figured greatly in this treatment. They are the natural variables for the description of this problem—an isolated body under the influence of potential fields. They will prove equally important in the quantum-mechanics treatment, where by the principle of correspondence we convert the Hamiltonian in equation (6.16) into an operator.

Chapter 7

AN INTRODUCTION TO QUANTUM MECHANICS

7.1 Introduction

In order to describe atomic or molecular response to optical fields accurately, it is necessary to apply quantum mechanics. Unfortunately, there are no deterministic steps that can lead from classical physics (as used in previous chapters) to quantum mechanics, nor are there direct observations of subatomic behavior upon which theories can be built. In the words of Max Planck, the development of a quantum approach was "an act of desperation ... a theoretical interpretation had to be found at any cost, no matter how high" [57]. Niels Bohr said that "anyone who is not shocked by quantum theory has not understood it" [57]. These are strong words, but they reflect the fact that the postulates of quantum mechanics were (and usually are) difficult to accept, because they depart from the classical style of theoretical development.

Classical mechanics has always relied upon direct observation to develop and test theories. This is so ingrained that we often use anthropomorphic imagery, saying that systems "behave" in certain ways or "want to do" something, as though they do it of their own volition. In classical mechanics, the behavior of a macroscopic system is caused by observable conditions at known locations and/or a specified time. The system behavior can be described exactly if the system and these causes are known. It is a completely deterministic approach. Research in the early part of the twentieth century proved, however, that it is not possible to transfer notions from the observed, macroscopic world into the subatomic realm. An entirely new formalism is required. In the 1920's especially, developments in the new theory of quantum mechanics came rapidly but they were driven by flashes of insight that even their hosts found difficult to accept, because quantum mechanics necessarily deals with phenomena that are not directly observable. Eventually, even the notion of causality was abandoned in favor of a more probabilistic approach. On a philosophical level, Albert Einstein found this departure especially troubling, arguing that "(God) does not play dice" [57].

Engineers are typically trained just up to this point of departure from classical mechanics. A baccalaureate-level engineer who needs to understand quantum mechanics can actually find himself or herself at the same philosophical break point that Einstein, Bohr, Heisenberg, Schrödingier and the others found difficult. These historic figures understood that the foundations for quantum mechanics were a significant departure. They discussed this dilemma into the late hours, holding spontaneous technical sessions even in hotel lobbies. At times the arguments became corrosive. Despite these conceptual difficulties, however, quantum theory continues to describe what indirect observations we can make regarding the subatomic realm. In addition, one can move from quantum mechanics into accepted classical mechanics by taking the quantum expressions to the macroscopic limit. To use a theory that has proven successful time and again, despite a few philosophical questions, is quite practical. Engineers often do this and ought to feel comfortable with it, so long as the foundations and limitations of the approach are well understood.

This presentation will begin with a short description of this historic departure from classical mechanics (following Baggott [57], Schwabl [58] and van der Waerden [59]), because it is very well matched to an engineering background. When the opportunity presents itself, we demonstrate that the theory is internally consistent and mathematically rigorous and that, when extended to the macroscopic scale, it reproduces classical mechanics. Quantum mechanics is actually a very large field. Many good texts on the subject exist. The books by Baggott [57], van der Waerden [59], Levine [60], Schwabl [58], Sakurai [61], Boyd [22], and Brandt and Dahmen [62] are especially useful. What is presented here is actually a minute (and dated) subset of the field.

In order to contain the size of the presentation, it will be necessary at times to omit details. Because the subject is not deterministic, the following presentation will seem somewhat disjointed. In truth, it is a collection of justifications, viewpoints, and useful ideas. The goal underlying this presentation style is to make first-time users sufficiently comfortable with the topic that they can use the outcomes relevant to spectroscopic techniques with confidence.

7.2 Historical Perspective

Some of the more compelling discoveries and developments that fostered what became known as the "Copenhagen interpretation" of quantum theory are briefly described here. The Copenhagen interpretation is the conventional view of quantum mechanics one typically finds in textbooks. Our purpose in this presentation is to justify the need for a new, non-classical formalism, and to provide some description of the foundations of quantum theory that are used in spectroscopy.

The Planck distribution

Perhaps the best place to start is with Planck's spectral distribution for electromagnetic radiation from a black body, a topic understood by engineers trained in heat transfer. At the beginning of the twentieth century, physicists were struggling to develop a fundamental description of the spectral distribution for radiated energy from thermal sources. In the year 1900, the German physicist Max Planck found a mathematical form that could describe the black-body spectrum over all of the temperatures and wavelengths necessary. His first description was, however, purely phenomenological. In order to put this description on a theoretical foundation, he was forced to use the statistical description proposed by Ludwig Boltzmann. Planck did not agree with Boltzmann's statistical approach, but it was the only avenue that would lead to the appropriate form (number density of photons at specific energy levels in this case, see Section 3.5 in this book for a description of how statistical mechanics leads to the correct form). His distaste for statistics led to Planck's "act of desperation" statement. The requirement for a statistical description of discrete optical quanta (photons) then reintroduced the notion of a "corpuscular" form of light.

Later, in 1917, Albert Einstein demonstrated that the Planck distribution can be obtained by enforcing LTE for a gas in a radiation field [59] (just the inverse of Planck's treatment). He also showed that the emission process for a single emitter could not be omnidirectional, as depicted in Figure 5.4. In order to maintain thermodynamic equilibrium, the individual emission process has to be unidirectional. The atom or molecule emits a photon with a defined direction of propagation, but the aggregate emission from a statistical ensemble of emitters must in fact have the pattern depicted in Figure 5.4. The pattern in Figure 5.4 is the probability density distribution for direction of single-photon emission. Time and again we will find that the single photon/atom event can seem inconsistent with classical expectations, but a statistical sample will produce a result that matches classical expectations.

The photoelectric and Compton effects

It was Einstein who successfully used the notion of photons in 1905 to explain the photoelectric effect, using Planck's assertion that each photon had an energy given by $h\nu$. In the photoelectric effect, wavelength, not irradiance, controls the threshold of photoelectron emission, and this photon energy argument successfully explained the process. The quantization of light was further demonstrated in 1923, in an experiment by Arthur Compton during which photons collided with electrons. Particle-based collisional conservation of momentum was used successfully to explain the outcome.

Through these developments (the Planck distribution, the photoelectric effect, and the Compton experiment), the photon theory had reached the point where it was undeniable, but it did not invalidate the wave nature of light. Too many discoveries had clearly identified a wave phenomenon to abandon such an idea. The notions of waves and particles were therefore required to coexist.

Orbital theories and spectra

In 1912 Niels Bohr adapted an orbital geometry previously proposed by Ernest Rutherford, to describe electron energy levels in a hydrogen atom. This formalism was able to describe the Balmer series of emission lines in the H spectrum, because he allowed seemingly unphysical quantum jumps between specific energy orbitals. These orbitals were delineated by assigned quantum numbers that corresponded to sequences in the observed hydrogen spectrum. The emission spectrum was then determined by differences between energy levels in the assigned orbitals. Bohr equated this energy level difference to the photon energy $h\nu$, which was a new assertion. Prior to this, the optical emission frequency was thought to correspond in some way to a mechanical oscillation frequency within the atom. While the Bohr theory failed to describe the spectra of multi-electron atoms, it accurately reproduced the relative spectral spacings in the Balmer series, and was therefore provisionally accepted.

deBroglie waves

In 1923, building upon Planck's theory and the Compton experiment, and upon Einstein's new theory of relativity, Louis de Broglie suggested that if photons possess momentum, then they must have a finite rest mass. The logical extension of this is that primary particles of mass (e.g. an electron) are likely to have an intrinsic wavelength. This wavelength would then be given by

$$ h\nu = pc, \quad \text{or } \lambda = \frac{h}{p} \tag{7.1}$$

where p is the particle momentum. Note that de Broglie used the relativistic total energy mc^2, not the kinetic energy. The reason that wave-like behavior is not observable in macroscopic masses has to do with the fact that h is very small. Furthermore, de Broglie suggested that one could do an electron scattering experiment and observe diffraction patterns. In 1925, Clinton Davisson and Lester Germer obtained a diffraction pattern from an electron scattering experiment, although that was not their original intent. A contemporary electron diffraction pattern can be found in Figure 7.1, which was originally published by Tonomura *et al.* [63]. The image was generated by electron diffraction through a double slit apparatus (similar to Young's optical diffraction experiment). The first figure in the series of frames contains the spot images created by just a few scattered electrons. As the number of particles increases with each successive frame, one can clearly see recognizable Young's fringes building up via coherent addition. The

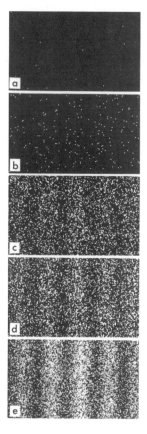

Figure 7.1: Electron diffraction from a double-slit apparatus (from Tonomura *et al.* [63]). From frame "a" to frame "e", electrons add coherently to give recognizable Young's fringes, conclusively demonstrating the wave behavior of matter (reproduced with permission).

text by Brandt and Dahmen [62] contains a complementary image of electron diffraction patterns from a sharp edge, displayed alongside a corresponding optical pattern.

To summarize, undeniable experimental evidence for wave behavior in particles and for particle behavior in light had been developed by this point in our historical review.

Schrödinger's equation

In 1925, Erwin Schrödinger gave a presentation on de Broglie's wave/ particle hypothesis. Peter Debye attended the seminar, and afterwards he suggested that a wave description of matter requires a wave equation—where was the wave equation? From this one audience question, the Schrödinger equation evolved. It is quite simple to use the Helmholtz equation [equation (4.46)] to infer the form for Schrödinger's time-independent equation. The Helmholtz (or stationary-wave) equation can be written as

$$\nabla^2 \psi = -k^2 \psi \tag{7.2}$$

where k is the wave vector and ψ is a wave function, analogous to the harmonic electric field in equation (4.46). We know from de Broglie that

$$\lambda = \frac{h}{p} \tag{7.3}$$

and we can say that $p = m_\mathrm{o} \mathrm{v}$, where m_o is the rest mass and v is the non-relativistic velocity. Then

$$k = \frac{2\pi}{\lambda} = \frac{2\pi m_\mathrm{o} \mathrm{v}}{h} \tag{7.4}$$

and

$$\nabla^2 \psi = -\frac{4\pi^2 m_\mathrm{o}^2 \mathrm{v}^2}{h^2} \psi \tag{7.5}$$

The total system energy is

$$E = T + V = \frac{1}{2} m_\mathrm{o} \mathrm{v}^2 + V \tag{7.6}$$

which produces

$$m_\mathrm{o} \mathrm{v}^2 = 2(E - V) \tag{7.7}$$

and

$$\nabla^2 \psi = -\frac{8\pi^2 m_\mathrm{o}}{h^2}(E - V)\psi \tag{7.8}$$

This can be rearranged to give

$$\boxed{-\frac{\hbar^2}{2m_\mathrm{o}} \nabla^2 \psi + V \psi = E \psi} \tag{7.9}$$

where:

$$\hbar \equiv h/2\pi \tag{7.10}$$

Equation (7.9) is the time-independent Schrödinger equation. The development just presented is by no means a derivation of the Schrödinger equation from first principles. Herein lies the problem; it was not clear what the "first principles" ought to be. Furthermore, equation (7.9) does not prove that the Schrödinger equation is appropriate for description of atomic or molecular phenomena. It simply combines a stationary wave equation (to satisfy Debye) with de Broglie's hypothesis, nothing more. Historically, it is also one of several approaches actually used by Schrödinger.

Schrödinger realized that such a simple development would convince nobody. For this reason, he wrote equation (7.9) in spherical coordinates, inserted a spherical potential field for V, and solved the hydrogen atom problem. Equation (7.9) is an eigenvalue problem. It is a linear, homogeneous partial differential equation, and for the one-electron atom, the boundary conditions are linear and homogeneous. Schrödinger was simply required to assume that the wave functions were single-valued, finite and continuous. The solutions were a series of eigenfunctions (the wave functions, ψ) with naturally occuring eigenvalues (energy levels, E). Buried in the eigenvalues were quantum numbers which determined the succession of energy levels in the atom. Quantum jumps between these energy levels then determined energy differences, and thus the emission spectrum. Schrödinger's original paper on this subject (published in 1926) was entitled "Quantization is an eigenvalue problem". Where Bohr had been forced to insert empirically-based quantum numbers, Schrödinger was able to generate them as a natural outcome of his theory. This wave theory was quickly applied to other spectroscopic problems where Bohr's theory had failed, and it succeeded (sometimes with logical modification).

Born's probability hypothesis

As astounding as this development was, it simply shifted the conundrum to another question—what was the physical meaning of these wave functions (It had been possible for Schrödinger to solve for the eigenvalues without explaining the eigenfunctions)? To solve the mathematics, one necessarily had to allow complex eigenfunctions. How can this be explained in terms of physical particles? Max Born suggested a controversial notion—that wave functions are probability amplitudes. Once squared (using the wave function and its complex conjugate to generate a real outcome), they become probability densities.

The probabilistic interpretation for absorption and emission catalyzed Einstein's assertion that *"He* does not play dice". Einstein was convinced that a fundamental, deterministic approach could be found. As an example, he questioned whether hidden variables underlying the observed phenomena did not actually control processes we now describe statistically. If found, these hidden variables could be described (giving new "first principles") and a deterministic approach developed. In the mid-1960s, Bell's theorem (after John Bell) showed that, if they exist, these hidden variables cannot form an underpinning to quantum mechanics. If they exist, quantum mechanics will have to be discarded. Einstein searched for deterministic, hidden variables for the remainder of his life and did not succeed. These ideas are discussed further in the work by Baggott [57].

Heisenberg uncertainty

In 1926, using a "thought experiment", Werner Heisenberg proposed his well-known uncertainty relationship. He suggested that it is not possible to know both position and momentum of a primary particle to arbitrary accuracy at the same time. This is expressed mathematically by stating that the uncertainty in position times the uncertainty in momentum has a lower limit:

$$\boxed{\Delta x \Delta p_x \geq \frac{h}{4\pi} = \frac{\hbar}{2}} \tag{7.11}$$

where x represents a Cartesian position along axis x, p_x is the momentum along that axis, and Δ indicates uncertainty. From a modern perspective, the simplicity of the experimental arguments used by Heisenberg is surprising. Much more exacting experiments than Heisenberg imagined are now possible, and his original thought process seems outdated. All the same, relation (7.11) can be proven easily using the Schwartz inequality and the properties of Hermitian operators. We will not prove it here because we have not yet discussed Hermitian operators and their relationship to momentum and position. Interested readers are referred to Schwabl [58] or Sakurai [61].

An introduction to commutators; with a connection between commutation, Heisenberg, and Schrödinger

At the limit of quantum mechanics the following holds true:

$$xp_x - p_x x = i\hbar \neq 0 \qquad (7.12)$$

In classical mechanics, the left-hand-side of equation (7.12) does equal zero. In fact, h is very small. As x and p_x become macroscopic, the size of the left hand side of equation (7.12) makes the right hand side appear to be zero (in a relative sense), and we obtain the classical solution. In the case under discussion, position and momentum are termed "observables", quantities that we could indeed observe in the laboratory (although not simultaneously, according to Heisenberg). In quantum mechanics, observables are related to a mathematical operator. These operators are not necessarily the observable quantity itself, although that is not ruled out. Since these are mathematical operators, we need a function upon which they operate. It is probably not a surprise that we choose the wave function ψ. It is common to denote an operator with a carat, so equation (7.12) is written more correctly as

$$\hat{x}\hat{p}_x\psi - \hat{p}_x\hat{x}\psi = i\hbar\psi \qquad (7.13)$$

This is one of the "canonical commutation relations". Paul Dirac called it one of the fundamentals of quantum mechanics. It is necessary, however, to ascertain the form of the operators. We seek functions that will prove consistent with other forms in quantum mechanics. We therefore assume (knowing now that this works) that $\hat{x}$ corresponds to multiplication by position "x", just as it does in classical mechanics. Next, we assume (knowing this works) that

$$\hat{p}_x = a\frac{\partial}{\partial x} \qquad (7.14)$$

where a is a dummy variable. Inserting both operators into equation (7.13), we find

$$
\begin{aligned}
xa\frac{\partial}{\partial x}\psi - a\frac{\partial}{\partial x}(x\psi) &= i\hbar\psi \\
ax\frac{\partial\psi}{\partial x} - ax\frac{\partial\psi}{\partial x} - a\psi &= i\hbar\psi \\
\text{and therefore } a &= -i\hbar \qquad (7.15)
\end{aligned}
$$

In three-dimensional Euclidean space, therefore, the momentum operator can be written as

$$\hat{p} = -i\hbar\nabla \tag{7.16}$$

Now consider energy conservation:

$$T + V = E, \quad \text{or} \quad \frac{p^2}{2m} + V = E \tag{7.17}$$

Replace the classical terms with the quantum-mechanical operators to obtain

$$-\frac{\hbar^2}{2m}\nabla^2\psi + \hat{V}\psi = E\psi \tag{7.18}$$

Equation (7.18) is identical to equation (7.9), demonstrating that our choice of the operator for momentum was consistent with the overall formalism. In addition, the Schrödinger equation is clearly identified with energy conservation. The classical version has already been encountered in equation (6.16), which defines the Hamiltonian. In fact, kinetic energy T in the Hamiltonian was written the same way, as $p^2/2m$. Equation (7.18) is often written in shorthand as

$$\boxed{\hat{H}\psi = E\psi} \tag{7.19}$$

$$\boxed{\hat{H} \equiv -\frac{\hbar^2}{2m}\nabla^2 + \hat{V}} \tag{7.20}$$

where we clearly mean the quantum-mechanical Hamiltonian. This is a demonstration of the correspondence principle in quantum mechanics, wherein we replace certain classical observables with quantum operators and we recover the quantum-mechanically correct formalism.

We will discuss that notion further, but for now we return to the Heisenberg uncertainty principle and the commutation relation. If two operators (say $\hat{A}$ and $\hat{B}$) commute, then

$$\hat{A}\hat{B}\psi - \hat{B}\hat{A}\psi = 0 \tag{7.21}$$

and if they do not commute

$$\hat{A}\hat{B}\psi - \hat{B}\hat{A}\psi \neq 0 \tag{7.22}$$

In shorthand, the "commutator" is written as

$$\boxed{[\hat{A}, \hat{B}] \equiv \hat{A}\hat{B} - \hat{B}\hat{A}} \tag{7.23}$$

Howard Robertson (see reference [57]) has demonstrated that, in general, the product of the standard deviations of two non-commuting operators is:

$$\Delta A \Delta B \geq \frac{1}{2}|\bar{b}| \tag{7.24}$$

where b is related to the commutator by

$$[\hat{A}, \hat{B}] = ib \tag{7.25}$$

According to Heisenberg

$$\Delta x \Delta p_x \geq \frac{1}{2}\hbar \tag{7.26}$$

and therefore

$$[\hat{x}, \hat{p}_x] = i\hbar \tag{7.27}$$

which recovers the commutation relation (7.12). Levine [60] develops the same point using an integral-based approach. This collection of findings could certainly be viewed as nothing more than a detailed consistency check. What we have done, however, is consistent with mathematical principles and with experimental findings.

Conclusion

In the past, some have taken Schrödinger's equation as a basic postulate and used that to derive the Heisenberg uncertainty principle, while others have taken the uncertainty principle as a basic postulate and worked the other way. Either way, it is necessary to begin with one of these postulates. In what follows, we describe selected portions of the full quantum mechanics formalism. Following that, the formal postulates of quantum mechanics are presented.

7.3   Additional Components of Quantum Mechanics

There are many approaches to this subject, each one enlightening in its own way. In this section we develop some specific, selected results using an approach similar to that taken by Schwabl [58]. This is done for two reasons. First, it is likely to overlap well with the background of a modern engineer. This material, therefore, can cast quantum mechanics in a more familiar context. Second, this section allows remaining gaps to be filled.

Material plane waves

As mentioned above, freely propagating electrons have been shown to diffract from slits or edges, just as light does. That necessarily means that electrons have wave properties. It should therefore be possible to describe the behavior of an electron using the plane wave formalism of equations (4.38) and (4.39). We therefore write

$$\Psi(\vec{r}, t) = C e^{i(\vec{k} \cdot \vec{r} - \omega t)} \tag{7.28}$$

where Ψ is the total wave function (including time dependence), and C is an amplitude term. This form is a solution to the Schrödinger equation, as demonstrated in a later section. For a free electron (i.e. in the absence of a potential field) the kinetic energy is related to momentum via

$$\frac{E}{\hbar} = \omega = \frac{p^2}{2m\hbar} \tag{7.29}$$

In addition, we can take the magnitude of the wave vector:

$$k = \frac{2\pi}{\lambda} \tag{7.30}$$

and use the de Broglie material wavelength $\lambda = h/p$:

$$k = \frac{p}{\hbar}, \quad \text{and} \quad \vec{k} = \frac{\vec{p}}{\hbar} \tag{7.31}$$

The wave function is then

$$\Psi(\vec{r}, t) = C e^{\frac{i}{\hbar}(\vec{p} \cdot \vec{r} - \frac{p^2}{2m} t)} \tag{7.32}$$

This is an interesting outcome. In the discussion on optical plane waves, we stated that "This description forms the basis for the field of Fourier optics, in which the individual $\vec{k}$ vectors are the Fourier spatial frequencies of the composite field". Here, the momentum vector performs that same function. In short, the basis momentum vectors are the spatial Fourier frequency components of the wave function. Schwabl is then able to use the notion of transformation between wave space and momentum space, together with Parseval's theorem of transforms, to develop a special case of the Heisenberg uncertainty principle. It is quite similar to the Fourier transform limitation on the time/bandwidth product given in equation (3.29).

The optical double-slit problem generates an interference pattern from an intensity distribution via [48]

$$I \propto |E_1 + E_2|^2 \qquad (7.33)$$

and thus

$$I \propto I_1 + I_2 + 2 \operatorname{Re}|E_1^* E_2| \qquad (7.34)$$

The third term gives rise to the interference pattern. One would expect, then, that the probability distribution for particles in a double-slit experiment would have to be

$$|\Psi_1 + \Psi_2|^2 \qquad (7.35)$$

which is, in fact, Born's probability assertion. This is certainly no proof of Born's interpretation, but, if particles can behave as plane waves, then this notion is certainly a logical hypothesis.

The time-dependent Schrödinger equation

To put time dependence into Schrödinger's equation is not difficult using this plane wave approach. If we begin with a free particle we can avoid description of a potential field for the time being. One can then simply state the requirements of such an equation. Following Schwabl [58]:

1. The expression should be first order in time, so that $\Psi(\vec{r}, t)$ can evolve from $\Psi(\vec{r}, 0)$.

2. The expression should be linear in $\Psi(\vec{r}, t)$, so that superposition can be used to describe the wave-like phenomena that are observed. For the same reason, constants in the equation cannot depend upon the state (e.g. energy, momentum, etc.) of the particle.

3. The equation needs to be homogeneous so that the probabilistic explanation can be used.

We begin by describing plane waves using equation (7.32):

$$\Psi(\vec{r}, t) = C e^{\frac{i}{\hbar}(\vec{p}\cdot\vec{r} - \frac{p^2}{2m}t)} \tag{7.36}$$

The time derivative of the plane wave expression is then

$$\frac{\partial}{\partial t}\Psi(\vec{r}, t) = -\frac{ip^2}{2m\hbar} C e^{\frac{i}{\hbar}(\vec{p}\cdot\vec{r} - \frac{p^2}{2m}t)} = -\frac{ip^2}{2m\hbar}\Psi(\vec{r}, t) \tag{7.37}$$

Multiplying this equation by $i\hbar$, and using equation (7.29), we obtain

$$i\hbar\frac{\partial}{\partial t}\Psi(\vec{r}, t) = E\Psi(\vec{r}, t) \tag{7.38}$$

The time-independent Schrödinger equation [equations (7.9) and (7.18)] is

$$-\frac{\hbar^2}{2m}\nabla^2\psi = E\psi \tag{7.39}$$

where the potential field interaction has been left off in order to describe a free particle. The right-hand side of the Schrödinger equation makes the next move obvious. We combine equation (7.38) with the time-independent Schrödinger equation to obtain

$$-\frac{\hbar^2}{2m}\nabla^2\Psi(\vec{r}, t) = i\hbar\frac{\partial}{\partial t}\Psi(\vec{r}, t) \tag{7.40}$$

Now we reinsert the potential energy operator to obtain a generalized Schrödinger equation:

$$-\frac{\hbar^2}{2m}\nabla^2\Psi(\vec{r}, t) + \hat{V}\Psi(\vec{r}, t) = i\hbar\frac{\partial}{\partial t}\Psi(\vec{r}, t) \tag{7.41}$$

or

$$\boxed{\hat{H}\Psi(\vec{r}, t) = i\hbar\frac{\partial}{\partial t}\Psi(\vec{r}, t)} \tag{7.42}$$

This is the full time-dependent Schrödinger equation. Once again, this is not a proof based upon first principles. It was introduced to justify the form of the time evolution operator in a heuristic way. In fact, if one takes time to consider what we have done, it was simply reliance upon superposition to reverse a separation of variables process. It does, however, use the concept of a material plane wave to infer a time dependence that is consistent with the requirements given above.

For the sake of completeness, we will separate equation (7.42) before continuing. We begin by assuming that the Hamiltonian is not time dependent. This really means that we assume that the potential $\hat{V}$ is not time dependent. The subject of time-dependent forms for $\hat{V}$ will be explored in Chapter 14. We therefore assert that Ψ is separable via

$$\Psi(\vec{r}, t) = f(t)\psi(\vec{r}) \tag{7.43}$$

Using this in equation (7.42), we obtain

$$\frac{1}{f(t)} i\hbar \frac{\partial}{\partial t} f(t) = \frac{1}{\psi(\vec{r})} \hat{H}\psi(\vec{r}) \tag{7.44}$$

As usual with a separation of variables problem, the left-hand side is dependent upon one variable (time) and the right-hand side is dependent upon another (spatial variables). The two sides must therefore equal the same constant, call it "E" here. Not surprisingly, we then find that

$$f(t) = Ce^{-\frac{iEt}{\hbar}} \tag{7.45}$$

where C is an integration constant. One can also find the energy represented by $E = h\nu = \hbar\omega$ so that

$$f(t) = Ce^{-i\omega t} \tag{7.46}$$

and

$$\hat{H}\psi(\vec{r}) = E\psi(\vec{r}) \tag{7.47}$$

which is the eigenvalue problem discussed previously.

The states defined by

$$\Psi(\vec{r}, t) = e^{-\frac{iEt}{\hbar}} \psi(\vec{r}) \tag{7.48}$$

are called "stationary states" because the probability densities are time independent:

$$|\Psi(\vec{r}, t)|^2 = |\psi(\vec{r})|^2 \tag{7.49}$$

Some additional mathematics

It turns out that classical, three-dimensional Euclidean space is insufficient to describe the full quantum mechanics problem. As an example, quantum mechanics must keep track of an electron "spin coordinate".

Spin is not literally a rotational motion of the electron, otherwise it would be possible to use Euclidean space to describe it. Spin is a new internal coordinate for the electron. It is therefore necessary to describe the quantum-mechanics problem in a space that occupies more than three dimensions. This multidimensional parameter space is called a "Hilbert space", and Euclidean space is a subset of it. Hilbert space is normally considered infinite in extent, and it can be occupied by constants, vectors, functions and so forth.

The "inner product" behaves in Hilbert space the same way that the dot (or "scalar") product behaves in Euclidean space. We use the inner product to find the components of a function in a specified Hilbert space just as one uses the dot product to find vector components in Euclidean space. Here, the inner product between two functions (e.g. φ and ψ) is defined by

$$(\varphi, \psi) \equiv \int \varphi^*(\vec{r})\psi(\vec{r}) \, d\tau \tag{7.50}$$

where $d\tau$ is a differential volume element in configuration space (it could be a space in Cartesian coordinates, momentum, energy and so forth). It is simple to show that normalized, orthogonal functions φ and ψ generate the same outcome from equation (7.50) as do Cartesian unit vectors and the dot product. Fourier sine and cosine functions are good examples. The argument is a little circular, however, because we use this analogy to define orthogonality for functions. We will use the inner product to express functions, averages, and so forth in terms of basis wave functions, using the expansion theorem given below.

The integral notation we have been using can be cumbersome. In response to this, Dirac developed a specialized notation that simplifies the mathematics considerably. It may seem equally cumbersome to use a new notation, but it can be learned quickly and the benefits of a simple system become obvious even more quickly. As one example, the expression given in (7.50) can be written in Dirac notation as

$$\int \varphi^*(\vec{r})\psi(\vec{r}) \, d\tau = \langle \varphi(\vec{r})|\psi(\vec{r}) \rangle \tag{7.51}$$

The notation $\langle \, | \, \rangle$ is called a bracket, while $\langle \, |$ is a "bra" and $| \, \rangle$ is a "ket".

The wave functions are clearly related to the state (via the quantum number) of an atomic or molecular system. The quantum states of an atom or molecule are therefore represented by Dirac "state vectors". The ket $|\psi_n\rangle$ is thus the state vector for the quantum number n (it is sometimes written as $|n\rangle$). The bra then represents the complex conjugate. The Dirac ket $|n\rangle$ is not literally equal to the wave function ψ_n, although they can perform the same way in parallel mathematical systems. The state ket vectors can be related to the wave functions, as shown by Sakurai [61].

In the section on correspondence to follow, we will use both the integral representation and the Dirac notation to smooth the transition to Dirac notation. Following that, we will use whichever system best illustrates the point we wish to make. Both inner products and Dirac notation are described in a much more structured, mathematical context in most quantum texts.

Correspondence

In quantum mechanics, quantities that can be experimentally observed ("observables") are represented by operators. We have encountered this notion already. For example, let the operator $-i\hbar\nabla$ act on the material plane wave given by equation (7.36).

$$
\begin{aligned}
-i\hbar\nabla\Psi &= -i\hbar\left(\frac{i}{\hbar}p_x\hat{x} + \frac{i}{\hbar}p_y\hat{y} + \frac{i}{\hbar}p_z\hat{z}\right) \\
&= \vec{p}
\end{aligned}
\tag{7.52}
$$

Therefore, the operator $-i\hbar\nabla$ is the momentum operator. Whenever we wish to describe momentum in quantum mechanics, we use this operator. Similarly, when $i\hbar\partial/\partial t$ acts on the plane wave:

$$
\begin{aligned}
i\hbar\frac{\partial}{\partial t}\Psi &= \frac{p^2}{2m}\Psi \\
&= E\Psi
\end{aligned}
\tag{7.53}
$$

This demonstrates that $i\hbar\partial/\partial t$ could be (and often is) called an energy operator. Time, however, is not an observable in quantum mechanics, it is a parameter. It is therefore more appropriate to use the Hamiltonian operator to represent energy. The position operator can then be taken as multiplication by position, as before. At this point, we have not bothered to define the potential energy operator, because it depends upon the specific problem one intends to analyze. An example of $\hat{V}$ and its application is given in Chapter 8.

The mean value of an observable A is the expectation value of its corresponding operator $(\hat{A})$:

$$\langle A_n \rangle = \frac{\int \Psi_n^* \hat{A} \Psi_n \, d\tau}{\int \Psi_n^* \Psi_n \, d\tau} = \frac{\langle \Psi_n | \hat{A} | \Psi_n \rangle}{\langle \Psi_n | \Psi_n \rangle} \tag{7.54}$$

This expression is a direct outcome of the notions we have already discussed (i.e. that the wave function is the probability amplitude), together with an application of basic probability calculus. Note that the "$\langle \; \rangle$" on $\langle A_n \rangle$ denotes the quantum-mechanical mean value, while the "$| \; \rangle$" on $| \Psi_n \rangle$ denotes a state vector. The Dirac formalism is designed to parallel the notation for averages in the imagery of the symbols [see, for example, equation (7.55) below].

The denominator of equation (7.54) is the inner product of the state vector $| \Psi_n \rangle$ with itself (analogous to the dot product between a unit vector and itself). Because the wave function is normalized the inner product is 1, giving

$$\boxed{\langle A_n \rangle = \int \Psi_n^* \hat{A} \Psi_n d\tau = \langle \Psi_n | \hat{A} | \Psi_n \rangle} \tag{7.55}$$

If the wave function is an eigenfunction of $\hat{A}$, then

$$\hat{A} \psi_n = a_n \psi_n \tag{7.56}$$

where a_n is the corresponding eigenvalue. Then equation (7.55) becomes

$$\boxed{\langle A_n \rangle = a_n \int \psi_n^* \psi_n \, d\tau = a_n \langle \psi_n | \psi_n \rangle = a_n} \tag{7.57}$$

where we have removed the time dependence because these are stationary states. This allows us to take full advantage of the properties of the eigenvalue problem. For example, when the wave function is an eigenfunction of the operator, then the mean value of an observable is the eigenvalue. In the time-independent Schrödinger equation [equations (7.9) and (7.18)], for example, the mean value of the Hamiltonian is the energy.

For this notion to make practical sense, the eigenvalue should be real because observables are real. Operators with real eigenvalues are termed "Hermitian" (to be more strict, Hermitian operators are self-adjoint):

$$\int \psi_n^* \hat{A} \psi_m \, d\tau = \left[\int \psi_n^* \hat{A} \psi_m d\tau \right]^*$$
$$\langle \psi_n | \hat{A} | \psi_m \rangle = \left[\langle \psi_n | \hat{A} | \psi_m \rangle \right]^* \tag{7.58}$$

where ψ_n and ψ_m are both eigenfunctions of $\hat{A}$. For eigenfunctions of Hermitian operators

$$\int \psi_n^* \psi_m d\tau = \langle \psi_n | \psi_m \rangle = 0 \tag{7.59}$$

for $n \neq m$. This can be shown without too much trouble. When the wave functions are normalized, then

$$\boxed{\int \psi_n^* \psi_m \, d\tau = \langle \psi_n | \psi_m \rangle = \delta_{nm}} \tag{7.60}$$

In general, the eigenfunctions of Hermitian operators are orthonormal, and in fact they form a complete basis set. This is proven in most quantum texts.

Superposition

Because the state vectors constitute a complete, orthonormal basis set, other functions can be expanded in terms of this set. It is therefore possible to solve difficult problems by finding expectation values in terms of state vectors we do know. Imagine, for example, that we want to find the mean value of an observable, $\langle A \rangle$, but the wave function (call it ψ for generality) is not an eigenfunction of the operator $\hat{A}$ (assumed Hermitian here). To say it another way, $|\psi\rangle$ is not an eigenstate of $\hat{A}$, but $|\psi\rangle$ is the state of the system.

Paraphrasing Baggott [57], the principle of superposition states that an arbitrary, well-behaved state vector can be expanded as a linear superposition of the complete set of orthonormal eigenstates of *any* Hermitian operator. We will therefore expand $|\psi\rangle$ in terms of the eigenstates of $\hat{A}$, using the orthonormality of those eigenstates. It should be noted that a state vector is "well behaved" when it has properties similar to the eigenstates we plan to use as the basis vectors: it should have the same physical interpretation, the same kind of boundary conditions, and so forth. Expanding a multi-electron atomic system in terms of single-electron atomic state vectors is a typical example.

In our example, we wish to find $\langle A \rangle$ for some state $|\psi\rangle$ which is not an eigenstate of $\hat{A}$, via

$$\langle A \rangle = \langle \psi | \hat{A} | \psi \rangle \tag{7.61}$$

(assuming that $|\psi\rangle$ is normalized). We then expand $|\psi\rangle$ in terms of the eigenstates of $\hat{A}$ (call them $|u_n\rangle$). This expansion is always possible because the $|u_n\rangle$ also form a complete basis set:

$$|\psi\rangle = \sum_n c_n |u_n\rangle \tag{7.62}$$

Consider that

$$\hat{A} |\psi\rangle = \sum_n c_n \hat{A} |u_n\rangle = \sum_n c_n\, a_n |u_n\rangle \tag{7.63}$$

where, as before, a_n is the eigenvalue of $\hat{A}$. By definition

$$\langle \psi | = \sum_m c_m^* \langle u_m | \tag{7.64}$$

where we use subscript m to emphasize that the bra and ket are not required to describe the same quantum state at all times. Then the term $\langle \psi | \hat{A} | \psi \rangle$ becomes

$$\langle \psi | \hat{A} | \psi \rangle = \sum_n |c_n|^2 \langle u_n | \hat{A} | u_n \rangle + \sum_m \sum_{n \neq m} c_m^* c_n \langle u_m | \hat{A} | u_n \rangle \tag{7.65}$$

Recall that

$$\langle u_n | \hat{A} | u_n \rangle = a_n \tag{7.66}$$

and

$$\langle u_n | \hat{A} | u_m \rangle = 0 \tag{7.67}$$

the outcome being

$$\boxed{\langle A \rangle = \langle \psi | \hat{A} | \psi \rangle = \sum_n |c_n|^2 a_n} \tag{7.68}$$

In short, $\langle A \rangle$ is a weighted average of the eigenvalues for the eigenstates that were used to expand $|\psi\rangle$, and the weighting constants are the square of the expansion coefficients. The weighting constants are the probabilities of finding state $|u_n\rangle$.

These expansion coefficients are just the projection amplitudes of $|\psi\rangle$ onto the basis $|u_n\rangle$. They can be found using the orthogonality property of the basis vectors via the inner product:

$$\langle u_n|\psi\rangle = \sum_m c_m \langle u_n|u_m\rangle = c_n \qquad (7.69)$$

This concept of superposition is used throughout quantum mechanics, and we will use it in this book for the development of the density matrix equations.

Measurements and compatible observables

In Dirac's words, "A measurement always causes the system to jump into an eigenstate of the dynamical variable that is being measured" [61]. If we wish to observe $\langle A_n\rangle$ to arbitrary precision and if ψ_n is an eigenfunction of $\hat{A}$, we have already shown that $\langle A_n\rangle = a_n$. Now if we wish to observe another quantity, call it $\langle B_n\rangle$, to equally arbitrary precision, then ψ_n must be an eigenfunction of $\hat{B}$ as well. Then the commutator becomes

$$[\hat{A}, \hat{B}] = 0 \qquad (7.70)$$

This is also a straightforward result demonstrated in most quantum texts. For two observables to be simultaneously measurable to arbitrary precision, their commutator must be zero. They are then called "compatible observables", and they can share a basis set of ket vectors. In this case, it is possible to find a true expectation value, call it $\langle A_n\rangle$, for the operator $\hat{A}$. It is simply the eigenvalue for that operator (a_n in this case), as shown in equation (7.57). A good example would be the quantized energy levels of a one-electron atom. The wave function is an eigenfunction of the Hamiltonian, and energy is the eigenvalue. Measurement at a particular energy level will return that specific energy level.

The commutator for the position and momentum operators is not zero, and the Heisenberg uncertainty principle applies. Position and momentum are "incompatible observables". Incompatible observables do not share a basis set of ket vectors. In this case, if the state wave function is not an eigenfunction of $\hat{A}$, we will still obtain an eigenvalue of $\hat{A}$ when we measure A, but we will not be able to tell in advance exactly which one. It is only possible to find a weighted average value $\langle A \rangle$ for the eigenvalues in question [e.g. equation (7.68)]. Our ability to predict specific observables exactly is thus controlled by the commutation relations.

The complete set of canonical commutation relations for position and momentum is

$$[x_i, x_j] = 0$$
$$[p_i, p_j] = 0$$
$$[x_i, p_j] = i\hbar\delta_{ij}$$

where the subscripts i and j indicate different Cartesian coordinates. These relations indicate that position and momentum cannot be assigned definite values simultaneously. If we detect one of them to arbitrary accuracy, then it is only possible to find a weighted average value for the other. As another example, we present the commutation relations for total angular momentum $\hat{L}$ and the Cartesian components of $\hat{L}$ ($\hat{L}_i$, using the same notation for Cartesian coordinates). The total angular momentum is related to the three components via

$$\hat{L}^2 = \sum_i \hat{L}_i^2 \tag{7.71}$$

The angular momentum commutation relations are

$$[\hat{L}_x, \hat{L}_y] = i\hbar\hat{L}_z$$
$$[\hat{L}_y, \hat{L}_z] = i\hbar\hat{L}_x$$
$$[\hat{L}_z, \hat{L}_x] = i\hbar\hat{L}_y$$
$$[\hat{L}^2, \hat{L}_i] = 0$$

which is why we use quantum numbers to specify the magnitude of the total orbital angular momentum (given by $\hat{L}^2$) and only one of its vector components. For a molecule, this is typically the component along the internuclear axis (given by $\hat{L}_z$).

Ehrenfest theorem

The Ehrenfest theorem states that quantum-mechanical average values will produce results from classical mechanics. We present it as a final topic in Section 7.3 because the outcome is quite convincing for those who remain skeptical.

We start with the Schrödinger equation:

$$i\hbar \frac{\partial}{\partial t}\Psi = \hat{H}\Psi \tag{7.72}$$

and the complex conjugate of the Schrödinger equation:

$$-i\hbar \frac{\partial}{\partial t}\Psi^* = \hat{H}\Psi^* \tag{7.73}$$

Now use the definition of an average:

$$\langle A \rangle = \int \Psi^*(\vec{r},t)\hat{A}\Psi(\vec{r},t)d\tau \tag{7.74}$$

and write the time derivative of that average:

$$\frac{\partial}{\partial t}\langle A \rangle = \int \left(\frac{\partial}{\partial t}\Psi^*(\vec{r},t)\right)\hat{A}\Psi(\vec{r},t) + \Psi^*(\vec{r},t)\frac{\partial \hat{A}}{\partial t}\Psi(\vec{r},t)$$
$$+ \Psi^*(\vec{r},t)\hat{A}\left(\frac{\partial}{\partial t}\Psi(\vec{r},t)\right)d\tau \tag{7.75}$$

Equations (7.72) and (7.73) contain the time derivatives of the wave function. Inserting these into the above and performing some algebraic manipulation, one can easily find

$$\frac{\partial}{\partial t}\langle A \rangle = \frac{i}{\hbar}\langle [H,A] \rangle + \langle \frac{\partial A}{\partial t} \rangle \tag{7.76}$$

Compare this to the classical equation (6.28). There is a clear correspondence between the classical Hamiltonian expression (6.28), written in terms of a Poisson bracket, and the quantum-mechanical expression (7.76), written in terms of the commutator.

That logically leads to the use of equation (7.76) as a demonstration of Ehrenfest's theorem. We have two purposes for following this line of thought. First, we will generate Newton's law using quantum-mechanical averages, which is a fairly compelling result. Second, the process will allow us to demonstrate how commutator and operator algebra can be used.

Consider the case where the operator is simply multiplication by position, and in this case it will be positioned along one Cartesian coordinate. We will need, therefore, to evaluate the commutator $[H, x_i]$ where the Hamiltonian is

$$
\begin{aligned}
H &= -\frac{\hbar^2}{2m}\nabla^2 + V(\vec{r}) \\
&= -\frac{\hbar^2}{2m}\sum_j \frac{\partial^2}{\partial x_j^2} + V(\vec{r}) \\
&= -\frac{\hbar^2}{2m}\sum_j \frac{\partial}{\partial x_j}\frac{\partial}{\partial x_j} + V(\vec{r})
\end{aligned}
$$

Again, the subscripts indicate the Cartesian coordinates. In order to keep track of operators, it is helpful to write them as they operate upon some function:

$$
\begin{aligned}
[H, x_i]\Psi &= \left[\left(-\frac{\hbar^2}{2m}\sum_j \frac{\partial}{\partial x_j}\frac{\partial}{\partial x_j} + V(\vec{r})\right), x_i\right]\Psi \\
&= \left(-\frac{\hbar^2}{2m}\sum_j \frac{\partial}{\partial x_j}\frac{\partial}{\partial x_j} + V(\vec{r})\right)x_i\Psi \\
&\quad - x_i\left(-\frac{\hbar^2}{2m}\sum_j \frac{\partial}{\partial x_j}\frac{\partial}{\partial x_j} + V(\vec{r})\right)\Psi \quad\quad (7.77)
\end{aligned}
$$

Rearranging gives

$$
\begin{aligned}
[H, x_i]\Psi &= \left(-\frac{\hbar^2}{2m}\sum_j \frac{\partial}{\partial x_j}\frac{\partial}{\partial x_j}\right)x_i\Psi + V(\vec{r})x_i\Psi \\
&\quad - x_i\left(-\frac{\hbar^2}{2m}\sum_j \frac{\partial}{\partial x_j}\frac{\partial}{\partial x_j}\right)\Psi - x_iV(\vec{r})\Psi \quad (7.78)
\end{aligned}
$$

The two terms involving $V(\vec{r})$ cancel, and

$$
[H, x_i]\Psi = \left[\left(-\frac{\hbar^2}{2m}\sum_j \frac{\partial}{\partial x_j}\frac{\partial}{\partial x_j}\right), x_i\right]\Psi \quad\quad (7.79)
$$

One theorem in the algebra of commutators provides

$$
[AB, C] = A[B, C] + [A, C]B \quad\quad (7.80)
$$

Using this, we can write equation (7.79) as

$$[H, x_i]\Psi = -\frac{\hbar^2}{2m}\sum_j \left(\frac{\partial}{\partial x_j}\left[\frac{\partial}{\partial x_j}, x_i\right]\Psi + \left[\frac{\partial}{\partial x_j}, x_i\right]\frac{\partial}{\partial x_j}\Psi\right) \quad (7.81)$$

Consider the first term inside the parentheses on the right-hand side of equation (7.81). It can be simplified as follows:

$$\frac{\partial}{\partial x_j}\left[\frac{\partial}{\partial x_j}, x_i\right]\Psi = \frac{\partial}{\partial x_j}\left(\frac{\partial}{\partial x_j}(x_i\Psi) - x_i\frac{\partial}{\partial x_j}\Psi\right)$$

$$= \frac{\partial}{\partial x_j}\Psi\delta_{ij} \quad (7.82)$$

The second term on the right-hand side of equation (7.81) can be handled with the same approach, and it generates the same result. Noting that the j component of the momentum is given by the operator $p_j = -i\hbar\frac{\partial}{\partial x_j}$, we obtain the following commutation relation:

$$[H, x_i] = -\frac{i\hbar p_i}{m} \quad (7.83)$$

Using a similar approach, one can show that:

$$[H, p_i] = \left[V(\vec{r}), \frac{\hbar}{i}\frac{\partial}{\partial x_i}\right] = i\hbar\frac{\partial V}{\partial x_i} \quad (7.84)$$

Now combine equations (7.76) and (7.83):

$$\frac{\partial}{\partial t}\langle\vec{r}\rangle = \frac{i}{\hbar}\langle-\frac{i\hbar\vec{p}}{m}\rangle = \frac{1}{m}\langle\vec{p}\rangle \quad (7.85)$$

Here we have assumed that the material is isotropic so that $\langle\partial\vec{r}/\partial t\rangle = 0$. In short, the average of the velocity vectors is zero because the medium is quiescent and in equilibrium. Combine equations (7.76) and (7.84) in the same way to obtain

$$\frac{\partial}{\partial t}\langle\vec{p}\rangle = -\langle\nabla V(\vec{r})\rangle \quad (7.86)$$

For a conservative force, $\vec{F}(\vec{r}) = -\nabla V(\vec{r})$. Combine this fact with equations (7.85) and (7.86) to obtain Newton's equation of motion in terms of quantum-mechanical averages:

$$m\frac{d^2}{dt^2}\langle\vec{r}\rangle = \langle\vec{F}(\vec{r})\rangle \quad (7.87)$$

This becomes the full Newtonian equation when $\vec{F}(\vec{r})$ changes slowly within the range of motion. At the macroscopic limit, quantum mechanics generates the expected macroscopic result. We did not work the complimentary classical example in order to conserve space. Marion and Thornton [56] show, however, that the same version of Newton's law can be developed using purely classical Hamiltonian dynamics.

7.4 Postulates of Quantum Mechanics

As mentioned above, a key question is what one could consider "first principles" in a description of atomic and subatomic phenomena? There are several very basic notions in quantum theory from which other results can be generated, and these have become the postulates of the formalism. It is not terribly complex in hindsight. Undeniable wave behavior (e.g. Figure 7.1) requires a wave expression, but time decay requires an exponential decay or growth expression. That generates the Schrödinger equation. Stationary states generate eigenvalue problems with known mathematics. The analogy with optics and electromagnetic waves (or probability theory) generates a justification for use of the wave function as a probability amplitude.

In what follows, we quickly present the postulates of quantum mechanics for the sake of completeness.

Postulate 1. The state of a quantum-mechanical system is completely described by the wave function Ψ. It contains all of the information that can be known about the system. According to Born's interpretation, wave functions are probability amplitudes. Then the probability density is given by

$$\boxed{|\Psi|^2 = \Psi^*\Psi}$$
(7.88)

For those who have not considered such an expression in the past, use of the conjugate makes sense. We seek the modulus squared of the wave function (in a vector, or Pythagorean sense). By using the complex conjugate, we can cancel the cross-terms in the product and recover this modulus squared. Probability densities are normalized, so

$$\boxed{\int_{-\infty}^{+\infty} \Psi^*\Psi \; d\tau = 1}$$
(7.89)

Postulate 2. Observables are represented by linear Hermitian operators. These operators can be found using the correspondence principle.

Postulate 3. Measurement of the observable A will generate the eigenvalues a_n of the operator $\hat{A}$.

Postulate 4. If $\hat{A}$ is a linear Hermitian operator associated with an observable, then the eigenfunctions of $\hat{A}$ (call them u_n) form a complete set.

Postulate 5. If $\Psi(x_i, t)$ is the normalized state function of a system, then the average value of A is

$$\langle A \rangle = \int \Psi^* \hat{A} \Psi \; \mathrm{d}\tau \tag{7.90}$$

Note that this average is taken over space, not time.

Postulate 6. The time development of a state, if undisturbed (see Chapter 14 for the "disturbed" case), is given by equation (7.42), reproduced here:

$$\hat{H}\Psi(\vec{r}, t) = i\hbar \frac{\partial}{\partial t} \Psi(\vec{r}, t)$$

7.5 Conclusions

The Schrödinger equation [i.e. equation (7.42)] is the fundamental theoretical tool that we will apply to spectroscopy. It can be solved in various ways, to obtain various outcomes. The many usages of equation (7.42) found in texts and articles can confuse those who are new to quantum mechanics. The relationship can actually be used to solve for three things: eigenvalues (energy), quantum numbers (residing inside the solution for the eigenvalues), and wave functions (which can then be used to describe the expectation value; i.e. for electron position, which then leads to descriptions of orbitals). Moreover, the angular momentum operator squared is closely allied with the Hamiltonian because the two commute (share the same wave function). The momentum operator $\hat{L}^2$ consists of the angular components of $\hat{H}$ multiplied by $2m$ (to convert from kinetic energy to momentum) and by r^2 (to convert this into angular momentum squared). That equation is then solved, providing angular momentum squared (the eigenvalue), and the angular orbital momentum quantum numbers. That wave function is also shared with equation (7.42). Again, this works simply because the operators commute. Because one formalism can reap so many outcomes, the results one finds in standard sources can appear to disagree if they present only part of the total picture.

We have thus far described several of the remarkable discoveries made during the early part of the twentieth century. The foundations presented here are required to describe comfortably the resonance response of an atom or molecule and the spectroscopy of atoms or molecules, and to develop dynamic expressions for phenomena that change over time. These topics are discussed in the chapters to follow.

Chapter 8

ATOMIC SPECTROSCOPY

8.1 Introduction

Atoms can be important free radical intermediates in reacting flows. In combustion, for example, the important atomic species are H, O, N, and C, and they are usually measured with a spectroscopic technique. In this chapter the basic formalism for atomic spectroscopy is presented, starting with the one-electron atom. Significant detail is included because this topic has both didactic and historic importance.

This subject has didactic value because it includes almost every topic of importance to spectroscopy, including electronic energy levels, angular momentum, electron spin, and so forth. The only missing subject of importance to molecular spectroscopy is vibration. These fundamental atomic subjects will be covered in detail, therefore, because they have a broad impact.

Historically, the observation of emitted line spectra from hydrogen atoms in a flame was a driving force behind the development of quantum mechanics [15]. The Balmer series of emission lines in the H atom (spanning roughly from 656 to 101 nm) was the first spectrum to be studied in detail. It is so named because it was Balmer who originally found a sequence of numbers that reproduced the spectral locations of the lines. That series is

$$\frac{1}{\lambda} = R \left(\frac{1}{2^2} - \frac{1}{n_1^2} \right) \tag{8.1}$$

where $n_1 = 3, 4, 5...$ (i.e. $n_1 > 2$) and R is the Rydberg constant ($R = 1.097373177 \times 10^5$ cm^{-1}). Following this development, it was speculated that the "2" in the first term within the parentheses could possibly be replaced by another integer:

$$\frac{1}{\lambda} = R\left(\frac{1}{n_2^2} - \frac{1}{n_1^2}\right) \tag{8.2}$$

where $n_2 = 1, 2, 3, 4, 5...$ and $n_1 > n_2$. This expression predicted the existence of other line series that were indeed observed later on. These include the Lyman series ($n_2 = 1$) in the UV, Paschen ($n_2 = 3$) in the near IR, Brackett ($n_2 = 4$) in the IR, and Pfund ($n_2 = 5$) further into the IR. These series were related to differences in what were called "terms":

$$\frac{1}{\lambda} = T_2 - T_1 \tag{8.3}$$

In 1913, Bohr applied a modified Rutherford orbital model for an atom to the Balmer series. While rejected ultimately, the Bohr development included several conceptual breakthroughs that led towards quantum mechanics. Bohr realized that the spectroscopic terms were related to atomic structure, and so he decided to force the Rutherford planetary model to reproduce this term dependence. In order to do so, he was required to assume: (1) that only very specific orbits (orbital radii) were allowed; (2) that despite the fact that accelerating electrons emit radiation, electrons in these specific orbits would not do so; (3) that radiation was emitted when an electron jumped from a higher-energy orbit to a lower-energy orbit; and (4), as with Planck, that the energy of the emitted photon was related to optical frequency via the expression $h\nu = E_{n_2} - E_{n_1}$. While these assumptions violated accepted physics, the theory reproduced the line spectra of the hydrogen atom, including a reasonable value for the Rydberg constant. Indeed, it could reproduce the spectrum of any one-electron atom (e.g. He^+, Li^{++}, etc.). The theory failed to describe multi-electron atoms and molecules, and so the search for an improved theory commenced, but Bohr's initial assumptions have survived.

The spectroscopic terms are thus related to various quantized energy levels. The lowest of these levels is called a ground state, while those of higher energy are called excited states. These energy levels are associated with position of a negatively charged electron relative to a positively charged nucleus (potential energy arising from the field interaction between the two), together with the orbital kinetic energy of the electron. Line spectra of atoms at optical wavelengths are usually associated with loosely bound electrons, typically the outer one. Note that, as n tends to infinity in a series such as equation (8.2), the series will reach a limit (because $1/n \to 0$). This is the ionization limit. Finally, the zero of energy is usually defined to be the equilibrium state with infinite distance between the nucleus and all of the electrons. The various bound levels then have negative energy.

In what follows, we present portions of the one-electron atom problem in detail, and then discuss implications of the results. Following that we describe formalisms and present results for multi-electron atoms. In some cases, the treatment may appear to be overly detailed. Detail may be necessary, however, if the reader wishes fully to understand atomic quantum numbers and spectra. In other cases, details are omitted because they are not central to the focus of this chapter, or because they relate to specific atomic cases. The style of this chapter allows the discussion on molecular spectroscopy to move along quickly; this and the following chapter are meant to be read together. The material presented here follows the material presented in Hertzberg [15], Levine [60], Karplus and Porter [64], and Cowan [65].

8.2 The One-Electron Atom

The one-electron atom is depicted in Figure 8.1. It constitutes a two-particle central force problem. The electron has quantized energy levels determined by the spherical potential (a function of distance r from the central positive charge) and by electronic orbital kinetic energy (related to electronic orbital angular momentum: $mv^2/2 = p^2/2m$). The word "orbital" is used to identify this form of kinetic energy and momentum because we will also encounter descriptions of electron spin and nuclear rotation. Here, "electronic orbital angular momentum" specifies the orbit of electrons around the nucleus.

8.2.1 Definition of $\hat{V}$

We can define the atomic central force relative to a potential field (much as we have done in Chapter 4). Using classical mechanics, we can write

$$\vec{F} = -\vec{\nabla}V(x, y, z) \tag{8.4}$$

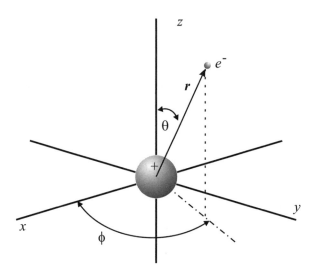

Figure 8.1: The one-electron atom problem, with coordinate definitions.

A simple example of this would be a mechanical force due to a grav-itational potential field. In such a case, $V = mgr$. F is necessarily a conservative force. The spherical coordinate form of ∇ is:

$$\nabla = \frac{\partial}{\partial r} + \frac{1}{r}\frac{\partial}{\partial \theta} + \frac{1}{r \sin \theta}\frac{\partial}{\partial \phi} \qquad (8.5)$$

We assume for the central force problem that $\hat{V}$ is not a function of θ or ϕ. Then $\hat{V} = V(r)$, and

$$\vec{F}(r) = -\frac{\partial V(r)}{\partial r}\hat{r} \qquad (8.6)$$

In the gravitational example, we would simply find that F$= -mg$. For the atomic problem, we use Coulomb's law for the force:

$$\vec{F}(r) = -\frac{1}{4\pi\epsilon_\circ}\frac{Ze^2}{r^2}\hat{r} \qquad (8.7)$$

where the nuclear charge is given by Ze (Z denotes the number of positive charges in the nucleus), the single electron charge is $-e$, and $\hat{r}$ is the unit vector acting along positive $\vec{r}$. The potential is then given by

$$\frac{\partial V(r)}{\partial r} = \frac{Ze^2}{4\pi\epsilon_\circ r^2} \qquad (8.8)$$

which leads to

$$V(r) = \frac{Ze^2}{4\pi\epsilon_\circ} \int_0^r \frac{dr}{r^2} = -\frac{Ze^2}{4\pi\epsilon_\circ r} \tag{8.9}$$

This is literally the spherical potential used in Schrödinger's equation to describe the one-electron atom.

8.2.2 Approach to the Schrödinger Equation

The Schrödinger equation can provide several solutions. It will generate values for E_n, the term energies that are the eigenvalues of the stationary equation. Embedded in these are the quantum numbers for the various states. In addition, Ψ gives the probability distribution, which can be used for orbital "locations", for example.

The time-dependent Schrödinger equation is presented in Chapter 7 as equation (7.42):

$$\hat{H}\Psi(\vec{r}, t) = i\hbar\frac{\partial}{\partial t}\Psi(\vec{r}, t)$$

In the discussion following the development of equation (7.42), it is separated into the product of a position-dependent, $\psi(\vec{r})$, and a time dependent, $f(t)$, function. The time-dependent function is then given by equation (7.46):

$$f(t) = Ce^{-\frac{iEt}{\hbar}} = Ce^{-i\omega t}$$

That leaves a time-independent part that is also equal to E, producing the stationary Schrödinger equation:

$$\left[-\frac{\hbar^2}{2m_a}\nabla^2 + V(\vec{r})\right]\psi(\vec{r}) = E\psi(\vec{r}) \tag{8.10}$$

In this particular development, m_a will denote mass in general, where the subscript a is attached because a magnetic quantum number m will be developed shortly, and this notation avoids confusion between the two. In addition, $\vec{r}$ is used to denote position in a three-dimensional space. Equation (8.10) is the eigenvalue equation that leads to quantization in the one-electron atom.

The problem is spherically symmetric, and in this case ∇^2 can be represented in one of two ways:

$$\nabla^2 = \frac{1}{r^2}\frac{\partial}{\partial r}\left(r^2\frac{\partial}{\partial r}\right) + \frac{1}{r^2\sin\theta}\frac{\partial}{\partial\theta}\left(\sin\theta\frac{\partial}{\partial\theta}\right) + \frac{1}{r^2\sin^2\theta}\frac{\partial^2}{\partial\phi^2} \tag{8.11}$$

$$\nabla^2 = \frac{\partial^2}{\partial r^2} + \frac{2}{r}\frac{\partial}{\partial r} + \frac{1}{r^2}\frac{\partial^2}{\partial \theta^2} + \frac{\cot\theta}{r^2}\frac{\partial}{\partial \theta} + \frac{1}{r^2\sin^2\theta}\frac{\partial^2}{\partial \phi^2} \qquad (8.12)$$

We can use either of these equivalent expressions for ∇^2 in equation (8.10). Equation (8.11) is a version commonly encountered in engineering, but we will usually adopt equation (8.12) here. In equations (8.11) and (8.12), the r is now missing vector notation. In spherical coordinates, r represents magnitude of the distance from the nucleus to the electron, as shown in Figure 8.1.

The frontal attack to this problem would be to assume that $\psi(\vec{r})$ is separable into functions of r, θ and ϕ such as

$$\psi(\vec{r}) = R(r)\Theta(\theta)\Phi(\phi) \qquad (8.13)$$

and this would be used to separate variables in equation (8.10). This approach does work, but there is a somewhat simpler path. The radial solution is fairly straightforward, and it will lead to the principal quantum number (the same n used in the Balmer series). The system does feature a radially symmetric potential, but there is also an orbital angular kinetic energy term that complicates the solution somewhat. Knowing that it is necessary to evaluate orbital motion, we begin with the orbital angular momentum.

It would be fair to ask how it is possible to redirect the problem statement to angular momentum. Consider the commutation relations

$$[\hat{H}, \hat{L}^2] = 0 \qquad (8.14)$$

where $\hat{H}$ is the Hamiltonian operator that is associated with energy and $\hat{L}$ is the orbital angular momentum operator. $\hat{H}$ and $\hat{L}^2$ commute; they share the same wave function. This has a classical analog; if we divide (momentum)^2 by $2m_a$, we find kinetic energy. In addition:

$$[\hat{H}, \hat{L}_x] = 0 \qquad (8.15)$$
$$[\hat{H}, \hat{L}_y] = 0$$
$$[\hat{H}, \hat{L}_z] = 0$$
$$[\hat{L}^2, \hat{L}_x] = 0$$
$$[\hat{L}^2, \hat{L}_y] = 0$$
$$[\hat{L}^2, \hat{L}_z] = 0$$
$$[\hat{L}^2, \hat{L}_z] = 0$$

but

$$[\hat{L}_x, \hat{L}_y] = i\hbar\hat{L}_z \qquad (8.16)$$
$$[\hat{L}_y, \hat{L}_z] = i\hbar\hat{L}_x$$
$$[\hat{L}_z, \hat{L}_x] = i\hbar\hat{L}_y$$

In short, $\hat{H}, \hat{L}^2$, and *one* of $\hat{L}_x, \hat{L}_y$, or $\hat{L}_z$ all commute (it is common to use $\hat{L}_z$ as the one component of orbital angular momentum that is solved). The same wave function in r, θ, ϕ space is an eigenfunction of $\hat{H}, \hat{L}^2$, and $\hat{L}_z$. It is thus at least acceptable practice to begin with $\hat{L}^2$ instead of $\hat{H}$.

For the angular momentum problem, there is no r component. We write

$$\hat{L}^2 Y(\theta, \phi) = cY(\theta, \phi) \qquad (8.17)$$
$$\hat{L}_z Y(\theta, \phi) = bY(\theta, \phi) \qquad (8.18)$$

where $\psi(r, \theta, \phi) = R(r)Y(\theta, \phi)$.

z-Component of angular momentum

The z momentum problem is the simplest to solve, and it is embedded in the other solutions, so we work it first. The z-directed component of momentum is depicted in Figure 8.2. In classical mechanics, $\vec{L} = \vec{r} \times \vec{p} = \vec{r} \times m\vec{v}$ (using these variables in a generic sense). Then L_z acts along z according to the right-hand rule. The angular momentum denoted by L_z thus confines its motion to the x/y plane. The linear momentum within the x/y plane can be written as mv. The angular momentum is then mv times a lever arm, given by $r\sin\theta$ in this case.

We can rely upon correspondence to apply these ideas to the quantum problem:

$$\hat{L}_z = (r\sin\theta)\,\hat{p}_{(x/y\ plane)} \qquad (8.19)$$

where the carats (ˆ) indicate operators, not unit vectors. In quantum mechanics, $\hat{p}$ is given by $-i\hbar\nabla$, but for momentum in the x/y plane only

$$\nabla = \frac{1}{r\sin\theta}\frac{\partial}{\partial\phi} = \frac{1}{r\sin\theta}\frac{d}{d\phi} \qquad (8.20)$$

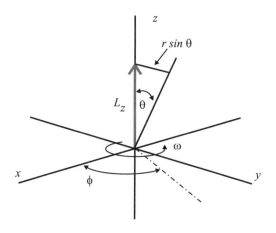

Figure 8.2: z-directed momentum.

There is no θ or r dependence to L_z (other than the parametric dependence of the cross-term $r\sin\theta$), and the differential becomes ordinary
because there is only ϕ dependence; thus

$$\hat{L}_z = r\sin\theta\left(-i\hbar\frac{1}{r\sin\theta}\frac{d}{d\phi}\right) \tag{8.21}$$

The $r\sin\theta$ terms cancel and the z-momentum equation is

$$-i\hbar\frac{d}{d\phi}Y(\theta,\ \phi) = bY(\theta,\ \phi) \tag{8.22}$$

Separating Y by $Y(\theta,\ \phi) = \Theta(\theta)\ \Phi(\phi)$, the equation is rewritten as

$$-i\hbar\frac{d}{d\phi}\left[\Theta(\theta)\ \Phi(\phi)\right] = b\left[\Theta(\theta)\ \Phi(\phi)\right] \tag{8.23}$$

Because there is no θ dependence, $\Theta(\theta)$ cancels on both sides and we
find

$$-i\hbar\frac{d}{d\phi}\Phi(\phi) = b\Phi(\phi) \tag{8.24}$$

or

$$-i\hbar\frac{d\Phi(\phi)}{\Phi(\phi)} = b\ d\phi \tag{8.25}$$

This is directly integrable:

$$\Phi(\phi) = A\ e^{\frac{ib\phi}{\hbar}} \tag{8.26}$$

(the minus sign has been removed via multiplication by i/i). It is necessary to evaluate both A and b. First, since the wave function must close upon itself after one 2π rotation, it can not be multi valued anywhere. There is therefore a periodicity requirement represented by

$$\Phi(\phi + 2\pi) = \Phi(\phi) \tag{8.27}$$

Therefore

$$2\pi \frac{b}{\hbar} = 2\pi m, \quad \text{where } m = 0, \pm 1, \pm 2, ... \tag{8.28}$$

or

$$\boxed{b = m\hbar} \tag{8.29}$$

Note that, by knowing the value of b, we know the value of the component of orbital angular momentum along z, because b is the eigenvalue of the $\hat{L}_z$ problem.

The wave function is thus given by

$$\Phi(\phi) = A e^{im\phi} \tag{8.30}$$

and A can be evaluated by normalizing $| \Phi |^2$:

$$\int_0^{2\pi} \left(A e^{im\phi}\right)^* \left(A e^{im\phi}\right) d\tau = 1 \tag{8.31}$$

In Cartesian space, $d\tau = r^2 \sin\theta \, dr \, d\theta \, d\phi$, but here we have only ϕ dependence, so

$$| A |^2 \int_0^{2\pi} e^{-im\phi} e^{im\phi} \, d\phi = 1 \tag{8.32}$$

The integrand is 1, and

$$| A |^2 (2\pi) = 1 \tag{8.33}$$

Thus, Φ becomes

$$\boxed{\Phi(\phi) = \frac{1}{\sqrt{2\pi}} e^{im\phi}, \quad m = 0, \pm 1, \pm 2, ...} \tag{8.34}$$

The integer m is the first quantum number we have encountered in this solution. It is usually called a "magnetic" quantum number, for reasons that will become clear.

Magnitude of the total orbital angular momentum

The orbital angular wave function is now given by

$$Y(\theta, \phi) = \Theta(\theta)\Phi(\phi) = \frac{\Theta(\theta)}{\sqrt{2\pi}}e^{im\phi} \tag{8.35}$$

and the next eigenvalue problem to be solved is equation (8.17):

$$\hat{L}^2 Y(\theta, \phi) = cY(\theta, \phi)$$

Again, we have to convert linear momentum to angular momentum, but it is no longer confined to the x/y plane. Therefore, $\hat{p} = -i\hbar\nabla$, $\hat{p}^2 = -\hbar^2\nabla^2$, and

$$\hat{L}^2 = r^2(-\hbar^2\nabla^2) \tag{8.36}$$

For ∇ we use only the angular portions, as there is no r dependence to angular momentum. Multiplying through by r^2, we find

$$\hat{L}^2 = -\hbar^2 \left[\frac{\partial^2}{\partial\theta^2} + \cot\theta\frac{\partial}{\partial\theta} + \frac{1}{\sin^2\theta}\frac{\partial^2}{\partial\phi^2} \right] \tag{8.37}$$

The $\hat{L}^2$ eigenvalue equation is then

$$
\begin{aligned}
\hat{L}^2 Y(\theta, \phi) &= -\hbar^2 \left[\frac{\partial^2}{\partial\theta^2} + \cot\theta\frac{\partial}{\partial\theta} + \frac{1}{\sin^2\theta}\frac{\partial^2}{\partial\phi^2} \right] \frac{\Theta(\theta)}{\sqrt{2\pi}}e^{im\phi} \\
&= cY(\theta, \phi) = c\frac{\Theta(\theta)}{\sqrt{2\pi}}e^{im\phi}
\end{aligned} \tag{8.38}
$$

The $1/\sqrt{2\pi}$ term can be canceled on both sides. Then divide by $-\hbar^2$, reorganize, convert to an ordinary differential, and eliminate $e^{im\phi}$:

$$
\begin{aligned}
e^{im\phi}\frac{d^2\Theta}{d\theta^2} + e^{im\phi}\cot\theta\frac{d\Theta}{d\theta} - \frac{m^2}{\sin^2\theta}e^{im\phi}\Theta &= -\frac{c}{\hbar^2}\Theta e^{im\phi} \\
\frac{d^2\Theta}{d\theta^2} + \cot\theta\frac{d\Theta}{d\theta} - \frac{m^2}{\sin^2\theta}\Theta &= -\frac{c}{\hbar^2}\Theta
\end{aligned} \tag{8.39}
$$

Now we perform the first of many substitutions. The goal is to put the equation in a form that is more amenable to solution:

$$
\begin{aligned}
x &\equiv \cos\theta \quad \text{and therefore } dx = -\sin\theta \, d\theta \\
\Theta(\theta) &\equiv G(x)
\end{aligned} \tag{8.40}
$$

Converting the differentials

$$\frac{d\Theta}{d\theta} = \frac{dG}{dx}\frac{dx}{d\theta} = -\sin\theta\frac{dG}{dx} \tag{8.41}$$

Furthermore, $\sin^2\theta + \cos^2\theta = 1$, so $\sin\theta = \sqrt{1-x^2}$. Then

$$\frac{d}{d\theta} = -\sqrt{1-x^2}\frac{d}{dx} \tag{8.42}$$

The second derivative is then found by operating twice:

$$
\begin{aligned}
\frac{d^2}{d\theta^2} &= \frac{d}{d\theta}\left(\frac{d}{d\theta}\right) \\
&= \left(-\sqrt{1-x^2}\frac{d}{dx}\right)\left(-\sqrt{1-x^2}\frac{d}{dx}\right) \\
&= -\sqrt{1-x^2}\left[-\sqrt{1-x^2}\frac{d^2}{dx^2} + \left(-\frac{1}{2}\frac{1}{\sqrt{1-x^2}}(-2x)\right)\frac{d}{dx}\right] \\
&= (1-x^2)\frac{d^2}{dx^2} - x\frac{d}{dx} \tag{8.43}
\end{aligned}
$$

Using these forms for $d/d\theta$ and $d^2/d\theta^2$ to reformulate the $\hat{L}^2$ equation, we find

$$(1-x^2)\frac{d^2G}{dx^2} - x\frac{dG}{dx} + \cot\theta\left(-\sqrt{1-x^2}\frac{dG}{dx}\right) - \frac{m^2}{1-x^2}G = -\frac{c}{\hbar^2}G \tag{8.44}$$

Using the definition of x, it is easy to show that $\cot\theta = \cos\theta/\sin\theta = x/\sqrt{1-x^2}$. This can be used in the above equation, and with some rearrangement we obtain

$$(1-x^2)\frac{d^2G}{dx^2} - 2x\frac{dG}{dx} + \left(\frac{c}{\hbar^2} - \frac{m^2}{1-x^2}\right)G = 0 \tag{8.45}$$

This form of the orbital angular momentum equation is a Legendre differential equation. The canonical form for a Legendre equation is

$$(1-x^2)\frac{d^2G}{dx^2} - 2x\frac{dG}{dx} + \left(\nu(\nu+1) - \frac{\mu^2}{1-x^2}\right)G = 0 \tag{8.46}$$

Very little would be required to make equation (8.45) match equation (8.46). We could then examine the properties of Legendre equations and find relations that lead to quantum numbers, the eigenvalue (providing the orbital angular momentum), the wave function and so on. We will not do that, however, for several reasons. First, once into the Legendre equations we must spend time discussing their properties. There are several faster paths to quantum numbers and the orbital angular momentum. The Legendre equation path does provide the wave functions as well, but since this chapter is on spectroscopy, we have chosen to focus on quantum numbers and eigenvalues, not so much on the eigenfunctions. We therefore perform another substitution that leads to conclusions about the solution to equation (8.45). It is important to point out that we will not find an outcome that is different from what we would find if we explored the properties of Legendre functions in a handbook. We will in fact be finding these properties ourselves without reliance upon a mathematical handbook. Therefore, we define a new form for the wave function:

$$G(x) \equiv (1 - x^2)^{\frac{|m|}{2}} H(x) \tag{8.47}$$

In the end, m will take positive and negative values, which is why we use $\mid m \mid$ in what follows. We also streamline the conversions by the use of H' to indicate $\mathrm{d}H/\mathrm{d}x$, and so forth:

$$\frac{\mathrm{d}G}{\mathrm{d}x} = (1 - x^2)^{\frac{|m|}{2}} H' - \mid m \mid x(1 - x^2)^{\frac{|m|}{2}-1} H \tag{8.48}$$

$$\frac{\mathrm{d}^2 G}{\mathrm{d}x^2} = (1 - x^2)^{\frac{|m|}{2}} \left[H'' - 2 \mid m \mid x(1 - x^2)^{-1} H' \right.$$

$$\left. + \left(2 \mid m \mid x^2 \left(\frac{\mid m \mid}{2} - 1 \right)(1 - x^2)^{-2} - \mid m \mid (1 - x^2)^{-1} \right) H \right]$$

When these are used in equation (8.45), with a good deal of rearrangement and simplification, we obtain

$$(1 - x^2)H'' - 2x(\mid m \mid +1)H' + \left[\frac{c}{\hbar^2} - \mid m \mid (1+ \mid m \mid) \right] H = 0 \tag{8.49}$$

Next we assume a power series solution to H which will provide a recursion relation that generates quantum numbers and eigenvalues:

$$H(x) \equiv \sum_{j=0}^{\infty} a_j x^j \tag{8.50}$$

Then

$$H' = \sum_{j=1}^{\infty} j \, a_j \, x^{j-1}$$

$$H'' = \sum_{j=2}^{\infty} j(j-1) \, a_j \, x^{j-2} \qquad (8.51)$$

The lower indices on the summations in equations (8.51) are set by the fact that for lower values of j, the individual terms in the summation are equal to zero. We can rearrange the indices in useful ways that allow us to simplify the ODE and reach the recursion relation we seek. As an example, for H' in equation (8.51), we set $k = j - 1$ (i.e. $j = k + 1$) and

$$H' = \sum_{k=0}^{\infty} (k+1) \, a_{k+1} \, x^k \qquad (8.52)$$

but, in fact, we do not care if we use k or j, so the same expression can be written as

$$H' = \sum_{j=0}^{\infty} (j+1) \, a_{j+1} \, x^j \qquad (8.53)$$

We can do the same for H'' to obtain

$$H'' = \sum_{j=0}^{\infty} (j+1)(j+2) \, a_{j+2} \, x^j \qquad (8.54)$$

Note that in equation (8.51) we have an x^{j-2} and in equation (8.54) we have x^j, but they both represent H''. Note also in equation (8.49) that there is one term with 1 times H'' and one term with $-x^2$ times H''. If we use equation (8.54) to represent H'' in the first case, and equation (8.51) to represent H'' in the second, the final summation for H'' will contain nothing but terms in x^j. That is what this exercise is about, representation of equation (8.49) in terms of a summation using terms in x^j only. In the same line of thought, note for example that

$$H' = \sum_{j=1}^{\infty} j \, a_j \, x^{j-1} = \sum_{j=0}^{\infty} j \, a_j \, x^{j-1} \qquad (8.55)$$

because that just adds a zero. Now, by judiciously choosing which expansion to substitute into equation (8.49), we obtain

$$\sum_{j=0}^{\infty} \Bigg[(j+2)(j+1)a_{j+2} \tag{8.56}$$

$$+\Big(-j^2 - j - 2 \mid m \mid j + \frac{c}{\hbar^2} - \mid m \mid^2 - \mid m \mid \Big)a_j \Bigg]x^j = 0$$

We can either set $x^j = 0$, which is not useful, or set the terms inside the square braces equal to zero, which leads to

$$
\begin{aligned}
a_{j+2} &= \frac{j^2 + j + 2 \mid m \mid j - \frac{c}{\hbar^2} + \mid m \mid^2 + \mid m \mid}{(j+2)(j+1)} a_j \\
&= \frac{(j+ \mid m \mid) (j+ \mid m \mid +1) - \frac{c}{\hbar^2}}{(j+2)(j+1)} a_j
\end{aligned}
\tag{8.57}
$$

which is the desired recursion relation. This outcome generates an even series for a if we start with a_0 and an odd series for a if we start with a_1. Since it is an infinite series, we can just set either a_0 or a_1 equal to zero, and it is typical to choose $a_1 = 0$.

It turns out that the series in equation (8.57) blows up for large j. Some say this happens because there is a singularity at $x = \pm 1$ in equation (8.45), while others have more complicated physical arguments. In either case, this series for a is meant to describe H, which is a substitution for G, which is itself a substitution for the wave function Θ. We cannot let the wave function blow up, so we restrict the expansion by making a equal to zero at some value of $j = k$, where k is an integer. We can do that by setting the numerator of equation (8.57) equal to zero at $j = k$. Once $a = 0$, it remains zero at higher values of j. This limitation to the recursion relation leads to

$$(k + \mid m \mid) (k + \mid m \mid +1) = \frac{c}{\hbar^2} \tag{8.58}$$

This can be true for various integer values of k. The quantum number m takes the values $0, \pm 1, \pm 2 ...$ while $\mid m \mid = 0, 1, 2 ...$ Then $k + \mid m \mid = 0, 1, 2 ...$ Now define a new quantum number:

$$\ell \equiv k + \mid m \mid \tag{8.59}$$

meaning that

$$\ell \geq \mid m \mid \tag{8.60}$$

m is thus confined to the values $-\ell, -\ell + 1, \ldots - 1, 0, 1 \ldots, \ell - 1, \ell$. In fact, m is usually designated as m_ℓ to make this relationship clear. The letter ℓ is called the azimuthal quantum number.

Then, from equation (8.58):

$$c = \ell(\ell + 1)\hbar^2 \quad \text{for } \ell = 0, 1, 2, \ldots \tag{8.61}$$

Because c is the eigenvalue for the orbital momentum equation, this relationship provides the magnitude of the orbital angular momentum squared.

To recap, so far we have solved the following expressions for their eigenvalues:

$$\hat{L}^2 \, Y(\theta, \phi) = c \, Y(\theta, \phi)$$
$$\hat{L}_z \, Y(\theta, \phi) = b \, Y(\theta, \phi)$$

The solutions have provided two sets of quantum numbers. Because $\hat{L}^2$ and $\hat{L}_z$ commute, they share the eigenfunction $Y(\theta, \phi)$. Because they are both eigenvalue problems they generate observables. The expression

$$c = \ell(\ell + 1)\hbar^2, \quad \ell = 0, 1, 2, \ldots$$

provides the value of the orbital angular momentum squared, for example, and then

$$|\, c\,|^{1/2} = \hbar\sqrt{\ell(\ell + 1)}$$

is the magnitude of the orbital angular momentum. The eigenvalue b is then

$$b = m_\ell \hbar, \quad |\, m_\ell\,| \leq \ell$$

which is the component of $\hat{L}$ along z.

We have already solved for Φ, but not Θ. It is possible to sketch the solution quickly. Going backwards, we have

$$x = \cos\theta \text{ and } \Theta(\theta) = G(x)$$
$$G(x) = (1 - x^2)^{\frac{|m|}{2}} H(x)$$
$$H(x) = \sum_{j=0}^{\infty} a_j x^j$$

combining

$$\Theta(\theta) = (1 - \cos^2\theta)^{\frac{|m|}{2}} \sum_{j=0}^{\infty} a_j \cos^j\theta$$

Since j is limited by k, we can write this as

$$\boxed{\Theta_{\ell,m}(\theta) = \sin^{|m|}\theta \sum_{j=0}^{\ell-|m|} a_j \cos^j\theta} \qquad (8.62)$$

These $\Theta_{\ell,m}(\theta)$ are a form of associated Legendre functions. The same equation is presented in terms of Legendre polynomials in references [60] and [65]. The orbital angular wave function can then be written as

$$\boxed{Y(\theta,\phi) = \sin^{|m|}\theta \left[\sum_{j=0}^{\ell-|m|} a_j \cos^j\theta \right] \frac{e^{im\phi}}{\sqrt{2\pi}}} \qquad (8.63)$$

The central point to this discussion is that, although somewhat messy, the wave functions can be found.

The angular momentum problem can actually be solved in a much quicker and neater way using "ladder operators" [60, 58, 66]. Ladder operators were not used here because they take some time to understand fully. To solve the radial problem will require a series solution. It therefore seemed best to fix on a single mathematical approach for the entire Chapter.

Energy and the full Hamiltonian

Now we return to the full one-electron Schrödinger equation (8.10):

$$\left[-\frac{\hbar^2}{2m_a} \nabla^2 + V(r) \right] \psi(\vec{r}) = E\psi(\vec{r})$$

where

$$\nabla^2 = \frac{\partial^2}{\partial r^2} + \frac{2}{r}\frac{\partial}{\partial r} + \frac{1}{r^2}\frac{\partial^2}{\partial\theta^2} + \frac{\cot\theta}{r^2}\frac{\partial}{\partial\theta} + \frac{1}{r^2\sin^2\theta}\frac{\partial^2}{\partial\phi^2}$$

Note, however, that

$$\hat{L}^2 = -\hbar^2 \left[\frac{\partial^2}{\partial \theta^2} + \cot \theta \frac{\partial}{\partial \theta} + \frac{1}{\sin^2 \theta} \frac{\partial^2}{\partial \phi^2} \right] \qquad (8.64)$$

so we can recast ∇^2 as

$$\nabla^2 = \frac{\partial^2}{\partial r^2} + \frac{2}{r} \frac{\partial}{\partial r} - \frac{1}{r^2 \hbar^2} \hat{L}^2 \qquad (8.65)$$

The Hamiltonian operator is thus

$$-\frac{\hbar^2}{2m_a} \left[\frac{\partial^2}{\partial r^2} + \frac{2}{r} \frac{\partial}{\partial r} \right] + \left(\frac{1}{2m_a r^2} \right) \hat{L}^2 + V(r) \qquad (8.66)$$

Note that $1/2m_a r^2$ is an appropriate conversion from angular momentum squared to kinetic energy. In addition, we already know that

$$\hat{L}^2 \psi = \ell(\ell + 1)\hbar^2 \psi \qquad (8.67)$$

the point being that we have already solved the $\hat{L}^2$ eigenvalue problem. When $\hat{L}^2$ operates on ψ, using equation (8.66), we do not have to worry about the operator because $\hat{L}^2$ and $\hat{H}$ commute. We just insert $\ell(\ell + 1)\hbar^2 \psi$. The one-electron atom Schrödinger equation is then

$$\left[-\frac{\hbar^2}{2m_a} \left(\frac{\partial^2}{\partial r^2} + \frac{2}{r} \frac{\partial}{\partial r} \right) + \frac{\ell(\ell + 1)\hbar^2}{2m_a r^2} + V(r) \right] \psi = E\psi \qquad (8.68)$$

Next we assume that $\psi(r, \theta, \phi)$ is separable into $R(r)Y(\theta, \psi)$, and we have already solved for $Y(\theta, \psi)$. Insert $\psi(r, \theta, \phi) = R(r)Y(\theta, \psi)$ into equation (8.68) and divide by $Y(\theta, \psi)$ to obtain

$$\left[-\frac{\hbar^2}{2m_a} \left(\frac{d^2}{dr^2} + \frac{2}{r} \frac{d}{dr} \right) + \frac{\ell(\ell + 1)\hbar^2}{2m_a r^2} + V(r) \right] R(r) = ER(r) \qquad (8.69)$$

At this point, it is necessary to be careful about how we write the terms in equation (8.69). The atom is actually two bodies. The group translates through space (somewhat like the free particle problem) and there is relative motion within the center of mass reference frame (internal energy states). We wish to solve the problem in the center of mass frame because we are solving for the states of an electron that is bound by the potential field $V(r)$. This requires a diversion that will cast the problem in the correct frame.

For simplicity, we assume for now that the axis connecting the electron and the nucleus is aligned along the x axis. This is no doubt a simplification because the probability density given by $|\psi|^2$ is clearly three-dimensional. All the same, the outcome of this simplified approach can be directly applied to the full problem. There are now two particles with masses m_n and m_e (designating the nuclear and electronic masses respectively), located at positions x_n and x_e and the potential is written as $V(x_n - x_e)$. We write a two-particle Hamiltonian operating on ψ:

$$-\left(\frac{\hbar^2}{2m_n}\frac{\partial^2}{\partial x_n^2}\right)\psi(x_n, x_e) - \left(\frac{\hbar^2}{2m_e}\frac{\partial^2}{\partial x_e^2}\right)\psi(x_n, x_e) + V(x_n - x_e)\psi(x_n, x_e)$$

$$(8.70)$$

We now define new variables by

$$\text{Relative displacement } x \equiv (x_n - x_e) \qquad (8.71)$$

$$\text{Location of the center of mass } X \equiv \frac{m_n x_n + m_e x_e}{m_n + m_e} \qquad (8.72)$$

We will need to transform variables, starting with

$$\frac{\partial x}{\partial x_n} = 1$$

$$\frac{\partial x}{\partial x_e} = -1$$

$$\frac{\partial X}{\partial x_n} = \frac{m_n}{m_n + m_e}$$

$$\frac{\partial X}{\partial x_e} = \frac{m_e}{m_n + m_e} \qquad (8.73)$$

Then, for any function $f(x, X)$

$$\frac{\partial f(x, X)}{\partial x_n} = \frac{\partial f(x, X)}{\partial x}\frac{\partial x}{\partial x_n} + \frac{\partial f(x, X)}{\partial X}\frac{\partial X}{\partial x_n}$$

$$= \frac{\partial f(x, X)}{\partial x} + \left(\frac{m_n}{m_n + m_e}\right)\frac{\partial f(x, X)}{\partial X} \qquad (8.74)$$

Similarly

$$\frac{\partial f(x, X)}{\partial x_e} = -\frac{\partial f(x, X)}{\partial x} + \left(\frac{m_e}{m_n + m_e}\right)\frac{\partial f(x, X)}{\partial X} \qquad (8.75)$$

or, in shorthand,

$$\frac{\partial}{\partial x_n} = \frac{\partial}{\partial x} + \left(\frac{m_n}{m_n + m_e}\right)\frac{\partial}{\partial X}$$

$$\frac{\partial}{\partial x_e} = -\frac{\partial}{\partial x} + \left(\frac{m_e}{m_n + m_e}\right)\frac{\partial}{\partial X} \tag{8.76}$$

The second derivatives are then found via

$$\frac{\partial^2}{\partial x_n^2} = \left[\frac{\partial}{\partial x} + \left(\frac{m_n}{m_n + m_e}\right)\frac{\partial}{\partial X}\right]\left[\frac{\partial}{\partial x} + \left(\frac{m_n}{m_n + m_e}\right)\frac{\partial}{\partial X}\right]$$

$$= \frac{\partial^2}{\partial x^2} + \left(\frac{m_n}{m_n + m_e}\right)\frac{\partial}{\partial x}\frac{\partial}{\partial X} + \left(\frac{m_n}{m_n + m_e}\right)\frac{\partial}{\partial X}\frac{\partial}{\partial x}$$

$$+ \left(\frac{m_n}{m_n + m_e}\right)^2\frac{\partial^2}{\partial X^2} \tag{8.77}$$

If we can assume smooth functions in which the order of differentiation does not matter, this reduces to

$$\frac{\partial^2}{\partial x_n^2} = \frac{\partial^2}{\partial x^2} + \left(\frac{2m_n}{m_n + m_e}\right)\frac{\partial}{\partial x}\frac{\partial}{\partial X} + \left(\frac{m_n}{m_n + m_e}\right)^2\frac{\partial^2}{\partial X^2} \tag{8.78}$$

Similarly

$$\frac{\partial^2}{\partial x_e^2} = \frac{\partial^2}{\partial x^2} - \left(\frac{2m_e}{m_n + m_e}\right)\frac{\partial}{\partial x}\frac{\partial}{\partial X} + \left(\frac{m_e}{m_n + m_e}\right)^2\frac{\partial^2}{\partial X^2} \tag{8.79}$$

Expression (8.70) then becomes

$$-\frac{\hbar^2}{2m_n}\left[\frac{\partial^2}{\partial x^2} + \left(\frac{2m_n}{m_n + m_e}\right)\frac{\partial}{\partial x}\frac{\partial}{\partial X} + \left(\frac{m_n}{m_n + m_e}\right)^2\frac{\partial^2}{\partial X^2}\right]\psi(x, X)$$

$$-\frac{\hbar^2}{2m_e}\left[\frac{\partial^2}{\partial x^2} - \left(\frac{2m_e}{m_n + m_e}\right)\frac{\partial}{\partial x}\frac{\partial}{\partial X} + \left(\frac{m_e}{m_n + m_e}\right)^2\frac{\partial^2}{\partial X^2}\right]\psi(x, X)$$

$$+ V(x)\psi(x, X) \tag{8.80}$$

We then reorganize in terms of $\partial^2/\partial x^2$, $(\partial/\partial x)(\partial/\partial X)$, and $\partial^2/\partial X^2$. After some algebraic manipulation, the full stationary Schrödinger equation becomes

$$-\frac{\hbar^2}{2}\left(\frac{m_n + m_e}{m_n m_e}\right)\frac{\partial^2}{\partial x^2}\psi(x, X) - \frac{\hbar^2}{2}\left(\frac{1}{m_n + m_e}\right)\frac{\partial^2}{\partial X^2}\psi(x, X)$$

$$+ V(x)\,\psi(x, X) = E\,\psi(x, X) \tag{8.81}$$

We can now separate this wave function via $\psi(x, X) = \epsilon(x)\eta(X)$:

$$-\frac{\hbar^2}{2}\eta(X)\left(\frac{m_n + m_e}{m_n m_e}\right)\frac{\partial^2\epsilon(x)}{\partial x^2} - \frac{\hbar^2}{2}\epsilon(x)\left(\frac{1}{m_n + m_e}\right)\frac{\partial^2\eta(X)}{\partial X^2}$$
$$+V(x)\,\epsilon(x)\eta(X) = E\,\epsilon(x)\eta(X) \tag{8.82}$$

Then divide by $\epsilon(x)\eta(X)$ and rearrange to obtain

$$-\frac{\hbar^2}{2}\frac{1}{\epsilon(x)}\left(\frac{m_n + m_e}{m_n m_e}\right)\frac{\partial^2\epsilon(x)}{\partial x^2} + V(x) - E$$
$$= \frac{\hbar^2}{2}\frac{1}{\eta(X)}\left(\frac{1}{m_n + m_e}\right)\frac{\partial^2\eta(X)}{\partial X^2} \tag{8.83}$$

As usual, everything on the left is a function of one variable, x, and everything on the right is a function of another, X, and so they must be equal to a constant. Here, we choose $-E_{cm}$, where cm denotes the center of mass as it moves through space. Therefore

$$-\frac{\hbar^2}{2(m_n + m_e)}\frac{d^2\eta(X)}{dX^2} = \eta(X)E_{cm} \tag{8.84}$$

This equation describes a free-space (no potential field is involved), two-particle problem for the motion of the center of mass (with a total mass given by $m_n + m_e$), and η is the free-space wave function. We are not interested in the solution to this problem. The left-hand portion of equation (8.83), however, is of interest. Using the same separation we find

$$-\frac{\hbar^2}{2}\left(\frac{m_n + m_e}{m_n m_e}\right)\frac{d^2\epsilon(x)}{dx^2} + V(x)\epsilon(x) = (E - E_{cm})\epsilon(x) \tag{8.85}$$

Compare this with equation (8.69). They are not exactly alike because equation (8.85) was written in simplified coordinates and it ignores angular momentum, while equation (8.69) ignored the two-particle details we have just discussed. Note, however, that the term we called m_a in equation (8.69) is represented here by a "reduced mass" μ defined by

$$\boxed{\mu \equiv \frac{m_n m_e}{m_n + m_e}} \tag{8.86}$$

Moreover, $\epsilon(x)$ in equation (8.85) is functionally the same as $R(r)$ in equation (8.69). Here, both x and r represent the relative distance between the electron and the nucleus. Furthermore, the E in equation (8.69) is actually the internal energy represented explicitly as $E - E_{cm}$ above. In light of these developments, we can recast equation (8.69) more accurately in terms of internal dynamics as

$$\left[-\frac{\hbar^2}{2\mu} \left(\frac{d^2}{dr^2} + \frac{2}{r} \frac{d}{dr} \right) + \frac{\ell(\ell+1)\hbar^2}{2\mu r^2} + V(r) \right] R(r) = ER(r) \qquad (8.87)$$

where we recognize that E [in equation (8.87)] is functionally the same as $E - E_{cm}$ [in equation (8.85)]. Rearrange this ODE, simplify the differentials, and insert the expression for V from equation (8.9) to give

$$R'' + \frac{2}{r} R' + \left[\frac{2\mu}{\hbar^2} E + \frac{2\mu Z e^2}{\hbar^2 4 \pi \epsilon_o r} - \frac{\ell(\ell+1)}{r^2} \right] R = 0 \qquad (8.88)$$

As with the angular momentum problem, we can apply a series solution. In order to adopt an appropriate solution, however, we analyze the case for very large r. The series solution will have to approach this case as r grows. The assumption of large r will allow us to neglect the complicating terms $(2\mu Z e^2)/(\hbar^2 4 \pi \epsilon_o r)$, $\ell(\ell+1)]/(r^2)$, and $(2/r)R'$ in equation (8.88). Equation (8.88) is thus reduced to the following:

$$R'' + \left(\frac{2\mu}{\hbar^2} E \right) R = 0 \qquad (8.89)$$

which is solved by

$$R = e^{\pm i \sqrt{\frac{2\mu}{\hbar^2} E} \, r} \qquad (8.90)$$

This is a spherical harmonic wave. It is not quantized and it describes the system at large r, which is an ionized state. For a bound electron, then, we assume

$$R(r) = K(r) e^{-Cr} \qquad (8.91)$$

where $C = \sqrt{-(2\mu E)/\hbar^2}$. Note that the "i" in the exponent has been converted to $\sqrt{-1}$. The derivatives are then

$$
\begin{aligned}
R' &= K' e^{-Cr} - CK e^{-Cr} \\
R'' &= K'' e^{-Cr} - 2CK' e^{-Cr} + C^2 K e^{-Cr}
\end{aligned} \qquad (8.92)
$$

These can be inserted into equation (8.88). The e^{-Cr} terms then fall out and are canceled. In addition, the term $(2\mu/\hbar^2)E$ in equation (8.88) can be set equal to $-C^2$, and the entire equation can be multiplied by r^2 to produce

$$r^2 K'' + (2r - 2Cr^2)K' + \left[\frac{2\mu Z e^2 r}{\hbar^2 4\pi\epsilon_\circ} - 2Cr - \ell(\ell+1)\right] K = 0 \quad (8.93)$$

Assume a series solution of the form

$$K \equiv \sum_{k=0}^{\infty} c_k r^k \qquad (8.94)$$

If this form is substituted into equation (8.93) and the equation is evaluated, the first few terms in the expansion will be zero. This would require some reorganization of the expansion. As an alternative, we rewrite the expansion as

$$K = \sum_{k=s}^{\infty} c_k r^k \qquad (8.95)$$

where s is the first value of k that produces a nonzero c_k. We can reorder the expansion as before by setting $j = k - s$ so that

$$K = \sum_{j=0}^{\infty} c_{j+s} r^{j+s} = \sum_{j=0}^{\infty} b_j r^{j+s} \qquad (8.96)$$

where $b_j = c_{j+s}$. Reorganizing gives

$$K = r^s \sum_{j=0}^{\infty} b_j r^j = r^s M(r) \qquad (8.97)$$

where $M(r)$ is defined as

$$M(r) \equiv \sum_{j=0}^{\infty} b_j r^j \qquad (8.98)$$

for notational convenience. Then the derivatives are

$$\begin{aligned} K' &= r^s M' + s r^{s-1} M \\ K'' &= r^s M'' + 2s r^{s-1} M' + s(s-1) r^{s-2} M \end{aligned} \qquad (8.99)$$

If these are inserted into equation (8.93), it is then rearranged, and if we divide by r^s, the outcome is

$$r^2 M'' + [(2s+2)r - 2Cr^2]M' \tag{8.100}$$

$$+ \left[s^2 + s + \left(\frac{2\mu Ze^2}{\hbar^2 4\pi\epsilon_\circ} - 2Cs - 2C \right)r - \ell(\ell+1) \right] M = 0$$

The series for M can be written as

$$M(r) \;=\; \sum_{j=0}^{\infty} b_j r^j = b_\circ + b_1 r + b_2 r^2 + \cdots$$

$$M'(r) \;=\; \sum_{j=0}^{\infty} j b_j r^{j-1} = 0 + b_1 + b_2 2r + \cdots$$

$$M''(r) \;=\; \sum_{j=0}^{\infty} j(j-1)b_j r^{j-2} = 0 + 0 + 2b_2 + \cdots \tag{8.101}$$

At $r = 0$ these are simply

$$M(0) \;=\; b_\circ$$
$$M'(0) \;=\; b_1$$
$$M''(0) \;=\; 2b_2 \tag{8.102}$$

Insert these into equation (8.100), under the condition that $r = 0$, to produce

$$0 + 0 + [s^2 + s - \ell(\ell+1)]b_\circ = 0 \tag{8.103}$$

Since $b_\circ \neq 0$, then $[s^2 + s - \ell(\ell+1)] = 0$, and this has two roots: $s = \ell$ and $s = -\ell - 1$. In short, we have located the index where the series stops producing zero, and we can use that fact to generate an appropriate series solution, but first we need to decide which of the two values for s to use. This decision is based upon which value produces a physically reasonable outcome. One way to check that is to evaluate $|R|^2$. Recall that

$$R(r) = K(r)e^{-Cr} = e^{-Cr}r^s \sum_{j=0}^{\infty} b_j r^j \tag{8.104}$$

The normalization integral is

$$\int |R|^2 r^2 \, dr \tag{8.105}$$

where $d\tau = r^2\,dr$ for the r component in spherical coordinates. It turns out that this integral (which ought to equal 1) blows up when $s = -\ell - 1$, but not when $s = \ell$. In other words, the quantum approach fails unless we assume that $s = \ell$. We therefore set

$$R(r) = \mathrm{e}^{-Cr} r^\ell M(r) \tag{8.106}$$

Use this result in equation (8.100), simplify and divide by r to obtain

$$rM'' + [(2\ell + 2) - 2Cr]M' + \left[\frac{2\mu Ze^2}{\hbar^2 4\pi\epsilon_\circ} - 2C\ell - 2C\right]M = 0 \tag{8.107}$$

As before, we reorganize the indices in the expansions in order to make judicious choices about what to insert into equation (8.107):

$$M(r) = \sum_{j=0}^{\infty} b_j r^j$$

$$M'(r) = \sum_{j=0}^{\infty} j b_j r^{j-1} = \sum_{j=0}^{\infty} (j+1) b_{j+1} r^j$$

$$M''(r) = \sum_{j=0}^{\infty} j(j-1) b_j r^{j-2} = \sum_{j=0}^{\infty} (j+1) j b_{j+1} r^{j-1} \tag{8.108}$$

Now we insert into equation (8.107) whatever versions of these will return r^j, our goal being to generate another recursion relation. After rearrangement, we find

$$\sum_{j=0}^{\infty} \Bigg[[(j+1)j + 2(\ell+1)(j+1)]b_{j+1}$$

$$+ \left(\frac{2\mu Ze^2}{\hbar^2 4\pi\epsilon_\circ} - 2C\ell - 2C - 2Cj\right)b_j\Bigg]r^j = 0 \tag{8.109}$$

Our choice is either to set $r = 0$, or to set the terms in square braces equal to zero, and for practical reasons we choose the latter. This generates the recursion relation

$$b_{j+1} = -\left[\frac{\frac{2\mu Ze^2}{\hbar^2 4\pi\epsilon_\circ} - 2C\ell - 2C - 2Cj}{(j+1)j + 2(\ell+1)(j+1)}\right]b_j \tag{8.110}$$

As before, $R(r)$ will blow up unless we limit the recursion relation to some value of j, call it k. Therefore $b_{k+1} = b_{k+2} = \cdots = 0$. To do this, we set the numerator of equation (8.110) equal to zero when $j = k$:

$$\frac{\mu Z e^2}{\hbar^2 4\pi\epsilon_\circ} = C(k + \ell + 1) \tag{8.111}$$

Both k and ℓ are integers. We therefore define

$$\boxed{n \equiv k + \ell + 1, \qquad n = 1, 2, 3...} \tag{8.112}$$

where n is the principal quantum number. Because $k = 0, 1, 2...$

$$\boxed{\ell \leq n - 1 \quad \text{or} \quad \ell = 0, 1, 2..., n - 1} \tag{8.113}$$

Applying the definition of C to equation (8.111) produces

$$n\sqrt{\frac{-2\mu E}{\hbar^2}} = \frac{\mu Z e^2}{\hbar^2 4\pi\epsilon_\circ} \tag{8.114}$$

or

$$\boxed{E_n = -\left(\frac{Z e^2}{h\epsilon_\circ}\right)^2 \frac{\mu}{8n^2}} \tag{8.115}$$

We can then describe the hydrogen-like line series via

$$\frac{1}{\lambda} = \frac{\Delta E}{hc} = \left(\frac{Z e^2}{h\epsilon_\circ}\right)^2 \frac{\mu}{8hc}\left(\frac{1}{n_2^2} - \frac{1}{n_1^2}\right) \tag{8.116}$$

which clearly provides a quantum mechanical value for the Rydberg constant.

Recall that

$$R(r) = e^{-\sqrt{\frac{-2\mu E}{\hbar^2}}r} r^\ell \sum_{j=0}^{\infty} b_j r^j \tag{8.117}$$

where b_j can be found from equation (8.110). Thus, we also have a set of expressions for R. Since $\psi = R(r)Y(\theta, \phi)$, we now have the full wave function as well. These wave functions are clearly expressed in terms of expansions and special functions. They can be found tabulated in various sources (e.g. [65, 64]).

One-electron, or "hydrogen-like", atoms have three quantum numbers. The principal quantum number, n $(= 1, 2, 3, ...)$, is related to energy levels of the atom via equation (8.115). As it turns out, the number of nodes in ψ is given by $n-1$. The azimuthal quantum number is ℓ $(= 0, 1, 2..., n-1)$, and it is related to the magnitude of the orbital angular momentum squared via equation (8.61). ℓ specifies the number of angular nodes, so the number of radial nodes is then $n-\ell-1$. Finally, the magnetic quantum number is m_ℓ $(= 0, \pm 1, \pm 2, ..., \pm \ell)$, and it provides the magnitude of the orbital angular momentum along direction z via equation (8.29).

8.2.3   Introduction to Selection Rules and Notation

Transition moments are described elsewhere in this book (e.g. Chapter 10). Here, we will briefly mention that not all transitions are significant. This has to do with the transition moment [defined in equation (10.9)], which is determined by the inner product:

$$\langle \psi_n | \hat{\mu} | \psi_m \rangle$$

Selection rules determine which transitions are spectroscopically active [e.g. can absorb or emit (a resonant process) or inelastically scatter light (a nonresonant process)]. When the transition moment is effectively zero, the transition is "not allowed" by selection rules. These "forbidden transitions", with negligible transition, moment have a very low probability of occurrence and are rarely observed spectroscopically. Selection rules for one-electron atoms are as follows:

$$\begin{aligned}
\Delta n &= 0, 1, 2, ... \text{ can be any integer} \\
\Delta \ell &= \pm 1 \\
\Delta m_\ell &= 0, \pm 1
\end{aligned}$$

(8.118)

This list is not yet complete because we have not discussed spin.

Spectroscopic notation for atoms is fairly standardized, although it is common to encounter variations. The term symbols for individual atoms consist of a series of numbers and letters that indicate the atomic structure. The first symbol in a total electronic term symbol is the principal quantum number n. The second is a letter that designates the value of ℓ. For $\ell = 0, 1, 2, 3, 4, 5, ...$ we use the letters s, p, d, f, g, h, ... respectively. A 2s orbital, therefore, has $n = 2$ and $\ell = 0$.

Figure 8.3 contains a simplified image of several states and allowed transitions in a one-electron atom.

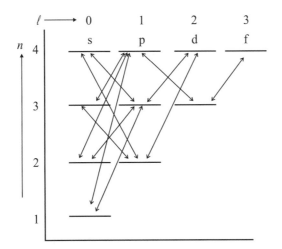

Figure 8.3: States and allowed transitions in a one-electron atom.

8.2.4 Magnetic Moment

Zeeman effect

From the Θ solution for L^2 [equation (8.61)], it is clear that the value of L^2 is quantized, and the magnitude can therefore have discrete values only. One component of the momentum vector is also quantized, thanks to the Φ solution for L_z [equation (8.29)]. This is referred to as "space quantization". Space quantization is not obvious from spectra unless the measurement is made in a strong external magnetic field.

This phenomenon can be explained using semi-classical mechanics. A moving charge q with an angular momentum vector $\vec{L}$ has a magnetic dipole moment $\vec{\mu}_{mag}$ given by

$$\vec{\mu}_{mag} = \gamma \vec{L} \qquad (8.119)$$

where $\gamma = q/(2m_q c)$, m_q being the mass of the charge. An external magnetic field can couple into such a magnetic moment and add or extract energy. This effect can be described by

$$\Delta E_{mag} = -\vec{\mu}_{mag} \cdot \vec{B} \qquad (8.120)$$

If $\vec{B}$ is directed along z, then:

$$\Delta E_{mag} = -\gamma L_z B \qquad (8.121)$$

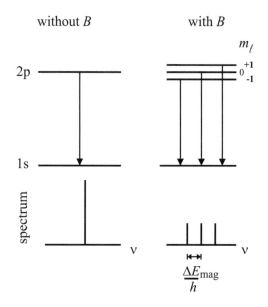

Figure 8.4: Magnetic splitting of a 2p state.

For a one-electron atom, q is equal to $-e$ and

$$\Delta E_{mag} = m_\ell \hbar \frac{e}{2m_e c} B = m_\ell \mu_B B \qquad (8.122)$$

where μ_B is the Bohr magneton given by

$$\mu_B \equiv \frac{\hbar e}{2m_e c} \qquad (8.123)$$

A single emission line from a one-electron atom originally in a 2p state and in the absence of a magnetic field will generate three separate lines in the presence of a strong magnetic field (see Figure 8.4). There is a magnetic splitting that occurs (called the Zeeman effect). The central line of these three will be at the wavelength of the original unsplit line because in that case $\Delta E_{mag} = 0$ ($m_\ell = 0$). The other two lines will be spaced away from that line by $\Delta \nu = \Delta E_{mag}/h$.

This process can be explained using the image in Figure 8.5. For a p orbital, $\ell = 1$, which means that m_ℓ can take the values $-1, 0, +1$, as shown. The magnitude but not the direction of L is known, but we do know the magnitude of the z component. For a d orbital, $\ell = 2$, which means that m_ℓ can take the values $-2, -1, 0, +1, +2$ and the picture

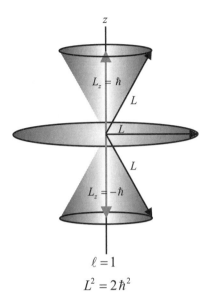

Figure 8.5: Magnetic splitting of a 2p state.

becomes more complicated with further options for L_z (two additional "cones" are added to Figure 8.5).

Spin

The external magnetic field we have discussed actually induces further line splitting, even in the simplest atomic structures. The conclusion is that the electron has some form of intrinsic magnetic moment. This notion was originally attributed to the idea that the electron spins on an axis. If this were literally true, then normal wave mechanics in Cartesian coordinates could be used to describe this behavior. This turns out to be impossible. Although spin does behave in some ways like orbital angular momentum, it does not in other ways (such as a restriction in the allowed quantum numbers). It is thus inappropriate to think of spin as the mechanical rotation of an electron. Paul Dirac developed a relativistic quantum mechanical description of the electron in 1928, from which spin was a natural result. We have not dealt with relativistic quantum mechanics in this book, however, and his treatment is therefore outside the scope of the text.

Here, we can treat spin with the same approach used for the orbital angular momentum, while making concessions to observation. We postulate a spin angular momentum described by

$$S^2 = s(s+1)\hbar^2 \tag{8.124}$$

with a z component given by

$$S_z = m_s\hbar \tag{8.125}$$

where $m_s = 0, \pm s$. The only experimentally observed value of s for an electron is $1/2$, giving

$$S^2 = \frac{3}{4}\hbar^2 \tag{8.126}$$

$$S_z = \pm\frac{1}{2}\hbar$$

As an aside, photons also have spin (right or left circular polarization), with $s = 1$.

Returning to the magnetic moment, we can combine the splitting effect of both the orbital angular momentum and spin by defining

$$(\mu_s)_z = m_s\gamma_s\hbar \tag{8.127}$$

with $\gamma_s = 2\gamma = -e/(m_e c)$. Then

$$\Delta E_{mag} = \mu_B B(m_\ell + 2m_s) \tag{8.128}$$

Splitting of the 2p $\rightarrow$ 1s emission line is then more complex, as shown in Figure 8.6. Note that there are six different energy levels in the 2p state, two of which are degenerate even in the presence of a magnetic field because the spin contribution cancels the orbital contribution.

The arrows in Figure 8.6 are meant to designate just a few of the possible transitions. These transitions are further limited, however, by the spin selection rule which is that the spin quantum number m_s cannot change during a transition:

$$\Delta m_s = 0 \tag{8.129}$$

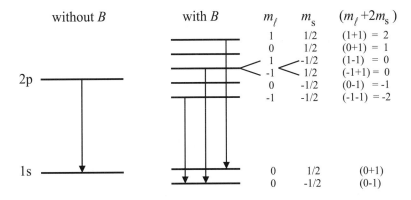

Figure 8.6: Magnetic splitting of a 2p state, including the effect of spin splitting.

Only six transitions from the 2p to the 1s state are thus allowed

1. $(m_\ell = +1, m_s = +1/2) \to (m_\ell = 0, m_s = +1/2)$
2. $(m_\ell = +1, m_s = -1/2) \to (m_\ell = 0, m_s = -1/2)$
3. $(m_\ell = 0, m_s = +1/2) \to (m_\ell = 0, m_s = +1/2)$
4. $(m_\ell = 0, m_s = -1/2) \to (m_\ell = 0, m_s = -1/2)$
5. $(m_\ell = -1, m_s = +1/2) \to (m_\ell = 0, m_s = +1/2)$
6. $(m_\ell = -1, m_s = -1/2) \to (m_\ell = 0, m_s = -1/2)$

8.2.5 Selection Rules, Degeneracy, and Notation

Final selection rules for one-electron atoms are as follows:

$$\begin{aligned}
\Delta n &= 0, 1, 2, \dots \text{ can be any integer} \\
\Delta \ell &= \pm 1 \\
\Delta m_\ell &= 0, \pm 1 \\
\Delta m_s &= 0
\end{aligned} \qquad (8.130)$$

The energy of a one-electron atom is dependent upon n, but not upon ℓ, m_ℓ, or m_s in the absence of an external magnetic field. Each possible value of ℓ, m_ℓ, and m_s (for the same n) thus has the same energy, and the atomic energy states are degenerate. Consider the allowed levels for the various quantum numbers: $n = 1, 2, 3, ...; \ell = 0, 1, 2, 3, ..., n-1$; and $m_\ell = -\ell, -\ell+1, ..., 0, ..., \ell-1, \ell$. For a particular value of n, there are n possible values of ℓ, and for each ℓ there are $2\ell+1$ values for m_ℓ. Therefore, the degeneracy due to ℓ and m_ℓ is the summation over all ℓ of each $2\ell+1$ value:

$$g_{(\ell \; \& \; m_\ell)} = \sum_{\ell=0}^{n-1}(2\ell+1) = \sum_{\ell=0}^{n-1}(2\ell) + \sum_{\ell=0}^{n-1}1 \qquad (8.131)$$

The final summation on the right-hand side of this expression is simply n. The other summation can be manipulated to give

$$\sum_{\ell=0}^{n-1}(2\ell) = 2\sum_{\ell=1}^{n-1}\ell \qquad (8.132)$$

where the lower summation index is changed because the first term was equal to zero. Note that ℓ is simply an integer. The summation of integers from 1 to $n-1$ is $\frac{1}{2}(n-1)n$ [60], so

$$g_{(\ell \; \& \; m_\ell)} = 2\frac{1}{2}(n-1)n + n = n^2 \qquad (8.133)$$

So far spin has been ignored. m_s has two allowed values, so $g_s = 2$ and the total one-electron atom degeneracy is

$$g_{(\text{one-electron atom})} = 2n^2 \qquad (8.134)$$

The total degeneracy of a one-electron atomic state with principal quantum number n is simply $2n^2$.

The notation discussed previously still applies, with one addition. Thanks to spin, even a 1s state can have more than one electron; they would have mismatched spin. It is common therefore, to add a succeeding superscript to indicate the number of electrons. For example, a 1s^2 state has two electrons with $n = 1$, $\ell = 0$, and mismatched spin. The same state will sometimes be described by the notation 1s(2).

8.3 Multi-Electron Atoms

For atoms with more than one electron (and usually more complex nuclei), the full Hamiltonian for N_e electrons is written as

$$\hat{H} = -\frac{\hbar^2}{2m_e}\sum_{i=1}^{N_e}\nabla_i^2 - \sum_{i=1}^{N_e}\frac{Ze^2}{4\pi\epsilon_\circ r_i} + \sum_{i=1}^{N_e}\sum_{j=i+1}^{N_e}\frac{e^2}{4\pi\epsilon_\circ r_{ij}} \qquad (8.135)$$

Here, the mass of an electron, m_e has been used because the nucleus has a much larger mass than an individual electron, and hence the reduced mass becomes the electron mass. In addition, Z denotes the total charge number of the nucleus (often but not always equal to N_e), and e is the charge per electron.

The electron-electron interaction terms $\left(\frac{e^2}{4\pi\epsilon_\circ r_{ij}}\right)$ now make it impossible to separate the Schrödinger equation. Various approaches are applied to the solution of this problem. They are usually segregated into either a "hydrogen-like" problem (e.g. He, or ions with very few electrons), or a full multi-electron problem. Here we briefly sketch the approach to both cases.

8.3.1 Approximation Methods

Hydrogen-like atoms

In this case the problem is not too complex because Z is not large. Two approaches are used, depending upon what outcome is desired.

Variational methods can be used to estimate energy levels. If φ is a well-behaved wave function for a similar problem (e.g. the H atom), then the variational theorem states that

$$W \equiv \int \varphi^* \hat{H} \varphi \, d\tau \geq E_1 \qquad (8.136)$$

where E_1 is the ground state energy. The variational theorem has a very straightforward proof (see, for example, [60]). The variational problem is solved by varying φ in order to find a minimum W, and hence an approximation to E_1. For simple atoms, this minimization can be done by using an adjustable parameter in φ, such as

$$\varphi = e^{-cx^2} \qquad (8.137)$$

One can also use an expansion such as

$$\varphi = \sum_{j=1}^{n} c_j f_j \qquad (8.138)$$

where the f_j are trial functions that satisfy the problem boundary conditions. It is a simple process then to take a derivative or derivatives of W with respect to c or c_j to minimize W. This process provides an estimate for the ground state energy. Using this value for E_1, it is then possible to modify the variational theorem to approximate the next energy level, E_2, and so on.

Perturbation methods can be used to estimate both energy levels and wave functions. They can also be used to provide sequences of quantum numbers. Because they have more utility, we devote more time to their development here. In the case where a particular Schrödinger equation cannot be solved, we begin with a closely related system (denoted by °) that can be solved:

$$\hat{H}^\circ \psi_n^\circ = E_n^\circ \psi_n^\circ \qquad (8.139)$$

Again, this could be the one-electron atom problem. We then approximate the more complex solution by using:

$$\hat{H} = \hat{H}^\circ + \lambda \hat{H}' \qquad (8.140)$$

where $\hat{H}'$ is the "perturbation Hamiltonian" and $0 \le \lambda \le 1$ is used to adjust the impact of the perturbation. The Schrödinger equation then becomes:

$$[\hat{H}^\circ + \lambda \hat{H}']\psi_n = E_n \psi_n \qquad (8.141)$$

Now both ψ_n and E_n are functions of the perturbation coefficient λ. We therefore expand them both as a Taylor series about $\lambda = 0$:

$$\psi_n = \psi_n \left. \right|_0 + \frac{\partial \psi_n}{\partial \lambda} \left. \right|_0 \lambda + \frac{\partial^2 \psi_n}{\partial \lambda^2} \left. \right|_0 \frac{\lambda^2}{2!} + \cdots$$

$$E_n = E_n \left. \right|_0 + \frac{\partial E_n}{\partial \lambda} \left. \right|_0 \lambda + \frac{\partial^2 E_n}{\partial \lambda^2} \left. \right|_0 \frac{\lambda^2}{2!} + \cdots \qquad (8.142)$$

where the "0" on $|_0$ indicates evaluation at $\lambda = 0$. Clearly

$$\psi_n \left. \right|_0 = \psi_n^\circ$$

$$E_n \left. \right|_0 = E_n^\circ \qquad (8.143)$$

We can substitute the Taylor series expansions into the Schrödinger equation and then equate like powers of λ. The terms that do not have a λ multiplier provide the following:

$$\hat{H}^\circ \psi_n^\circ = E_n^\circ \psi_n^\circ \tag{8.144}$$

which is the unperturbed Schrödinger equation. Next, equating terms in λ produces

$$\hat{H}' \psi_n^\circ + \hat{H}^\circ \frac{\partial \psi_n}{\partial \lambda} \big|_0 = \frac{\partial E_n}{\partial \lambda} \big|_0 \psi_n^\circ + E_n^\circ \frac{\partial \psi_n}{\partial \lambda} \big|_0 \tag{8.145}$$

The term $\frac{\partial \psi_n}{\partial \lambda} \big|_0$ is also a wave function. For simplicity, we define it as a ket:

$$| \psi_n^{(1)} \rangle \equiv \frac{\partial \psi_n}{\partial \lambda} \big|_0 \tag{8.146}$$

where the superscript (1) indicates the first order in $\partial \lambda$. Now we switch to Dirac notation and rearrange equation (8.145):

$$\hat{H}^\circ | \psi_n^{(1)} \rangle - E_n^\circ | \psi_n^{(1)} \rangle = \frac{\partial E_n}{\partial \lambda} \big|_0 | \psi_n^\circ \rangle - \hat{H}' | \psi_n^\circ \rangle \tag{8.147}$$

Now form an inner product with $\langle \psi_m^\circ |$:

$$\langle \psi_m^\circ | \hat{H}^\circ | \psi_n^{(1)} \rangle - E_n^\circ \langle \psi_m^\circ | \psi_n^{(1)} \rangle$$
$$= \frac{\partial E_n}{\partial \lambda} \big|_0 \langle \psi_m^\circ | \psi_n^\circ \rangle - \langle \psi_m^\circ | \hat{H}' | \psi_n^\circ \rangle \tag{8.148}$$

which is

$$\langle \psi_m^\circ | \hat{H}^\circ | \psi_n^{(1)} \rangle - E_n^\circ \langle \psi_m^\circ | \psi_n^{(1)} \rangle$$
$$= \frac{\partial E_n}{\partial \lambda} \big|_0 \delta_{mn} - \langle \psi_m^\circ | \hat{H}' | \psi_n^\circ \rangle \tag{8.149}$$

Consider the first term on the left-hand side, and apply a property of Hermitian operators:

$$
\begin{aligned}
\langle \psi_m^{\circ} \mid \hat{H}^{\circ} \mid \psi_n^{(1)} \rangle &= \langle \psi_n^{(1)} \mid \hat{H}^{\circ} \mid \psi_m^{\circ} \rangle^* \\
&= \langle \psi_n^{(1)} \mid E_m^{\circ} \mid \psi_m^{\circ} \rangle^* \\
&= E_m^{\circ *} \langle \psi_n^{(1)} \mid \psi_m^{\circ} \rangle^* \\
&= E_m^{\circ} \langle \psi_m^{\circ} \mid \psi_n^{(1)} \rangle
\end{aligned}
\qquad (8.150)
$$

Here we have used the fact that the eigenvalues of Hermitian operators are real. Equation (8.149) is then

$$
(E_m^{\circ} - E_n^{\circ}) \langle \psi_m^{\circ} \mid \psi_n^{(1)} \rangle = \frac{\partial E_n}{\partial \lambda} \mid_0 \delta_{mn} - \langle \psi_m^{\circ} \mid \hat{H}' \mid \psi_n^{\circ} \rangle
\qquad (8.151)
$$

We wish to find the energy of an individual level, so we now set $n = m$, which means that $(E_m^{\circ} - E_n^{\circ}) = 0$ and $\delta_{mn} = 1$:

$$
\frac{\partial E_n}{\partial \lambda} \mid_0 = \langle \psi_n^{\circ} \mid \hat{H}' \mid \psi_n^{\circ} \rangle
\qquad (8.152)
$$

Recall that the first two terms of the expansion for E give

$$
E_n = E_n^{\circ} + \frac{\partial E_n}{\partial \lambda} \mid_0 \lambda
\qquad (8.153)
$$

Now take the full perturbation (set $\lambda = 1$) and use equation (8.152):

$$
E_n = E_n^{\circ} + \langle \psi_n^{\circ} \mid \hat{H}' \mid \psi_n^{\circ} \rangle = E_n^{\circ} + \int \psi_n^{\circ *} \hat{H}' \psi_n^{\circ} \, d\tau
\qquad (8.154)
$$

Note in equation (8.154) that the total energy is the unperturbed energy plus the expectation value for the perturbation Hamiltonian (the energy operator) using the unperturbed wave functions. This seems like a reasonable first-order energy correction. It is possible to find a perturbation term for the wave function as well, but we stop here because energy determines spectral locations.

The exact nature of the solution to either the variational method or the perturbation technique depends upon exactly which atom is analyzed. Typically, published data on the spectroscopy of an atom contains both experimental data and the outcome from models that have been found to agree with trustworthy experimental data (see, for example, Alkemade et al. [67]). The models can then be used to extend beyond the range of experiments.

Hartree-Fock SCF

For more complex multi-electron atom problems, we return to equation (8.135). The Hartree-Fock self-consistent field (SCF) method is typically used. This technique is similar to the variational method described above, but much more detail regarding the electron-electron interaction is included [60]. A variation function that is a product of individual wave functions is chosen:

$$\psi = f_1(r_1, \theta_1, \phi_1) f_2(r_2, \theta_2, \phi_2)...f_n(r_n, \theta_n, \phi_n) \qquad (8.155)$$

where the f are given by $f = R_{n\ell} Y_\ell^m(\theta, \phi)$. If we were to apply the one-electron Hamiltonian to each f, it would produce a very rough approximation. As one example, it would apply a full nuclear charge to each electron, and the effect of other electrons in the same proximity would be ignored. Instead, in accordance with the variational principle, we will find the f values that minimize W:

$$W \equiv \int \psi^* \hat{H} \psi \, d\tau \qquad (8.156)$$

Alternatively, if the wave functions ψ are not normalized it is necessary to use

$$W \equiv \frac{\int \psi^* \hat{H} \psi \, d\tau}{\int \psi^* \psi \, d\tau} \qquad (8.157)$$

At this point we do not have easily-manipulated constants c or c_j which we can use for minimization. The problem is too complex for that. It is necessary to begin with a guess that is probably close. As an example, it is valid to assert that the charge-related problems described above will affect the radial portion of the wave function much more than the orbital portions. In such a case, one can assume that

$$f_i = h_i(r_i) Y_{\ell_i}^{m_i}(\theta_i, \phi_i) \qquad (8.158)$$

where $Y_{\ell_i}^{m_i}$ is an individual angular wave function from the one-electron atom problem. The term $h_i(r_i)$ is an approximation to the correct but unknown $R_i(r_i)$.

For electrons 1 and 2, the interelectron interaction term is written as

$$V_{12} = \frac{q_1 q_2}{4\pi\epsilon_\circ r_{12}} \tag{8.159}$$

where we have used q instead of e, with numerical labels, in order to describe the more complex details of an multielectron interaction. Equation (8.159) does not accurately represent the effect of other electrons on electron number 1. We therefore approximate the interaction between electrons 1 and 2 by assuming that q_2 represents a distributed charge, with fixed charge density ρ_2:

$$dq_2 \cong \rho_2 dv_2 \tag{8.160}$$

where v represents volume in this case, in order to avoid confusion with the potential V. The interelectron interaction term is then

$$V_{12} = \frac{q_1}{4\pi\epsilon_\circ} \int \frac{\rho_2}{r_{12}} dv_2 \tag{8.161}$$

An expression for ρ_2 can be generated by the approximation $\rho_2 \cong -e \mid f_2 \mid^2$, because $\mid f_2 \mid^2$ is the probability density as a function of position (and using $q = -e$ now because the charge is for one electron). The interaction term for electron number 2 acting on electron number 1 is then

$$V_{12} = \frac{e^2}{4\pi\epsilon_\circ} \int \frac{\mid f_2 \mid^2}{r_{12}} dv_2 \tag{8.162}$$

and the total interaction with electron number 1 is a sum of contributions from each electron:

$$V_{1-electrons} = \sum_{j=2}^{N_e} \frac{e^2}{4\pi\epsilon_\circ} \int \frac{\mid f_j \mid^2}{r_{1j}} dv_j \tag{8.163}$$

and the total potential (including the nucleus) is

$$V_1(r_1, \theta_1, \phi_1) = \sum_{j=2}^{N_e} \frac{e^2}{4\pi\epsilon_\circ} \int \frac{\mid f_j \mid^2}{r_{1j}} dv_j - \frac{Ze^2}{4\pi\epsilon_\circ r_1} \tag{8.164}$$

Z is the nuclear charge, but it could be reduced for electrons that are further out, in order to describe shielding. The central field approximation asserts that V should be a function of r only. This can be accommodated by

$$V_1(r_1, \theta_1, \phi_1) = V_1(r_1) = \frac{\int_0^{2\pi} \int_0^{\pi} V_1(r_1, \theta_1, \phi_1) \sin \theta_1 \, d\theta_1 \, d\phi_1}{\int_0^{2\pi} \int_0^{\pi} \sin \theta \, d\theta \, d\phi} \quad (8.165)$$

This potential can then be used in a pseudo-separated Schrödinger equation for electron number 1:

$$\left[-\frac{\hbar^2}{2m_e} \nabla^2 + V_1(r_1) \right] f_1^{(1)} = \epsilon_1 f_1^{(1)} \quad (8.166)$$

The superscript $^{(1)}$ on f_1 indicates that it is a solution to the first iteration of the Schrödinger equation for the actual f_1. $f_1^{(1)}$ represents an improvement over the original estimate for f_1 with which we started. Not surprisingly, the Y_ℓ^m term in $f_1^{(1)}$ is usually quite similar to the original, but the radial term h_1 may have changed considerably. It will not be similar to the one-electron $R(r)$. The solution will produce recognizable eigenvalue behavior with quantum numbers.

With $f_1^{(1)}$ in hand, the problem for electron number 2 is then worked, but the interelectron interaction is now written as

$$V_{2-electrons} = \frac{e^2}{4\pi\epsilon_\circ} \int \frac{|f_1^{(1)}|^2}{r_{21}} \, dv_1 + \sum_{j=3}^{N_e} \frac{e^2}{4\pi\epsilon_\circ} \int \frac{|f_j|^2}{r_{2j}} \, dv_j \quad (8.167)$$

This expression is used to solve for $f_2^{(1)}$, which is then used with $f_1^{(1)}$ and the original estimated for $f_j's$ (j> 4) to find $f_3^{(1)}$, and so forth. Once the first iteration has been completed for all N_e electrons, the full iterative cycle is begun again. The process is continued until there is no change (within an uncertainty bound) in the outcome.

The total energy is not a simple summation over all of the ϵ values from the various Schrödinger solutions, because that would double-count contributions from the interaction terms. Instead

$$E = \sum_{i=1}^{N_e} \epsilon_i - \sum_{i=1}^{N_e} \sum_{j=i+1}^{N_e} \int \int \frac{e^2}{4\pi\epsilon_\circ} \frac{|f_i|^2 |f_j|^2}{r_{ij}} \, dv_i \, dv_j \quad (8.168)$$

8.3.2 The Pauli Principle and Spin

In this section, we revisit electronic spin in more detail, briefly review nuclear spin, and discuss implications of the Pauli exclusion principle.

Electron spin

According to the second postulate of quantum mechanics, for every property there must be an associated operator. There must, therefore, be electron spin angular momentum operators that are analogous to $\hat{L}^2$, $\hat{L}_x$, $\hat{L}_y$, and $\hat{L}_z$. They are written as $\hat{S}^2$, $\hat{S}_x$, $\hat{S}_y$, and $\hat{S}_z$. These operators are postulated to be linear and Hermitian, just like the orbital operators. Therefore

$$[\hat{S}_x, \hat{S}_y] = i\hbar\hat{S}_z \tag{8.169}$$
$$[\hat{S}_y, \hat{S}_z] = i\hbar\hat{S}_x$$
$$[\hat{S}_z, \hat{S}_x] = i\hbar\hat{S}_y$$
$$[\hat{S}^2, \hat{S}_x] = [\hat{S}^2, \hat{S}_y] = [\hat{S}^2, \hat{S}_z] = 0$$

Following the orbital treatment, one can write the eigenvalues for $\hat{S}^2$ as

$$| \hat{S}^2 |= s(s+1)\hbar^2, \; s = 0, 1/2, 1, 3/2, ... \tag{8.170}$$

and the eigenvalues for $\hat{S}_z$ as

$$| \hat{S}_z |= m_s\hbar, \; m_s = -s, -s+1, ..., s-1, s \tag{8.171}$$

This has been said before. The experimentally observed result is that s is limited to $1/2$, which then leads to $m_s = \pm 1/2$ (see Figure 8.7 for a graphic depiction). The same result can be found using relativistic quantum mechanics.

The eigenfunctions for the $\hat{S}_z$ equation are typically labeled as α and β:

$$\hat{S}_z\alpha = +\frac{1}{2}\hbar\alpha$$
$$\hat{S}_z\beta = -\frac{1}{2}\hbar\beta \tag{8.172}$$

Since $\hat{S}^2$ and $\hat{S}_z$ commute,

$$\hat{S}^2\alpha = \frac{3}{4}\hbar^2\alpha$$
$$\hat{S}^2\beta = \frac{3}{4}\hbar^2\beta \tag{8.173}$$

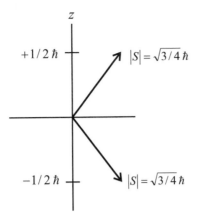

Figure 8.7: Depiction of spin angular momenta.

Because spin is not dependent upon position, the functional dependences of α and β are written as $\alpha = \alpha(m_s)$ and $\beta = \beta(m_s)$. Normalization is achieved via

$$\sum_{m_s=-1/2}^{m_s=1/2} |\alpha(m_s)|^2 = 1$$

$$\sum_{m_s=-1/2}^{m_s=1/2} |\beta(m_s)|^2 = 1 \qquad (8.174)$$

When normalization of the wave function includes electron spin it is actually written as

$$\sum_{m_s=-1/2}^{m_s=1/2} \int \int \int_{-\infty}^{+\infty} |\psi(x,y,z,m_s)|^2 \, dx \, dy \, dz = 1 \qquad (8.175)$$

and orthogonality is written as

$$\sum_{m_s=-1/2}^{m_s=1/2} \alpha^*(m_s)\beta(m_s) = 0 \qquad (8.176)$$

For simple atoms, the wave function can be represented by

$$\psi_{total} = \psi(x,y,z)g(m_s) \qquad (8.177)$$

where $\mathbf{g}(m_s)$ represents either $\alpha(m_s)$ or $\beta(m_s)$, whichever is appropriate. Then the Schrödinger equation is

$$\hat{H}\psi\mathbf{g} = \mathbf{g}\hat{H}\psi = \mathbf{g}E\psi \qquad (8.178)$$

$\mathbf{g}$ is not usually a function of position, it falls out of the Hamiltonian and the same term energy levels are calculated with or without spin. This produces a spin degeneracy of 2 ($+m_s = +1/2$, also called "spin up", and $-m_s = -1/2$, also called "spin down"). Spin coupling in certain atoms, however, can split the energy levels. This is an issue to be taken up in individual cases.

Because of the uncertainty principle, it is not possible to distinguish between identical particles. It is not possible to write relationships that specifically identify one electron in a specific orbital, for example. Helium, as an example, can not be described by 1s(1)1s(2) where the numbers in parentheses identify which electron is which. It is correct to specify the state as:

$$\frac{[1s(1)1s(2) \pm 1s(2)1s(1)]}{\sqrt{2}} \qquad (8.179)$$

where the $\sqrt{2}$ term renormalizes the expression.

This notion introduces the idea of interchange, and the question of what interchanges are allowed. If we define a system of variables for particle number 1 (x_1, y_1, z_1, m_{s1}) as q_1, then a system of N identical particles can be described in terms of

$$\psi = \psi(q_1, q_2, q_3, ..., q_N) \qquad (8.180)$$

One can define a permutation operator, $\hat{P}_{12}$, by

$$\hat{P}_{12}f(q_1, q_2, q_3, ...q_n) = f(q_2, q_1, q_3, ..., q_n) \qquad (8.181)$$

An important property of this operator is that $\hat{P}_{12}\hat{P}_{12} = \hat{P}_{12}^2 = 1$. If

$$\hat{P}_{12}f_i = c_i f_i \qquad (8.182)$$

then

$$\hat{P}_{12}^2 f_i = c_i \hat{P}_{12} f_i$$

$$1 f_i = c_i^2$$
$$c_i = \pm 1 \tag{8.183}$$

Hence the eigenvalues of the permutation operator are ± 1. Consider the eigenvalue $+1$ for an eigenfunction f:

$$\hat{P}_{12} f_+(q_1, q_2, q_3, ..., q_N) = +1 f_+(q_1, q_2, q_3, ..., q_N)$$
$$f_+(q_2, q_1, q_3, ..., q_N) = f_+(q_1, q_2, q_3, ..., q_N) \tag{8.184}$$

The function is unchanged when particles are switched, it is "symmetric with respect to interchange". Be aware that this is not the same thing as wave function symmetry with respect to wave function inversion in space. In addition

$$f_-(q_2, q_1, q_3, ..., q_N) = -f_-(q_1, q_2, q_3, ..., q_N) \tag{8.185}$$

which is antisymmetric with respect to interchange. Now consider wave functions. One cannot tell the difference between N particles. Thus the wave function $\psi(q_1, q_2, ..., q_i, ..., q_j, ..., q_N)$ is indistinguishable from $\psi(q_1, q_2, ..., q_j, ..., q_i, ..., q_N)$, and

$$\hat{P}_{ij} \psi(q_1, q_2, ..., q_i, ..., q_j, ..., q_N) = \text{constant } \psi(q_1, q_2, ..., q_i, ..., q_j, ..., q_N) \tag{8.186}$$

Wave functions are either symmetric or antisymmetric with respect to interchange of any two particles. Experimental evidence indicates that electrons will always follow the antisymmetric case. Relativistic quantum dictates that any particles with 1/2 integral spin will be antisymmetric, and these are called "fermions". Particles with integral spin will always be symmetric and they are called "bosons".

One outcome of this finding leads to the Pauli exclusion principle. Imagine there are two indistinguishable electrons, in the same coordinates ($q_1 = q_2$), then

$$\psi(q_1, q_2, ..., q_N) = -\psi(q_2, q_1, ..., q_N) = -\psi(q_1, q_2, ..., q_N)$$
$$2\psi = 0$$
$$\psi = 0 \tag{8.187}$$

where the minus sign is used because electrons are fermions. The meaning of this outcome is that there is zero probability that $q_1 = q_2$. Something must be different. If electrons have the same location, then they must have different values for m_s. For the He example, we then require

$$\frac{\alpha(1)\beta(2) \pm \beta(1)\alpha(2)}{\sqrt{2}} \tag{8.188}$$

to describe the two possible spin states. Indeed, it is very common to write a combined orbital and spin wave function as a "spin-orbital"; $1s(1)\alpha(1)$, for example. The Pauli exclusion principle states that no two electrons can occupy the same spin-orbital. An alternative statement is that no two electrons can have the same quantum numbers.

Nuclear spin

Nuclei also have an intrinsic angular momentum similar to electron spin, termed the nuclear spin. As before, there must be nuclear spin angular momentum operators that are analogous to $\hat{L}^2$, $\hat{L}_x$, $\hat{L}_y$, and $\hat{L}_z$. They are written as $\hat{I}^2$, $\hat{I}_x$, $\hat{I}_y$, and $\hat{I}_z$. The nuclear angular momentum is

$$| \hat{I}^2 | = I(I+1)\hbar^2 \tag{8.189}$$

where the value of the quantum number I is either integral (starting at zero) or half-integral, depending upon the makeup of the nucleus. Protons have spin of $1/2$, like an electron. Neutrons do as well. Nucleons have orbital angular momentum and spin that couple to each other (see, for example, Levine [68]). The resultant is called nucleon spin, and it can have integer values. The total spin of a nucleus, based upon coupling of all the possible spin values, cannot be described easily; each has its own behavior. As in equation (8.178), nuclear spin can have no effect upon energy. Often, however, the magnetic moment of the nucleus can split energy levels. This happens through angular momentum coupling (see Section 8.3.4 below).

8.3.3   The Periodic Table

The periodic table (presented in a truncated format in Figure 8.8) organizes ground electronic states of the atoms. The excited states

1 H																	2 He
3 Li	4 Be											5 B	6 C	7 N	8 O	9 F	10 Ne
11 Na	12 Mg											13 Al	14 Si	15 P	16 S	17 Cl	18 Ar
19 K	20 Ca	21 Sc	22 Ti	23 V	24 Cr	25 Mn	26 Fe	27 Co	28 Ni	29 Cu	30 Zn	31 Ga	32 Ge	33 As	34 Se	35 Br	36 Kr
37 Rb	38 Sr	39 Y	40 Zr	41 Nb	42 Mo	43 Tc	44 Ru	45 Rh	46 Pd	47 Ag	48 Cd	49 In	50 Sn	51 Sb	52 Te	53 I	54 Xe
55 Cs	56 Ba	57 La	....														

Figure 8.8: Truncated periodic chart containing only the first six periods.

are not populated at room temperature. The table consists of rows (periods) with 2, 8, 8, 18, 18, 32 and 32 elements. Elements in the same column (group) have similar chemical properties and spectra.

The first period is H and He, both in the 1s shells. The second period is for $n = 2$. There are now four orbitals: the 2s with $\ell = 0$; the 2p with $\ell = 1$ and $m_\ell = 0$; the 2p$_{+1}$ with $\ell = 1$ and $m_\ell = +1$; and the 2p$_{-1}$ with $\ell = 1$ and $m_\ell = -1$. In addition, $m_s = \pm 1/2$ for each orbital. The $n = 2$ shell can thus hold eight electrons. Add to that the two in the already filled 1s shell for a total of ten possible electron locations. The energy of the 2s level is less than the 2p, so the 2s is filled first. The shells in the second period are thus filled via

$$\underline{Li} : 1s^2 2s, \ \underline{Be} : 1s^2 2s^2, \ \underline{B} : 1s^2 2s^2 2p, \ \underline{C} : 1s^2 2s^2 2p^2$$
$$\underline{N} : 1s^2 2s^2 2p^3, \ \underline{O} : 1s^2 2s^2 2p^4, \ \underline{F} : 1s^2 2s^2 2p^5, \ \underline{Ne} : 1s^2 2s^2 2p^6$$

The six electrons in the 2p^6 of Ne account for all three values of m_ℓ $(0, \pm 1)$, times 2 for $m_s = \pm 1/2$.

The next period is for $n = 3$. The energy level ordering is 3s < 3p < 3d, so the 3s and 3p levels fill with eight electrons first. These all start with a "Ne-like" core of ten electrons in the $n = 1$ and $n = 2$ shells $(1s^2 2s^2 2p^6)$. A shorthand notation writes these with a preceeding "[Ne]". The $n = 3$ configurations are then

$$\underline{Na} : [Ne]3s, \ \underline{Mg} : [Ne]3s^2, \ \underline{Al} : [Ne]3s^2 3p, \ \underline{Si} : [Ne]3s^2 3p^2$$
$$\underline{P} : [Ne]3s^2 3p^3, \ \underline{S} : [Ne]3s^2 3p^4, \ \underline{Cl} : [Ne]3s^2 3p^5, \ \underline{Ar} : [Ne]3s^2 3p^6$$

For $n = 4$, the 4s orbital is more stable than the 3d. The elements past Ar begin with 4s shells, not 3d. That period has a $1s^2 2s^2 2p^6 3s^2 3p^6$

"Ar-like" core. We can use the same shorthand to write:

$K : [Ar]4s$, $\underline{Ca} : [Ar]4s^2$, $\underline{Sc} : [Ar]4s^23d$, $\underline{Ti} : [Ar]4s^23d^2$, $V : [Ar]4s^23d$
$\underline{Cr} : [Ar]4s^13d^5$, $\underline{Mn} : [Ar]4s^23d^5$, $\underline{Fe} : [Ar]4s^23d^6$, $\underline{Co} : [Ar]4s^23d^7$,
$\underline{Ni} : [Ar]4s^23d^8$, $\underline{Cu} : [Ar]4s^13d^{10}$, $\underline{Zn} : [Ar]4s^23d^{10}$, $\underline{Ga} : [Ar]4s^23d^{10}4p$
$\underline{Ge} : [Ar]4s^23d^{10}4p^2$, $\underline{As} : [Ar]4s^23d^{10}4p^3$, $\underline{Se} : [Ar]4s^23d^{10}4p^4$,
$\underline{Br} : [Ar]4s^23d^{10}4p^5$, $\underline{Kr} : [Ar]4s^23d^{10}4p^6$

The fourth period thus corresponds to the filling of the 4s (2), 3d (10), and 4p (6) orbitals, for a total of 18 elements in that period.

Period 5 now corresponds to the filling of the 5s, 4d, and 5p orbitals, period 6 is the 6s, 5d, 6p, and 4f orbitals, period 7 fills the 7s, 6d, 7p, and 5f orbitals, and so forth.

8.3.4   Angular Momentum Coupling

For closed shells and subshells, all the electron configurations are fully specified because every available quantum number has been taken. For partially filled shells, however, it is not immediately clear how an electron with a given n and ℓ will be distributed among the possible values for m_ℓ and m_s. The concern, therefore, lies with several electrons and the orientations of their orbital angular momenta and spin angular momenta. Because of the electron-electron repulsion term and electron shielding, each of these states has a slightly different energy. This also affects level degeneracies. By specifying the way in which the various forms of angular momentum couple to each other, therefore, one is specifying term values and degeneracies.

Consider an atom with k electrons in an open shell. For each electron i there is an orbital angular momentum $\vec{\ell}_i$ and electron spin angular momentum $\vec{s}_i$ (and perhaps nuclear spin coupling, although this is a small effect that will be discussed parenthetically at the end of this section). These individual momenta will couple, primarily because there are electron-electron interactions, and the way in which they couple affects the energy of each state. It becomes necessary, therefore, to describe the way these angular momenta couple. Each atom has particular structural patterns that make it difficult to generalize. All the same, it is possible to classify various observed forms of coupling.

Russell-Saunders coupling (also called $L-S$ coupling) assumes that all of the $\vec{\ell}_i$ couple together to form a total orbital angular momentum vector $\vec{L}$, and all of the $\vec{s}_i$ couple to form a total spin angular momentum $\vec{S}$. Then $\vec{L}$ and $\vec{S}$ couple to form a total atomic angular momentum $\vec{J}$.

In $j - j$ coupling, individual electronic orbital and spin angular momenta couple; $\vec{\ell}_i$ and $\vec{s}_i$ couple to give $\vec{j}_i$. Then all of the $\vec{j}_i$ couple to give $\vec{J}$. There is no definite L or S in $j - j$ coupling.

Obviously, one must decide whether an atom behaves according to Russell-Saunders or $j - j$ coupling. In general, atoms with $Z \leq 40$ are Russell-Saunders while atoms with $Z \geq 40$ are $j - j$. This is true because Russell-Saunders coupling is based upon electrostatics (more on this in a moment), while $j - j$ coupling is based upon coupling of magnetic moments induced by orbital and spin angular momenta (often called "spin-orbit" coupling). $j - j$ coupling becomes important when spin-orbit effects dominate over electrostatic. This happens only in very large atoms because the electronic velocities are greater at very large radii. Flowfield spectroscopists care more about smaller atoms, and about molecules formed from smaller atoms, so the following discussion will focus exclusively on Russell-Saunders coupling.

As a simple case, imagine that two electrons share a p orbital ($\ell = 1$ for both, and then m_ℓ can take $0, \pm 1$) in a Russell-Saunders atom. Imagine they have the same $m_\ell = 1$, but different values of m_s (see Figure 8.9). The two electrons have a large mutual repulsion (their orbitals are the same shape and are in the same space). As shown, the resultant value of L_z is $2\hbar$. Alternatively, if one electron is on the $m_{\ell 1} = +1$ orbital but the other is in $m_{\ell 2} = 0$, the repulsion is reduced while $L_z = \hbar$. If $m_{\ell 1} = m_{\ell 2} = 0$, repulsion is maximized (owing to the confined but matched orbital shapes), and $L_z = 0$. As it turns out, the weakest repulsion between the two occurs when $m_{\ell 1} = +1$ and $m_{\ell 2} = 0$, with parallel spins ($m_{s1} = m_{s2} = \pm 1/2$).

Now consider a group of k electrons. Possible values for the total orbital angular momentum (in Russell-Saunders coupling) include

$$
\begin{aligned}
L &= \ell_1 + \ell_2 + \ell_3 ... \ell_k, \\
&\quad \ell_1 + \ell_2 + \ell_3 ... \ell_k - 1, \\
&\quad \ell_1 + \ell_2 + \ell_3 ... \ell_k - 2, ...
\end{aligned}
\tag{8.190}
$$

There is a progression of possible L values because the possible values of ℓ include $0, 1, 2, ... n - 1$, making the smallest combination change -1. The minimum of L depends upon the ℓ_i. If all the ℓ_i are equal,

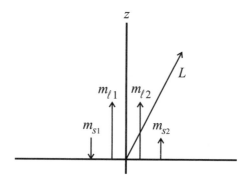

Figure 8.9: Cartoon of the coupling with $m_{\ell 1} = m_{\ell 2} = 1$. Magnitude of $L = \sqrt{\ell(\ell+1)}\hbar$, $| L_z |= (m_{\ell 1} + m_{\ell 2})\hbar = 2\hbar$.

$L_{min} = 0$, but if one ℓ_i is larger than the others, L_{min} is found by orienting all the others to oppose it, as long as $L \geq 0$.

Possible values for the total spin angular momentum are given by

$$\begin{aligned} S &= s_1 + s_2 + s_3...s_k, \\ &\quad s_1 + s_2 + s_3...s_k - 1, \\ &\quad s_1 + s_2 + s_3...s_k - 2, ... \end{aligned} \qquad (8.191)$$

In this case, the smallest combination change is -1 again. Because s can only take the values $\pm 1/2$, one combination change involves switching from $+1/2$ to $-1/2$, or -1. The minimum value for S is 0 for even k and $1/2$ for odd k.

As an example, for three p electrons ($\ell_1 = 1, \ell_2 = 1, \ell_3 = 1$) with $m_\ell = 0, \pm 1$, we find that $L = (1 + 1 + 1)$, $(1 + 1 + 1 - 1)$, $(1 + 1 + 1 - 2)$, $(1 + 1 + 1 - 3) = 3, 2, 1, 0$. In addition, $S = 3/2, 1/2$ (odd k). Or, for one f electron ($\ell_1 = 3$) and two p electrons ($\ell_2 = 1, \ell_3 = 1$), $L = 5, 4, 3, 2, 1$ and $S = 3/2, 1/2$ (same k). Knowing L and S, one can find J by $J = L + S, L + S - 1, L + S - 2, \quad ..., \quad | L - S |$.

The energy ordering of these states follow Hund's rules:

1.   The terms are ordered according to their S values. The term with maximum S (maximum spin multiplicity) is the most stable and is the ground state; stability decreases with decreasing S.

2.   For a given value of S, the state with the maximum value of L is the most stable.

3.   For given S and L, the term with the minimum value of J ($J =|$

$L - S$ |) is the most stable if the subshell is less than half-full; and the term with the maximum value of J ($J =| L + S$ |) is the most stable if the subshell is more than half-full.

When nuclear spin couples with other angular momenta, $\vec{J}$ and $\vec{I}$ couple to give a total angular momentum $\vec{F}$. The total angular momentum quantum number F can take the values $F = J+1, J+2, ... | J - I |$. The energy splitting introduced by nuclear spin coupling is roughly 10^{-3} that of electron spin coupling. This "hyperfine splitting" is often not observed in experiments. At such a point the nuclear spin contribution becomes a degeneracy. The total statistical weight is now $(2J + 1)(2I + 1)$. This factor of $(2I + 1)$ is the same for all states of an atom. Nuclear spin is discussed in more detail in reference [15].

8.3.5   Selection Rules, Degeneracy, and Notation

The selection rules for multi-electron atoms are not much different from those for single-electron atoms. They include

$$
\begin{aligned}
\Delta\, n &= \quad 0, 1, 2, 3...\text{can be any integer} \\
\Delta\, L &= \quad 0, \pm 1 \\
\Delta\, \ell &= \quad \pm 1 \\
\Delta\, J &= \quad 0, \pm 1; \; J=0 \nleftrightarrow J=0 \\
\Delta\, S &= \quad 0
\end{aligned}
$$

Degeneracy in multi-electron atoms is not as systematic as it was in the one-electron case. The term describing components of angular momentum along z will have a $(2J + 1)$ style of degeneracy (where this J is meant to describe any angular momentum in this particular expression). For Russell-Saunders coupling, therefore, the angular momentum degeneracy will be $(2L + 1)(2S + 1)$, and for $j - j$ coupling it will be $(2j + 1)(2j + 1)$ [65]. Nuclear degeneracy is the same for all states and it does not enter into calculations based upon Boltzmann statistics. These are really just guidelines, because electron-electron interactions can remove degeneracies. It becomes necessary to study the spectroscopy of each atom in detail (especially in heavier atoms), in order to understand when levels should be considered degenerate and when they are split. In practice, this question actually depends upon the spectral resolution of the instrument used to observe the spectrum. If the instrument cannot resolve the splitting, then it is common to consider the originating state degenerate. The words "resolve the splitting" themselves are open to interpretation. If, for example, a narrow-band spectrometer detects unusual broadening, it may be due to very weakly split lines that are both broadened. The instrument response might better be modeled as two broadened lines within one instrument profile, despite the fact that two discrete peaks were not detected.

For Russell-Saunders coupling, the atomic term symbol is written as

$$n^{(2S+1)}L_J \tag{8.192}$$

with $L = 0, 1, 2, 3$ being designated by S, P, D, F. Often the value of n is left off - it is assumed the reader knows n. The superscript preceeding L is called the spin multiplicity (it is a spin degeneracy). For $2S + 1 = 1, 2, 3$, these are called singlet, doublet and triplet atomic states.

8.4   Conclusion

This chapter may have contained much more about atomic spectroscopy than could be justified by interest in atomic species within flowfield research topics. As mentioned above, however, the approach taken here is the same approach taken in molecular spectroscopy. This chapter can therefore be considered an introduction to the next chapter, where molecular spectroscopy goes much faster than it would if this chapter had not been written.

For those interested in atomic spectroscopy itself, this chapter has developed basic concepts but it has not provided details related to any *particular* element. Each element has variations on the concepts presented here. Books such as Banwell and McCash [14], Herzberg [15], and Alkemade *et al.* [67] provide details on specific elements.

Chapter 9

MOLECULAR SPECTROSCOPY

9.1 Introduction

Molecules play the dominant role in the behavior of most chemically reacting flowfields, and therefore their concentrations (number densities) must be measured accurately in order to describe the flowfield. In combustion research, for example, the more stable reactants and final products (e.g. O_2, N_2, CH_4, CO_2, and H_2O) are usually measured using various analytical instruments (usually by probe extraction and analysis). Even here, the analytical instrument often relies upon spectroscopy. As an alternative, many major species can be detected *in situ* using Raman spectroscopy (see Chapter 13). The more reactive free-radical intermediates (e.g. OH, CH, CN, NH, and HCO) are most commonly measured *in situ* with a resonance spectroscopic technique. This is because such techniques can detect intermediate species without disturbing the flowfield, and this can usually be accomplished at high speed with good sensitivity.

Unfortunately, spectroscopic detection of free radicals is complicated. There are many molecules of interest, and they each have somewhat different spectroscopic behavior. Even among the diatomic molecules, it becomes important to collect a file on each particular molecule to be studied, so that individual attributes are accounted for. Polyatomic molecules are significantly more complex, because they are three-dimensional. They are also much more difficult to detect op-

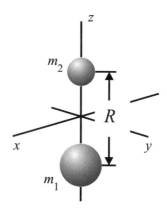

Figure 9.1: Diatomic molecule, note that the electron locations have been left out.

tically because they have so many closely spaced energy levels; the Boltzmann fraction for an individual energy level that would be addressed by a narrow-band laser is thus very low. Finally, because they are so reactive, these free radicals do not exist in large concentrations. Experimentalists often struggle near the edge of the lower detection limit for the spectroscopic instrument.

Most spectroscopic measurements in reacting flows are used to detect diatomic molecules. For this reason, we devote most of this chapter to diatomics. A short section introducing some issues associated with polyatomic species is included at the end of this chapter. Much of the material presented below draws heavily from Chapter 8; many shortcuts will in fact simply refer to a development from that treatment on atomic spectroscopy. If some point made in this chapter is not obvious, it may help to re-read the pertinent sections of Chapter 8.

9.2   Diatomic Molecules

Diatomic molecules are always linear, and the z coordinate is chosen to align with the internuclear axis (see Figure 9.1). The electronic structure of the molecule in Figure 9.1 has not been depicted, because both atoms are most likely multi-electron. The resulting molecular electron orbitals are complicated, and their image could obscure the nuclei. As depicted, the nuclei form a "dumbell" arrangement, with the internuclear spacing given by R. The spacing R can certainly change, but it is confined by a potential field that is developed by the molecular

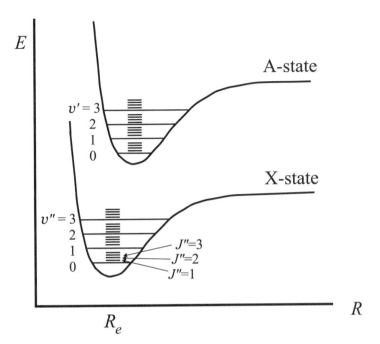

Figure 9.2: Schematic diagram of the potential wells for two electronic states in a diatomic molecule.

electronic orbitals. The value of R at the minimum of the potential well defines an equilibrium internuclear spacing R_e. The nuclei vibrate about R_e and rotate about the center of mass. To a first approximation, the impact of rotation about the internuclear axis (about z) is negligible relative to the rotation of that axis itself (i.e. about x or y). The value of R affects the electronic structure, entering as a parameter in the electronic solution. The electronic orbital structure then determines the potential field that sets the value of R_e. The two are thus interrelated.

Figure 9.2 includes a diagram of the potential fields established by electronic structure. The potential wells for the lowest (ground) electronic level ("X-state") and a first excited electronic level ("A-state") are shown. In addition, various vibrational energy levels, identified by their vibrational quantum numbers v, are depicted. Vibrational quantum numbers associated with energy levels in the lower electronic state are labeled with a double prime, v'', while those associated with the upper electronic state are labeled with a single prime, v'. While the vibrational energy level spacing is much less than the electronic spacing,

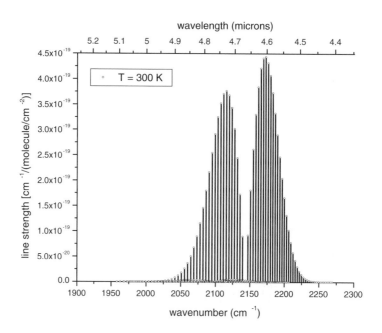

Figure 9.3: Absorption spectrum of the fundamental vibrational band of CO, within the ground electronic level (one hot band is barely visible as well). Temperature is 300 K and pressure is 1 atm. Spectrum was synthesized using the HITRAN database [69].

rotational spacings are even smaller. The rotational quantum numbers J'' for three rotational levels in the lowest vibrational level of the X-state are labeled. Again, a double prime indicates lower level while a single prime indicates upper level.

Because there is a hierarchy in energy spacing, transitions between various states are observed in different regions of the spectrum. Purely rotational transitions produce very small energy changes, and they are observed far into the infrared (IR), in the microwave regime. Vibrational transitions are observed in the IR while electronic transitions are observed in the visible and ultraviolet (UV). As an example, Figure 9.3 contains a synthesized portion of the spectrum for the diatomic molecule CO. It represents one vibrational transition (from $v'' = 0$ to $v' = 1$, e.g. absorption) within the ground electronic state. One can also see very weak lines near the base. These originate from the $v'' = 1$

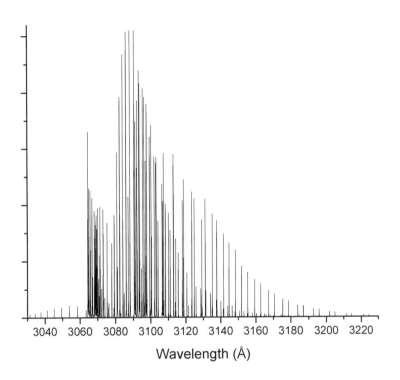

Figure 9.4: Absorption spectrum of OH in the lowest vibrational band (0,0) of the first electronic transition (A←X). Temperature is 2000 K and pressure is 1 atm, vertical units are relative. Spectrum was synthesized using LIFBASE [13].

to $v' = 2$ band (called a hot band). The figure depicts a large number of rotational transitions that can occur within this single vibrational transition.

Figure 9.4 contains another example, the absorption spectrum for the molecule OH, representing the first electronic transition (A←X), with no change in vibrational quantum number. Again, numerous rotational transitions can occur within this electronic/vibrational transition manifold. Note that this OH spectrum is not as neat as the CO example in Figure 9.3, for reasons that will become clear.

9.2.1   Approach to the Schrödinger Equation

For a molecule, a solution to the Schrödinger equation starts by separating nuclear and electronic motion. We use the Born-Oppenheimer approximation, which assumes that the electronic problem can be solved with stationary nuclei. This is justified by rate arguments; the electrons move much faster and in a relative sense the nuclei can be considered stationary. Nuclear rotation and vibration are then solved separately from the electronic solution. For a diatomic molecule made up of atoms number1 and 2 (as depicted in Figure 9.1), with N electrons, we write the Schrödinger equation as

$$\left(-\frac{\hbar^2}{2\mu}\nabla^2 - \frac{\hbar^2}{2m_e}\sum_{i=1}^{N}\nabla_i^2 + \hat{V}\right)\psi = E\psi \tag{9.1}$$

where μ is the reduced nuclear mass given by

$$\mu = \frac{m_1\, m_2}{m_1 + m_2}$$

In addition, ψ is the wave function for both nuclear and electronic states, and E is the total energy. The term $(\hbar^2/2\mu)\nabla^2$ describes the kinetic energy of the internal motion of the nuclei (note the use of a reduced mass based upon nuclear mass), while the term $(\hbar^2/(2m_e))\sum_{i=1}^{N}\nabla_i^2$ describes the total electronic kinetic energy. The electron mass, m_e, is used instead of a reduced mass, because the mass of the electron is so small relative to the nuclei that the reduced mass for electronic motion is essentially equal to the electronic mass.

The potential energy term can be written most generally as

$$\hat{V} = V_{ee} + V_{eN} + V_{NN} \tag{9.2}$$

where V_{ee} describes electron-electron repulsion:

$$V_{ee} = \sum_{i>j=1}^{N} \frac{e^2}{4\pi\epsilon_o r_{ij}} \tag{9.3}$$

V_{eN} describes electron-nuclear attraction:

$$V_{eN} = -\sum_{i=1}^{N} \frac{Z_1 e^2}{4\pi\epsilon_o r_{i1}} - \sum_{i=1}^{N} \frac{Z_2 e^2}{4\pi\epsilon_o r_{i2}} \tag{9.4}$$

and V_{NN} describes nuclear-nuclear repulsion:

$$V_{NN} = \frac{Z_1 Z_2 e^2}{4\pi\epsilon_\circ R} \tag{9.5}$$

The wave function is then a function of multiple position coordinates:

$$\psi = \psi(\vec{r}_1, \vec{r}_2, \vec{r}_3, ..., \vec{r}_n, \vec{R}) \tag{9.6}$$

Similar to the multi-electron problem, there are coupled interactions that make the Schrödinger equation inseparable. They include the former electron-electron $(e^2/4\pi\epsilon_\circ r)$ terms and the new $(Ze^2/4\pi\epsilon_\circ r)$ electron-nuclear terms. The electron-electron terms can be solved with the same techniques that were applied to the multi-electron atom.

There is a parametric linkage between the electronic and nuclear solutions for a diatomic molecule. We can show that by assuming the electronic motion occurs in the vicinity of quasi-stationary nuclei. Within this assumption, we can write an electronic Schrödinger equation:

$$\left(-\frac{\hbar^2}{2m_e} \sum_{i=1}^{N} \nabla_i^2 + V(\vec{r}, R) \right) \psi_e(\vec{r}, R) = E(R)\psi_e(\vec{r}, R) \tag{9.7}$$

(this equation will be developed via separation of variables below). In this equation, the nuclear terms have been extracted and the internuclear spacing R is a parameter in the calculation, not a variable. The equation is solved for multiple, fixed values of R. The energy term, $E(R)$, is the potential field within which the nuclei move, and it enters into the nuclear solution.

To separate variables, we assert that

$$\psi(\vec{r}, \vec{R}) = \psi_e(\vec{r}, R)\chi(\vec{R}) \tag{9.8}$$

where $\chi(\vec{R})$ is the nuclear wave function. Substitute this into the full Schrödinger equation to obtain

$$\left(-\frac{\hbar^2}{2\mu}\nabla^2 - \frac{\hbar^2}{2m_e} \sum_{i=1}^{N} \nabla_i^2 + V(\vec{r}, R) \right) \psi_e(\vec{r}, R)\chi(\vec{R})$$

$$= E_{total}\, \psi_e(\vec{r}, R)\chi(\vec{R}) \tag{9.9}$$

The terms containing ∇_i^2 act only on electronic terms; they have no effect upon $\chi(\vec{R})$, while the term containing ∇^2 differentiates with respect to nuclear components. It can act upon $\chi(\vec{R})$ and $\psi_e(\vec{r}, R)$. Recognizing that equation (9.7) is embedded in equation (9.9), we write

$$\left(-\frac{\hbar^2}{2\mu}\nabla^2 + E(R)\right)\psi_e(\vec{r}, R)\chi(\vec{R}) = E_{total}\ \psi_e(\vec{r}, R)\chi(\vec{R}) \qquad (9.10)$$

The assumption that the nuclear spacing is a parameter in the electronic solution is required to produce this result. Next, since $\psi_e(\vec{r}, R)$ changes more slowly with R than $\chi(\vec{R})$ does, one can simplify the nuclear differential term:

$$\nabla^2\psi_e(\vec{r}, R)\chi(\vec{R}) \cong \psi_e(\vec{r}, R)\nabla^2\chi(\vec{R}) \qquad (9.11)$$

This expression is a mathematical statement of the Born-Oppenheimer approximation. Equation (9.10) can be rearranged, the $\psi_e(\vec{r}, R)$ terms cancel, and we obtain

$$\left(-\frac{\hbar^2}{2\mu}\nabla^2 + E(R)\right)\chi(\vec{R}) = E_{total}\ \chi(\vec{R}) \qquad (9.12)$$

which is a molecular nuclear Schrödinger equation. It looks very much like the one-electron atom Schrödinger equation (8.10) where $E(R)$ provides a description of the potential field in equation (9.12). The following treatment will in fact closely parallel Section 8.2.2, allowing us to shorten this section significantly.

In spherical polar coordinates, equation (9.12) can be rewritten as

$$\left[-\frac{\hbar^2}{2\mu}\frac{1}{R^2}\left(\frac{\partial}{\partial R}\left(R^2\frac{\partial}{\partial R}\right) + \frac{1}{\sin\theta}\frac{\partial}{\partial\theta}\left(\sin\theta\frac{\partial}{\partial\theta}\right)\right.\right.$$

$$\left.\left.+\frac{1}{\sin^2\theta}\frac{\partial^2}{\partial\phi^2}\right) + E(R)\right]\chi(R,\theta,\phi) = E_{total}\ \chi(R,\theta,\phi) \quad (9.13)$$

Similar to the one-electron atom solution, it is possible to separate $\chi(\vec{R})$ into a radial component $\mathcal{R}(R)$ and an angular component $Y(\theta,\phi) = \Theta(\theta)\Phi(\phi)$. Inserting this separation into equation (9.13), and dividing by $\hbar^2/2\mu$, produces

$$
\left[-\frac{1}{R^2}\frac{\partial}{\partial R}\left(R^2\frac{\partial}{\partial R}\right) + \frac{1}{\sin\theta}\frac{\partial}{\partial\theta}\left(\sin\theta\frac{\partial}{\partial\theta}\right) + \frac{1}{\sin^2\theta}\frac{\partial^2}{\partial\phi^2} \right.
$$
$$
\left. +\frac{2\mu}{\hbar^2}E(R)\right]\mathcal{R}(R)\Theta(\theta)\Phi(\phi) = \frac{2\mu}{\hbar^2}E_{total}\,\mathcal{R}(R)\Theta(\theta)\Phi(\phi) \quad (9.14)
$$

Operating through, collecting terms in $\mathcal{R}(R)$, $\Theta(\theta)$ and $\Phi(\phi)$, and removing the explicit variable dependences [the $(R), (\theta)$, and (ϕ) notations] for compactness [except for the electronic potential $E(R)$], we obtain

$$
-\frac{\Theta\Phi}{R^2}\frac{\partial}{\partial R}\left(R^2\frac{\partial\mathcal{R}}{\partial R}\right) + \mathcal{R}\theta\Phi\frac{2\mu}{\hbar^2}E(R) - \frac{\mathcal{R}\Phi}{R^2\sin\theta}\frac{\partial}{\partial\theta}\sin\theta\frac{\partial\Theta}{\partial\theta}
$$
$$
-\frac{\mathcal{R}\Theta}{R^2\sin^2\theta}\frac{\partial^2\Phi}{\partial\phi^2} = \frac{2\mu}{\hbar^2}E_{total}\,\mathcal{R}\,\Theta\,\Phi \quad (9.15)
$$

The next step in separation of variables is to divide by χ, via multiplication by $-R^2/\mathcal{R}\Theta\Phi$ in this case:

$$
\frac{1}{\mathcal{R}}\frac{\partial}{\partial R}\left(R^2\frac{\partial\mathcal{R}}{\partial R}\right) - \frac{2\mu}{\hbar^2}R^2 E(R) + \frac{1}{\Theta\sin\theta}\frac{\partial}{\partial\theta}\sin\theta\frac{\partial\Theta}{\partial\theta}
$$
$$
+\frac{1}{\Phi\sin^2\theta}\frac{\partial^2\Phi}{\partial\phi^2} = -\frac{2\mu}{\hbar^2}R^2 E_{total} \quad (9.16)
$$

It is then convenient to multiply by $\sin^2\theta$, and to collect all the ϕ-dependent terms on one side:

$$
\frac{1}{\Phi}\frac{\partial^2\Phi}{\partial\phi^2} = fn(R,\theta) = \text{constant} \quad (9.17)
$$

which is mathematically identical to equation (8.24), the equation for ϕ dependence in the one-electron atom problem. It produces a similar result, including a magnetic quantum number $M = 0, \pm1, \pm2, ..., \pm J$ for the component of nuclear angular momentum along the nuclear axis z:

$$
\boxed{L_Z = M\hbar} \quad (9.18)
$$

Following the same treatment as in Section 8.2.2, this result could then be inserted into equation (9.16), the equation would be rearranged to isolate θ dependence, and we would find Legendre polynomial solutions. The square of the nuclear angular momentum is then

$$\boxed{L^2 = \mathcal{J}(\mathcal{J}+1)\hbar^2} \tag{9.19}$$

where $\mathcal{J}$ is the nuclear angular momentum quantum number. A different font is used here for this rotational quantum number because J will be used later on to describe coupled rotational states, and it would be best to avoid further confusion. Substituting these results back into equation (9.16) will produce

$$-\frac{\hbar^2}{2\mu R^2}\frac{\partial}{\partial R}\left(R^2\frac{\partial \mathcal{R}}{\partial R}\right) + \frac{\mathcal{J}(\mathcal{J}+1)\hbar^2}{2\mu R^2}\mathcal{R} + E(R)\mathcal{R} = E_{v,\mathcal{J}}\mathcal{R} \tag{9.20}$$

This is the radial Schrödinger equation for nuclear motion. The solutions for Φ and Θ in terms of $\mathcal{J}$ will provide rotational energy, while the solution for $\mathcal{R}$ will generate expressions for vibrational energy.

The Rigid Rotator

Here, we analyze the nuclear rotational kinetic energy separate from the electronic term. We begin by recognizing that there was a separated version of equation (9.16) (it was not shown) that was dependent only upon angular variables. If all of the separation steps had been presented, we would have found

$$-\hbar^2\left(\frac{1}{\sin\theta}\frac{\partial}{\partial \theta}\sin\theta\frac{\partial}{\partial \theta} + \frac{1}{\sin^2\theta}\frac{\partial^2}{\partial \phi^2}\right)\Theta\Phi$$
$$= \mathcal{J}(\mathcal{J}+1)\hbar^2\,\Theta\Phi \tag{9.21}$$

The operator $\hat{L}^2$, defined by

$$\hat{L}^2 \equiv -\hbar^2\left(\frac{1}{\sin\theta}\frac{\partial}{\partial \theta}\sin\theta\frac{\partial}{\partial \theta} + \frac{1}{\sin^2\theta}\frac{\partial^2}{\partial \phi^2}\right) \tag{9.22}$$

is therefore the operator for the square of the angular momentum. The eigenvalues (the observables) are $\mathcal{J}(\mathcal{J}+1)\hbar^2$. These are the values of the square of the angular momentum.

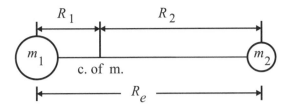

Figure 9.5: Classical interpretation for a diatomic molecule.

In classical mechanics, the magnitude of the angular momentum $| \vec{L} |$ is given by

$$| \vec{L} |= I_e \omega \qquad (9.23)$$

where ω is the angular velocity (not frequency in this case) given by $\omega = v/R_e$, and I_e is the moment of inertia. R_e is the equilibrium distance between the two masses (see Figure 9.5) Relative to the center of mass (labeled "c. of m." in the figure), we can define the moment of inertia by

$$I_e \equiv m_1 R_1^2 + m_2 R_2^2 \qquad (9.24)$$

and from the definition of the center of mass

$$R_1 = \frac{m_2}{m_1 + m_2} R_e$$
$$R_2 = \frac{m_1}{m_1 + m_2} R_e \qquad (9.25)$$

The moment of inertia can then be written as

$$I_e = \frac{m_1 m_2}{m_1 + m_2} R_e^2 = \mu R_e^2 \qquad (9.26)$$

Since there is no angular-dependent potential for an isolated diatomic nucleus:

$$E_{rotational} = \frac{1}{2} \mu v^2 = \frac{1}{2} I_e \omega^2 = \frac{(I_e \omega)^2}{2 I_e} = \frac{| \vec{L} |^2}{2 I_e} = \frac{| \vec{L} |^2}{2 \mu R_e^2} \qquad (9.27)$$

If we then set $| \vec{L} |^2 = \mathcal{J}(\mathcal{J} + 1)\hbar^2$ (equating classical and quantum mechanical angular momentum), we find

$$E_{\mathcal{J}} = \frac{\hbar^2}{2I_e}\mathcal{J}(\mathcal{J}+1) = \frac{\hbar^2}{2\mu R_e^2}\mathcal{J}(\mathcal{J}+1) \quad \text{for} \ \ \mathcal{J} = 0, 1, 2, ... \quad (9.28)$$

The selection rule for purely rotational transitions is $\Delta\mathcal{J} = \mathcal{J}_{\text{final}} - \mathcal{J}_{\text{initial}} = \pm 1$. For absorption, therefore:

$$
\begin{aligned}
\nu &= \frac{E_{\mathcal{J}'} - E_{\mathcal{J}''}}{h} = \frac{\hbar^2}{2I_e h}[(\mathcal{J}''+1)(\mathcal{J}''+2) - (\mathcal{J}''(\mathcal{J}''+1)] \\
&= 2B_e(\mathcal{J}''+1) \quad\quad\quad\quad\quad\quad\quad\quad\quad\quad\quad\quad (9.29)
\end{aligned}
$$

where B_e is a rotational constant:

$$B_e \equiv \hbar^2/2I_e h \quad\quad (9.30)$$

For emission

$$\nu = 2\mathcal{J}'B_e \quad\quad (9.31)$$

These formulas predict evenly spaced rotational lines. This is a first-order approximation because the bond is not actually rigid.

The Harmonic Oscillator

We now return to the nuclear Schrödinger equation [equation (9.20)]:

$$-\frac{\hbar^2}{2\mu R^2}\frac{\partial}{\partial R}\left(R^2\frac{\partial\mathcal{R}}{\partial R}\right) + E_{\mathcal{J}}\mathcal{R} + E(R)\mathcal{R} = E_{v,\mathcal{J}}\mathcal{R} \quad\quad (9.32)$$

If we define the zero of energy to be the minimum in the potential well, then electronic energy can be separated out and we write

$$E_{v,\mathcal{J}} = E_{nucleus} = E_{\mathcal{J}} + E_v \quad\quad (9.33)$$

and the nuclear Schrödinger equation becomes

$$-\frac{\hbar^2}{2\mu R^2}\frac{\partial}{\partial R}\left(R^2\frac{\partial\mathcal{R}}{\partial R}\right) + E(R)\mathcal{R} = E_v\mathcal{R} \quad\quad (9.34)$$

where, again, $E(R)$ is the solution to the electronic problem (the potential well depicted in Figure 9.2). Equation (9.34), therefore, describes vibrational dynamics within the potential well.

To simplify equation (9.34), define

$$\mathcal{R} \equiv \frac{1}{R} \, \chi(R) \tag{9.35}$$

and insert this into the equation. That generates an equation in χ:

$$-\frac{\hbar^2}{2\mu} \frac{d^2}{dR^2} \chi(R) + E(R) \, \chi(R) = E_v \, \chi(R) \tag{9.36}$$

For most diatomic molecules the potential field depicted in Figure 9.2 is appropriate. There are several functional forms that model such a potential (e.g. the Morse potential discussed below). For solution to the Schrödinger equation, $E(R)$ is typically written as an expansion in R about R_e:

$$E(R) \cong E(R_e) + \left(\frac{dE}{dR}\right)_{R_e} (R-R_e) + \frac{1}{2}\left(\frac{d^2E}{dR^2}\right)_{R_e} (R-R_e)^2 + \cdots \tag{9.37}$$

We have already chosen to set $E(R_e) = 0$, and, since the function reaches a minimum at R_e, $(dE/dR)_{R_e}$ is also zero. This leaves the third term as the first nonzero term. We define $\rho \equiv (R - R_e)$ and

$$E(R) = \frac{1}{2}k_e\rho^2 + \cdots \tag{9.38}$$

where $k_e \equiv (d^2E/dR^2)_{R_e}$. If we truncate this expression at the first term, we have a simple harmonic oscillator formalism for the potential, where k_e is an effective spring constant. The functional form is parabolic, which very nearly reproduces the bottom of the potential well depicted in Figure 9.2. The higher-order terms in ρ can be considered anharmonic corrections. For now, we take just the first term, and equation (9.36) becomes

$$-\frac{\hbar^2}{2\mu} \frac{d^2}{d\rho^2} \chi(\rho) + \frac{1}{2}k_e\rho^2\chi(\rho) = E_v\chi(\rho) \tag{9.39}$$

Now multiply by $-2\mu/\hbar^2$ to obtain

$$\frac{d^2}{d\rho^2}\chi(\rho) - \alpha^2\rho^2\chi(\rho) = -\frac{2\mu}{\hbar^2}E_v\chi(\rho) \qquad (9.40)$$

where $\alpha \equiv \sqrt{k_e\mu/\hbar^2}$. If we were to try a series solution at this point, we would find a three-term recursion relation. A simpler recursion relation can be found via the substitution:

$$\chi(\rho) \equiv e^{\frac{-\alpha\rho^2}{2}} f(\rho) \qquad (9.41)$$

Applying the chain rule, we find that

$$\frac{d^2\chi(\rho)}{d\rho^2} = e^{\frac{-\alpha\rho^2}{2}}\left[f'' - 2\alpha\rho f' - \alpha f + \alpha^2\rho^2\right] \qquad (9.42)$$

where the superscripts $''$ and $'$ denote second and first derivatives respectively (not lower and upper energy levels) in this case. Substituting this result into equation (9.40) produces

$$f'' - 2\alpha\rho f' + \left(\frac{2\mu}{\hbar^2}E_v - \alpha\right)f = 0 \qquad (9.43)$$

Now develop a series solution using

$$f(\rho) = \sum_{n=0}^{\infty} c_n\rho^n \qquad (9.44)$$

Term-by-term differentiation, with simplification, produces

$$f' = \sum_{n=0}^{\infty} nc_n\rho^{n-1}$$

$$f'' = \sum_{n=0}^{\infty}(n+2)(n+1)c_{n+2}\rho^n \qquad (9.45)$$

Using these series in equation (9.43) produces

$$\sum_{n=0}^{\infty}\left[(n+2)(n+1)c_{n+2} - 2\alpha n c_n + \left(\frac{2\mu}{\hbar^2}E_v - \alpha\right)c_n\right]\rho^n = 0 \qquad (9.46)$$

The coefficients for ρ^n should all be zero for this to hold true. Therefore, the recursion relation must be

$$C_{n+2} = \frac{\alpha + 2\alpha n - \frac{2\mu}{\hbar^2} E_v}{(n+2)(n+1)} C_n \qquad (9.47)$$

This relation contains two arbitrary constants. c_0 gives rise to an even series and c_1 generates an odd series. The most general solution is a linear combination:

$$\chi(\rho) = A e^{\frac{-\alpha\rho^2}{2}} \sum_{\ell=0}^{\infty} c_{(2\ell+1)} \, \rho^{(2\ell+1)} + B e^{\frac{-\alpha\rho^2}{2}} \sum_{\ell=0}^{\infty} c_{(2\ell)} \, \rho^{(2\ell)} \qquad (9.48)$$

where the values of c are derived from the recursion relation. As usual, it is necessary to explore regimes for this series in an effort to simplify it. Begin by dealing with the even series (setting $n = 2\ell$), and take the ratio of coefficients for $\rho^{(2\ell+1)}$ and for $\rho^{(2\ell)}$:

$$\frac{c_{(2\ell+1)}}{c_{(2\ell)}} = \frac{\alpha + 4\alpha\ell - 2\frac{\mu}{\hbar^2} E_v}{(2\ell+1)(2\ell+2)} \qquad (9.49)$$

For large ℓ, this becomes

$$\frac{c_{(2\ell+1)}}{c_{(2\ell)}} = \frac{4\alpha\ell}{(2\ell)(2\ell)} = \frac{\alpha}{\ell} \qquad (9.50)$$

As it turns out, the odd series produces the same outcome. This outcome, α/ℓ, is the same as the ratio of coefficients for a series expansion of $\exp(\alpha\rho^2)$ for large ρ, if the series is written as a sum over $\ell = 0 \to \infty$. This term ends up in equation (9.48), where this $\exp(\alpha\rho^2)$ is multiplied by $\exp(-\alpha\rho^2/2)$ to produce $\exp(\alpha\rho^2/2)$, which becomes very large at large ρ. That result is not a useful solution. If we break off the series in equation (9.48) after an interval, then the $\exp(-\alpha\rho^2/2)$ terms can generate a zero at large ρ. To do that, c_n must go to zero at some value of n, say $n = v$. Using the numerator of the recursion relation, we can write

$$\alpha + 2\alpha v - \frac{2\mu}{\hbar^2} E_v = 0 \qquad (9.51)$$

Rearranging and applying the definition of α ($\alpha = \sqrt{k_e\mu/\hbar^2}$) yields

$$E_v = \left(v + \frac{1}{2}\right) \frac{h}{2\pi} \sqrt{\frac{k_e}{\mu}} \qquad (9.52)$$

where v is clearly the vibrational quantum number, which must be some integer. Interestingly, the classical solution for a simple harmonic oscillator provides an oscillation frequency given by

$$\nu_{osc} \equiv \nu_e = \frac{1}{2\pi}\sqrt{\frac{k_e}{\mu}} \tag{9.53}$$

Therefore, we can write the quantum solution for vibrational energy as

$$E_v = \left(v + \frac{1}{2}\right)h\nu_e \quad \text{for } v = 0, 1, 2, ... \tag{9.54}$$

Note that even, when $v = 0$, there is some vibrational energy in the molecule.

There are fairly loose selection rules for vibrational transitions. If the harmonic oscillator assumption holds, then $\Delta v = \pm 1$ (note that $\Delta v = 0$ is allowed during an electronic transition). The resulting spectrum (including rotational structure) is called the "fundamental band". When $v'' = 1, 2, ...$ the resulting spectrum is called a "hot band" because higher vibrational levels are populated as temperature increases. The harmonic oscillator limit rarely holds. It is often necessary to include higher-order anharmonic terms, at which time $\Delta v \geq \pm 1$ is allowed. When $\Delta v > \pm 1$, the spectrum is called an "overtone".

9.2.2   Rotation-Vibration Spectra and Corrections to Simple Models

Changes in $\mathcal{J}$ can occur simultaneously with changes in v. In the absence of electronic change, this topic is generally called "IR spectroscopy". If we were to observe an IR spectrum with very low spectral resolution, we would simply see structures associated with changes in v. We would see a fundamental band, perhaps several hot bands (depending upon the temperature) and perhaps some overtones (depending upon instrument sensitivity). The spectral structures themselves would be somewhat indistinct, because they represent the convolution of the instrument spectral response with the rotational structure of each band. The poor instrument spectral response would smooth-out all of the rotational structure.

If we then improved the spectral resolution of the instrument, we could resolve structures such as those depicted in Figure 9.3. The more detailed structure is due to rotational transitions that occur during the one vibrational transition associated with that band.

If there is no change in electronic level, it is possible to focus entirely on the nuclear energy, given by

$$E_{nuclear} = E_{v,\mathcal{J}} = hB_e\,\mathcal{J}(\mathcal{J}+1) + h\nu_e\left(v+\frac{1}{2}\right) \tag{9.55}$$

Term energies (in units of cm^{-1}, or wave numbers) can also be used to describe the nuclear energy. They are found by dividing energy by hc. Equation (9.55) can then be recast in term energies by

$$\frac{E_{nuclear}}{hc} = G_v + F_{\mathcal{J}} = \frac{B_e}{c}\,\mathcal{J}(\mathcal{J}+1) + \frac{\nu_e}{c}\left(v+\frac{1}{2}\right) \tag{9.56}$$

where G_v typically denotes vibrational term energy and $F_{\mathcal{J}}$ denotes rotational term energy.

Note that it is very common to define B_e and ν_e in terms of wave numbers. The same notation is used, which can cause confusion. The terms B_e and ν_e defined here in terms of Hz will have been divided by c to obtain the wave number form [e.g. $B_e(\text{cm}^{-1}) = B_e(\text{Hz})/c$]. In the wave number case, we would write

$$\frac{E_{nuclear}}{hc} = G_v + F_{\mathcal{J}} = B_e\,\mathcal{J}(\mathcal{J}+1) + \nu_e\left(v+\frac{1}{2}\right) \tag{9.57}$$

where, again, this version uses a different definition for B_e and ν_e.

Continuing with our former frequency-based definitions for B_e and ν_e, the frequency of a "ro-vibrational" transition is found by dividing the energy change of the transition ($E_{v',\mathcal{J}'} - E_{v'',\mathcal{J}''}$) by h:

$$\nu = B_e[\mathcal{J}'(\mathcal{J}'+1) - \mathcal{J}''(\mathcal{J}''+1)] + \nu_e[v'-v''] \tag{9.58}$$

Selection rules then generate

$$\begin{aligned}\Delta v &= +1, \quad \Delta\mathcal{J} = \pm 1 \quad \text{absorption} \\ \Delta v &= -1, \quad \Delta\mathcal{J} = \pm 1 \quad \text{emission}\end{aligned} \tag{9.59}$$

where $\Delta v \equiv v' - v''$ and $\Delta\mathcal{J} \equiv \mathcal{J}' - \mathcal{J}''$. For every vibrational transition, there is a series of $\Delta\mathcal{J}$ values. For every transition in which

$\Delta \mathcal{J} = -1$, the collection of lines is called a "P branch" of the vibrational transition. Conversely, $\Delta \mathcal{J} = +1$ produces the "R branch". In Figure 9.3, the left-hand lobe is the P branch while the right-hand lobe is the R branch. For ro-vibrational transitions in an electronic state that is labeled "Σ", $\Delta \mathcal{J}$ cannot equal zero. For ro-vibrational transitions in an electronic state that is labeled "Π","Δ", etc., $\Delta \mathcal{J}$ can equal zero (this Σ, Π, Δ designation will be discussed in detail in material that follows; very briefly it is analogous to the S, P, D... in an atom; the total angular momentum along the internuclear axis is 0,1,2... for Σ, Π, Δ...). When $\Delta \mathcal{J} = 0$, this produces a "Q branch". There is no Q branch visible in Figure 9.3 because that spectrum occurs in a molecule that is in a Σ electronic state.

For a fundamental band, $v' - v'' = 1$, and for absorption in the R branch, $\mathcal{J}' = \mathcal{J}'' + 1$. Then the optical frequency associated with such a ro-vibrational transition can be written as

$$\nu = B_e[(\mathcal{J}''+1)(\mathcal{J}''+2) - \mathcal{J}''(\mathcal{J}''+1)] + \nu_e = 2B_e(\mathcal{J}''+1) + \nu_e \quad (9.60)$$

Note that ν_e locates the center of the vibrational transition. The rotational transitions are spaced $2B_e$ apart and they lie on the high-frequency (low-λ) side of ν_e. A similar expression can be developed for the P branch, and the lines will also be spaced by $2B_e$ but they will lie on the low-frequency side of ν_e. The first R branch line will be located at $\nu = \nu_e + 2B_e$ and the first P branch line will be at $\nu = \nu_e - 2B_e$.

By simply measuring the location of ν_e, one has measured an effective nuclear spring constant via equation (9.53). By measuring rotational line spacings, one has measured B_e and therefore the nuclear moment of inertia via equation (9.30).

Anharmonicity

To this point we have described vibration using the simple harmonic oscillator formalism. Molecules are somewhat anharmonic, however, so the transition moment for overtones is nonzero. Anharmonicity is also evidenced by the fact that the energy spacing between levels decreases for high levels of v, as shown in Figure 9.6.

To describe anharmonicity, we simply include more terms in the expansion for $E(\rho)$ [equation (9.38)] to produce

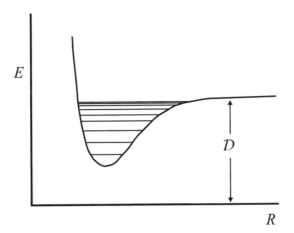

Figure 9.6: Anharmonicity in the vibrational levels.

$$E(\rho) = \frac{1}{2}k_e\rho^2 + \frac{1}{3!}\left(\frac{d^3 E}{d\rho^3}\right)_0 \rho^3 + \frac{1}{4!}\left(\frac{d^4 E}{d\rho^4}\right)_0 \rho^4 + \cdots \qquad (9.61)$$

where the subscript "0" indicates that the derivatives are evaluated at $\rho = 0$. To obtain these derivatives, it is necessary to assume a functional form for the potential $E(\rho)$. A common functional form is the Morse potential given by

$$E(\rho) = D\left(1 - e^{-\beta\rho}\right)^2 \qquad (9.62)$$

Here, D is the dissociation energy, which is defined as the energy difference between the minimum of the well (at R_e) and the asymptote at $\rho \to \infty$ (i.e. $R \to \infty$). The factor β is positive. For $\rho < 0$ (i.e. for $R < R_e$), therefore, $E(\rho)$ blows up, as expected. In addition, for very large ρ, $E(\rho) \to D$, which makes sense because we set the energy at the bottom of the well equal to zero.

Expanding $E(\rho)$ in a Maclaurin series gives

$$E(\rho) = D\left(\beta^2\rho^2 - \beta^3\rho^3 + \frac{7}{12}\beta^4\rho^4 - \cdots\right) \qquad (9.63)$$

Therefore, $D\beta^2 = (1/2)k_e$, which can be used to specify β. The entire potential curve is thus determined if D and k_e are known. The anharmonic terms in equation (9.61) are then

$$\frac{1}{3!}\left(\frac{\mathrm{d}^3 E}{\mathrm{d}\rho^3}\right)_\circ = -\beta^3 \mathcal{D}$$

$$\frac{1}{4!}\left(\frac{\mathrm{d}^4 E}{\mathrm{d}\rho^4}\right)_\circ = \frac{7}{12}\beta^4 \mathcal{D} \tag{9.64}$$

and so forth. The energy levels for the anharmonic case can then be found by substituting for $E(\rho)$ into the radial Schrödinger equation and solving the ODE. Alternatively, the additional anharmonic terms can be treated as energy perturbations. It is possible to use the wave functions from the harmonic oscillator solution to find the expectation value for the perturbation Hamiltonian. In both cases, the outcome is

$$E_v = h\nu_e\left(v + \frac{1}{2}\right) - h\nu_e x_e\left(v + \frac{1}{2}\right) \tag{9.65}$$

where x_e is an anharmonic correction term that is usually obtained from measured spectra. An alternative and commonly used formalism writes the same expression as

$$E_v = hc\omega_e\left(v + \frac{1}{2}\right) - hc\omega_e x_e\left(v + \frac{1}{2}\right) \tag{9.66}$$

where $\omega_e = \nu_e/c$. It is unfortunate that ω is used because there is no 2π in the conversion, which can potentially add confusion. Tabulations, however, usually contain $\omega_e x_e$. The term $\omega_e x_e$ is called the first anharmonicity constant, and x_e is given by

$$x_e = \frac{h\nu_e}{4\mathcal{D}} \tag{9.67}$$

Selection rules now allow

$$\Delta v = \pm 1,\ \pm 2,\ \pm 3,\ ... \tag{9.68}$$

and the vibrational overtone spacings provide a measure of $\mathcal{D}$.

Centrifugal distortion in rotation

Molecular rotators are not actually rigid; the value of the moment of inertia I can actually change with rotational level $\mathcal{J}$. This particular phenomenon does not include rotational interaction with vibration, it simply deals with centrifugal distortion.

Equation (9.28) presents the outcome of the rigid rotator solution, repeated here:

$$E_{\mathcal{J}} = \frac{\hbar^2}{2I_e}\mathcal{J}(\mathcal{J}+1) = \frac{\hbar^2}{2\mu R_e^2}\mathcal{J}(\mathcal{J}+1) = hB_e\mathcal{J}(\mathcal{J}+1)$$

To find R_e, it is necessary to use an expression for $E(R)$, or to use measurements of B_e. In order to include centrifugal distortion effects, it is necessary to include the internuclear potential given by the centrifugal stretching:

$$E_{\mathcal{J}}(R) = \frac{\hbar^2}{2\mu R_e^2}\mathcal{J}(\mathcal{J}+1) + \frac{1}{2}k_e(R - R_e)^2 \qquad (9.69)$$

An effective R_e is found by minimization of $E_{\mathcal{J}}(R)$ with respect to R, which produces

$$k_e(R_e' - R_e) = \frac{\hbar^2}{\mu R_e^3}\mathcal{J}(\mathcal{J}+1) \qquad (9.70)$$

where R_e is the minimum in the potential well for a case that ignores centrifugal distortion and R_e' is the minimum for a case that includes centrifugal distortion:

$$R_e' = R_e + \frac{\hbar^2}{k_e\mu R_e^3}\mathcal{J}(\mathcal{J}+1) \qquad (9.71)$$

We then use R_e' in the expressions for rotational energy. To simplify, that expression is expanded in powers of $\mathcal{J}(\mathcal{J}+1)$ to obtain

$$E_J = hB_e\,\mathcal{J}(\mathcal{J}+1) - hD_e\,[\mathcal{J}(\mathcal{J}+1)]^2 \qquad (9.72)$$

where $D_e = (4B_e^3)/\omega_e^2$.

Vibration/rotation interaction

As a molecule vibrates and rotates, the moment of inertia varies because R departs from R_e. The particle samples larger R for higher levels of v, for a realistic potential (see Figure 9.6). Instead of $1/R_e^2$ in the expression for B_e, it would be more correct to use the expectation value for R (e.g. $1/\langle R \rangle_v^2$) to generate a vibrationally dependent rotational "constant":

$$B_v = \frac{\hbar^2}{2h\mu \langle R \rangle_v^2} \tag{9.73}$$

Then

$$E_J = hB_v \, J(J+1) \tag{9.74}$$

To a first approximation

$$B_v = B_e - \alpha_e \left(v + \frac{1}{2} \right) \tag{9.75}$$

where

$$\alpha_e = \frac{6B_e^2}{\omega_e} \left[\sqrt{\frac{\omega_e x_e}{B_e}} - 1 \right] \tag{9.76}$$

Combined, corrected ro-vibrational energies

We have now presented corrected expressions for rotational and vibrational energy levels in diatomic molecules. The vibrational term energy G_v is most commonly written as

$$G_v = \frac{E_v}{hc} = \omega_e \left(v + \frac{1}{2} \right) - \omega_e x_e \left(v + \frac{1}{2} \right)^2 + \omega_e y_e \left(v + \frac{1}{2} \right)^3 \tag{9.77}$$

where the third anharmonicity constant ($\omega_e y_e$) has been included. It simply comes from a third term in the expansion for $E(R)$. The rotational term energy is commonly written as

$$F_J = \frac{E_J}{hc} = \frac{B_v}{c} \, J(J+1) - \frac{D_v}{c} \, J^2(J+1)^2 + \frac{H_v}{c} \, J^3(J+1)^3 \tag{9.78}$$

where a third rotational correction has also been included. As mentioned once already, the term constants used in other publications can have the $1/c$ built into them (i.e. they are expressed in cm^{-1}).

The total ro-vibrational term energy is the sum $G_v + F_J$. Each molecule will require a particular emphasis on the various corrections to the rigid rotator/harmonic oscillator solution. Spectral observations lead to conclusions regarding which corrections are important and which can be neglected.

Parity

The total wave function for an atom or molecule is a function of several position coordinates r_i and (if electronic terms are included) spin coordinates s_i. Then one can write the Schrödinger equation as

$$\hat{H}(\vec{r}_i, \vec{p}_i, \vec{\ell}_i, \vec{s}_i)\psi(\vec{r}_i, \vec{s}_i) = E\psi(\vec{r}_i, \vec{s}_i) \tag{9.79}$$

where $\vec{p}_i$ is a momentum vector (dependent upon $\vec{r}_i$), and $\vec{\ell}_i$ and $\vec{s}_i$ are orbital and spin vectors respectively. In addition

$$\hat{H}(-\vec{r}_i, -\vec{p}_i, \vec{\ell}_i, \vec{s}_i)\psi(-\vec{r}_i, \vec{s}_i) = E\psi(-\vec{r}_i, \vec{s}_i) \tag{9.80}$$

For an isolated atom or molecule

$$\hat{H}(\vec{r}_i, \vec{p}_i, \vec{\ell}_i, \vec{s}_i) = \vec{H}(-\vec{r}_i, -\vec{p}_i, \vec{\ell}_i, \vec{s}_i) \tag{9.81}$$

This means that

$$\psi(\vec{r}_i, \vec{s}_i) = c\psi(-\vec{r}_i, \vec{s}_i) = c^2\psi(\vec{r}_i, \vec{s}_i) \tag{9.82}$$

Clearly, $c = \pm 1$, where $+$ denotes "even parity" and $-$ denotes "odd parity". The total wave function is called "symmetric" if it is unchanged after it has been reflected through the origin (center of mass in this case). This reflection is indicated by $r_i \to -r_i$ and $p_i \to -p_i$ in the equations above. If a wave function is symmetric in this sense, it has even parity. It is anti-symmetric if it changes after reflection, an indication of odd parity. Note that wave function symmetry is not the same as interchange of particles, as discussed in Section 8.3.2. They are two different topics. As it turns out, there is a separate selection rule for parity.

9.2.3 A Review of Ro-Vibrational Molecular Selection Rules

The preceding text has already mentioned selection rules in several places. It seems best, however, to compile the ro-vibrational rules here so that the collection can be easily reviewed. Again, quantum mechanical selection rules determine which transitions are allowed. For allowed transitions, the matrix element (also called transition moment, $\vec{\mu}_{nm}$) is nonzero. For forbidden transitions the transition moment is zero (or nearly zero). This has to do with the overlap integral that defines the transition moment in equation (10.9).

Ro-vibrational ("IR active") transitions will occur only if the molecule has a permanent dipole moment, because the transition moment requires this. For homonuclear diatomic molecules, therefore, spectroscopically resonant ro-vibrational transitions are forbidden.

Purely rotational transitions can occur only between states of opposite parity. This is normally written as

$$+ \leftrightarrow - \quad + \not\leftrightarrow + \quad - \not\leftrightarrow - \tag{9.83}$$

where "$\leftrightarrow$" designates an allowed transition and "$\not\leftrightarrow$" designates a forbidden transition. The rotational selection rule for purely rotational transitions is

$$\Delta \mathcal{J} = \pm 1 \tag{9.84}$$

Note that issues like electron spin are ignored because this rule is for nuclear rotational changes alone.

The parity rule also holds for combined ro-vibrational transitions. For ro-vibrational transitions in a Σ electronic state [where Λ (the total electronic angular momentum vector along the internuclear axis) $= 0$], $\Delta \mathcal{J} = \pm 1$. For ro-vibrational transitions in any other electronic state (e.g. for $\Lambda \neq 0$), $\Delta \mathcal{J} = 0, \pm 1$. In short, for ro-vibrational transitions within a Σ electronic state there is no Q branch. There is no strict selection rule for vibrational transitions, but the transition moment is typically largest for $\Delta v = \pm 1$.

9.2.4 Electronic Transitions

To deal with electronic transitions, one must return to the original molecular Schrödinger equation (9.1)

$$\left(-\frac{\hbar^2}{2\mu}\nabla^2 - \frac{\hbar^2}{2m_e}\sum_{i=1}^{N}\nabla_i^2 + V \right)\psi = E\psi$$

As before, we rely upon the Born-Oppenheimer approximation to separate nuclear (vibration and rotation) motion from electronic. We set

$$\psi(\vec{r}, \vec{R}) = \psi_e(\vec{r}, R)\chi(\vec{R}) \tag{9.85}$$

and this separates an electronic equation of the form

$$\left(-\frac{\hbar^2}{2m_e}\sum_{i=1}^{N}\nabla_i^2 + V(\vec{r}, R)\right)\psi_e(\vec{r}, R) = E(R)\psi_e(\vec{r}, R) \tag{9.86}$$

The solution to this equation will, in part, specify the function $E(R)$, and it can change with each electronic level. Each electronic level, therefore, has a different potential well, which causes the vibrational structure to vary from one electronic state to the next. Rotation also interacts with the potential well, and rotational states are thus different from one electronic state to the next. The lowest value of the ground-state electronic potential well is usually set to zero. Then the first energy level above that is the $v''=0$ vibrational level with no rotation. From that level the various $v''=0$ state rotational levels can be populated, followed by further ro-vibrational levels in the ground electronic state. The energy difference between the ground electronic state and the first excited electronic state is typically (but not always) larger than ro-vibrational transitional energies.

The total energy is a sum, thanks to the Born-Oppenheimer approximation:

$$E = E_e + E_v + E_r \tag{9.87}$$

which can also be written in term energies (divide by hc):

$$T = \frac{E}{hc} = T_e + G_v + F_J \tag{9.88}$$

We have already developed expressions for G_v and F_J for ro-vibrational transitions in the previous section, but there are important distinctions now that electronic transitions have been included. The constants used to evaluate ro-vibrational term energies (ω_e, x_e, y_e, B_v, and D_v) are not the same in the ground and excited electronic states, owing to the fact that $E(R)$ changes from one electronic state to the next. In addition, F is now determined by the resultant angular momentum, including the portion of the electronic angular momentum that couples to the nuclear angular rotational momentum. It is thus necessary to describe electronic angular momentum coupling before one can evaluate the resultant value of F.

Electronic term energies are not easy to calculate. The potential wells are most commonly inferred using *ab initio* calculations, often relying upon comparison to measured spectroscopic parameters. One should be careful about the electronic term energy, however, because two forms are common. In the form used above, T_e is the energy spacing between the minimum points of two potential wells. Another commonly used form is to take the difference between the first vibrational levels (for both v' and $v'' = 0$). This form (often written T_o) is not the same, because the vibrational constants in the two electronic levels are not the same.

Building up electronic states

This subject is actually quite large. Here we provide a very brief summary of molecular bonding. Two approaches are used to model electronic states. The "united atom" approach is applied in the case when the two nuclei are relatively close to each other. In this limiting case the electronic states build up just as they do with a multi-electron atom. In the "separated atom" approach, individual atomic electron distributions are considered, and the two are added. This is also an asymptotic limit for molecular dissociation.

<u>United atom</u> Each electron within this molecule has a principal quantum number n and an orbital angular momentum quantum number ℓ. As before, ℓ is space quantized by $m_\ell = \ell$, $\ell - 1$, $... - \ell$. The quantum number λ is defined as $\lambda \equiv \mid m_\ell \mid$. Individual electrons are labeled σ, π, δ, $\phi, ...$ for $\lambda = 0$, 1, 2, 3, ... Each individual electron is labeled $n\ell\lambda$. Examples would be $1s\sigma$, or $3d\pi$ and so forth.

<u>Separated atom</u> Each electron within this molecule has a principal quantum number n and an orbital angular momentum quantum number ℓ. As before, ℓ is space quantized by $m_\ell = \ell$, $\ell - 1$, $..., -\ell$. The quantum number λ is defined as $\lambda \equiv \mid m_\ell \mid$. Individual electrons are labeled σ, π, δ, $\phi, ...$ for $\lambda = 0$, 1, 2, 3, ... Each individual electron is labeled $\lambda n\ell_i$, where "i" denotes the atom with which the electron is associated. Examples would be $\sigma 1 s_H$, or $\pi 2 p_O$, and so forth. If the

molecule is homonuclear, a slightly different system is used. Examples would include $\sigma_g 1s$ or $\pi_u 2p$. Here, the "g" denotes "gerade", meaning the orbital wave function is symmetric upon inversion, while "u" denotes "ungerade", meaning the orbital wave function is asymmetric upon inversion.

The electron shells add up (are populated) in a progression similar to atoms. σ orbitals hold only two electrons (two spin states), while π and δ have four ($\pm m_\ell$ and $\pm m_s$), and so forth. One assigns groups such as $1s\sigma^2$, $2s\sigma^2$, $2p\sigma^2$, $2p\pi^4$, ... It is not immediately clear which states have higher energy in a diatomic molecule, so there is not a clear formula for building up states. Many molecules build up in the same way, but some have a different order. As with atoms, the outer electron determines the electronic state designation for optical spectroscopy. In contrast, Auger and X-ray spectroscopy in the solid phase cause electronic transitions in inner electrons.

Electron spin

The individual values for each electronic spin state add to give a resultant:

$$\vec{S} = \sum_i \vec{s}_i \tag{9.89}$$

The component of $\vec{S}$ along the internuclear axis is space quantized and it is given by

$$\Sigma = S, S-1, ... -S \tag{9.90}$$

There are thus $2S+1$ possible values of Σ, or a multiplicity of $2S+1$.

Nuclear spin

Heteronuclear diatomic molecules have two nuclei that are distinguishable, which means that the Pauli exclusion principle is not relevant to this situation (inversion of the nuclei is never symmetric; the nuclei cannot be confused with each other). Each identifiable nucleus has a multiplicity (or degeneracy) of $2I+1$ (where I is the nuclear spin quantum number), so the total spin degeneracy for nuclei a and b is $g_I = (2I_a+1)(2I_b+1)$. Here, g_I is independent of J, and it is the same for all levels. For this reason, it divides out of the Boltzmann fraction expression [equation (2.69)] for number density in each energy level.

Conversely, homonuclear diatomic molecules have two nuclei that are indistinguishable. In order to satisfy the Pauli exclusion principle, therefore, it is necessary to track symmetry properties of the entire molecule carefully. Some nuclear spin states are not allowed, depending upon the behavior of the total molecular wave function. The rotational wave function (and therefore the rotational quantum number) thus affects what nuclear spin states are allowed. The allowed values for I and g_I are then dependent upon the value of J (see, for example, Levine [68]). This is why Table 2.1 contains values for g_I versus J only for homonuclear diatomics. Because homonuclear diatomic molecules are not IR active, this dependence of g_I on J can go unnoticed. Homonuclear diatomics are "Raman active", however, and this degeneracy is manifested in the Raman spectrum (see Chapter 13), via the Boltzmann fraction expression.

Electronic states and coupling

Just as in the case for a multi-electron atom, the electrons in a molecule can combine orbital angular momenta. This produces a total orbital angular momentum $\vec{L}$ for the electrons. In Russell-Saunders coupling, for example, $\vec{L}$ is the sum of individual orbital angular momenta $(\vec{\ell})$. In a molecule, $\vec{L}$ can combine with the nuclear orbital angular momentum (which was absent in the atomic case) to produce a resultant total orbital angular momentum. That term then determines the rotational energy level structure that appears in an electronic transition.

In the case of an atom, the component of $\vec{\ell}$ along z is $\vec{m}$. It is quantized and has a value of $\mid \vec{m} \mid = m_\ell \hbar$, where m_ℓ is the quantum number. The atomic problem provides no means by which to define the axis z, until a magnetic or electric field is applied. In the case of a diatomic molecule, there is a built-in electrostatic field along the internuclear axis that defines z. Electronic coupling then occurs along the diatomic internuclear axis. The nature of the coupling depends upon the nuclei, the electrons, and the field that they generate.

The diatomic molecular coupling problem is discussed in terms very much like the atomic problem (see Figure 9.7). A precession of $\vec{L}$ about the internuclear axis gives rise to a component along the axis,

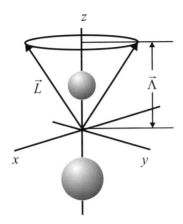

Figure 9.7: Cartoon of electronic coupling to the internuclear axis of a diatomic molecule.

called $\vec{\Lambda}$. The magnitude of $\vec{\Lambda}$ is given by $\mid M_L \mid \hbar$, where $M_L = L, L-1, L-2, ..., -L$, and L is the quantum number for $\vec{L}$. For a given orbital angular momentum, if the electron reverses direction, then the sign of M_L changes but energy is constant. There is thus a degeneracy of 2 in these couplings. Energy does change with $\mid \vec{\Lambda} \mid$, and since $\mid M_L \mid= 0, 1, 2, ..., L$, there are $L + 1$ distinct states with different energies for each value of L.

Electronic states are designated by Σ, Π, Δ, Φ, ... for $\Lambda = 0, 1, 2, 3, ...$ respectively. As just mentioned, for the Π, Δ, Φ, ... states, M_L can have two values, giving rise to a degeneracy of two. This is called "lambda doubling". It is possible for this degeneracy to be removed at high rotational quantum numbers.

Molecules have electronic spin multiplicity in the same way that atoms do. The resultant spin S will be integral if there are an even number of electrons, and half-integral if odd. The spin component along the internuclear axis is then $\Sigma = S, S - 1, S - 2, ..., -S$ (it is unfortunate, but the same symbol is used for the spin component along the axis and as a designation for the $\Lambda = 0$ electronic state). There are $2S + 1$ possible values of the spin component along the axis, Σ. Σ is not defined for a $\Lambda = 0$ (no orbital angular momentum coupling) state.

In an atom, the orbital and spin angular momenta add to give a composite value (we used the letter J). In a molecule, Λ and Σ add to give

$$\Omega \equiv \mid \Lambda + \Sigma \mid \qquad\qquad (9.91)$$

For a given $\Lambda(\neq 0)$, there are $2S+1$ values of Σ and therefore $2S+1$ values of Ω (in addition to the lambda doubling). Λ can produce a magnetic field, so the $2S+1$ states can possibly be split by the spin interaction with that field.

Angular momentum coupling now must account for the nuclear rotation as well. The resultant will always be called J, but the way it is generated depends upon the molecule. In the simplest case, if S and Λ are both zero, then the molecule is a nonrigid rotator. That is uncommon.

An extremely simple case related to coupling that can also arise in purely ro-vibrational transitions occurs when the electron orbital introduces moment of inertia about the internuclear axis that is not insignificant. This case is called a symmetric top (see Section 9.3 for a more detailed definition of a symmetric top molecule). Both diatomic and polyatomic molecules can be in the symmetric top form. In such a case, the rotational energy levels are described by

$$F_{\mathcal{J}} = \frac{E_{\mathcal{J}}}{hc} = \frac{B_v}{c}\,\mathcal{J}(\mathcal{J}+1) - \frac{D_v}{c}\,\mathcal{J}^2(\mathcal{J}+1)^2 + \frac{1}{c}(A-B_v)\Lambda^2 \quad (9.92)$$

Where the term $(A-B_v)\Lambda^2$ is the symmetric top correction. Here, Λ is the same component of electronic angular momentum along the internuclear axis. A and B_v are based upon equation (9.30), using the moments about the axis and normal to the axis respectively. The symmetric top correction describes a very specific case for diatomic molecules, where the electron orbit actually contributes momentum off the internuclear axis.

Geometries in which the electron orbit couples to the internuclear axis and thus affects the rotational spectrum of an electronic transition are typically described by one of Hund's coupling cases (see, for example, reference [17]). It is important to understand that these are idealized, limiting cases. It is not uncommon for a molecule to follow such a case fairly loosely, or in fact to change case as the rotational level increases. One simply chooses which case best describes the observed behavior. There are many of these cases, but we will briefly describe only Hund's cases a and b because most molecules encountered in flow-field diagnostics behave according to these two.

We have a small problem in that we called nuclear angular momentum "L", in order to draw a quick analogy to the electronic angular momentum, also called "L". It is now necessary to discuss both in the same block of text. For that reason, we now use "R" to designate the nuclear term. Note also that we have used J for the electronic angular momentum quantum number and for the nuclear angular momentum quantum number. In keeping with convention, it will be used here once again. It would appear that people simply like to use J for whichever version is under discussion.

Hund's case a In this case (see Figure 9.8) we assume the following: the interaction between nuclear rotation and electronic angular momentum is weak; spin orbit coupling is also weak; and electronic angular momentum is strongly coupled to the internuclear axis. $\vec{L}$ and $\vec{S}$ precess independently about the internuclear axis. Their projections, Λ and Σ, then add to give Ω, a well-defined quantum number that can be used in place of Λ as a representation for electronic motion. $\vec{R}$ represents nuclear angular momentum, and in Hund's case a the total angular momentum is

$$\vec{J} = \vec{R} + \vec{\Omega} \qquad (9.93)$$

with a quantum number $J = R + \Omega$.

States with different values of Λ will have slightly different energy levels. The rotational term energy for Hund's case a is usually written as

$$F_J = \frac{B_v}{c} J(J+1) - \frac{B_v}{c}\Omega^2 \qquad (9.94)$$

(see, for example, reference [17]). Others write this expression with an additional $(A/c)\Omega^2$ (the same term one would find in a symmetric top, see, for example, reference [28]), but that term is usually included in the electronic term energy [16]. It is important, however, to understand in detail where various energy terms are located and whether or not, by mixing references, one has neglected to include important terms.

Hund's case b In this case (see Figure 9.9) we assume the following: spin coupling to Λ is weak; as a result, spin couples to the nuclear rotational axis $\vec{R}$. Here, the magnetic field is weak, and S does not precess about the internuclear axis and therefore it remains fixed in

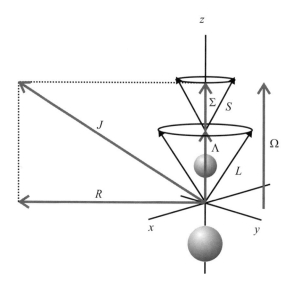

Figure 9.8: Schematic of one example for angular momentum coupling in Hund's case a.

space, adding vectorally to $\vec{R}$. A new angular momentum vector $\vec{N}$ and quantum number N are defined by

$$\vec{N} \equiv \vec{R} + \vec{\Lambda} \tag{9.95}$$

and then

$$\vec{J} = \vec{N} + \vec{S} \tag{9.96}$$

Hund's case b is very common in free radicals because it applies to Σ states, and to Π and Δ states in diatomics containing atoms with low atomic number. Rotational energy levels in Hund's case b are not described by one simple formula. Herzberg provides a more detailed discussion on the various forms used and their justification [17].

9.2.5   Electronic Spectroscopy

Because the energies add, spectral lines for electronic transitions are usually located at wave numbers given by

$$\frac{\nu}{c} = \frac{1}{\lambda_{\circ}} = \breve{\nu} = T' - T'' = (T'_e - T''_e) + (G'_v - G''_v) + (F'_J - F''_J) \quad \text{cm}^{-1} \tag{9.97}$$

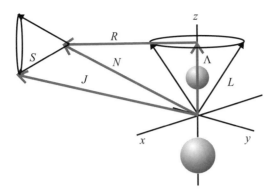

Figure 9.9: Schematic of one example for angular momentum coupling in Hund's case b.

The difference $T'_e - T''_e$ is set by measurement or *ab initio* calculation. Diagnostics users typically compile the data from various sources and use them. The differences $G'_v - G''_v$ and $F'_J - F''_J$ have already been discussed. The important distinction here is that the vibrational and rotational constants for the primed and double-primed states are different because they are associated with different electronic levels.

Consider the case where the rotational structure is ignored (e.g. when an instrument does not have the spectral resolution to detect rotational structure). Then

$$\breve{\nu} = \breve{\nu}_e + \breve{\nu}_v \equiv \breve{\nu}_\circ \tag{9.98}$$

Using results from the vibrational analysis, this can be written as

$$
\begin{aligned}
\breve{\nu}_\circ &= \breve{\nu}_e + \omega'_e \left(v' + \frac{1}{2} \right) - \omega'_e x'_e \left(v' + \frac{1}{2} \right)^2 \\
&\quad - \omega''_e \left(v'' + \frac{1}{2} \right) + \omega''_e x''_e \left(v'' + \frac{1}{2} \right)^2
\end{aligned}
\tag{9.99}
$$

This spectral location corresponds to the $J' = 0 \leftarrow J'' = 0$ rotational transition. In electronic transitions, there are no strict selection rules on Δv. Franck-Condon factors determine the strength of response [see, for example, equation (10.20) and the discussion preceeding it].

It is also possible to describe $\breve{\nu}_\circ$ in terms of the location of the (0,0) band center:

$$\check{\nu}_o = \check{\nu}_{0,0} + \omega_o' v' - \omega_o' x_o' v'^2 - \omega_o'' v'' + \omega_o'' x_o'' v''^2 \qquad (9.100)$$

where the location of the band center is

$$\check{\nu}_{0,0} \equiv \check{\nu}_e + \frac{1}{2}\omega_e' - \frac{1}{4}\omega_e' x_e' - \frac{1}{2}\omega_e'' + \frac{1}{4}\omega_e'' x_e'' \qquad (9.101)$$

Having established this formalism, we now add rotational structure:

$$\check{\nu} = \check{\nu}_e + \check{\nu}_v + \check{\nu}_J = \check{\nu}_o + \check{\nu}_J \qquad (9.102)$$

or, in terms of the term energies

$$\check{\nu} = \check{\nu}_o + F'(J') - F''(J'') \qquad (9.103)$$

In the symmetric top case, $F(J)$ is given by equation (9.92) (as just one example). Unfortunately, the symmetric top term $(A - B_v)\Lambda^2$ is sometimes bundled into $\check{\nu}_o$ for the purposes of "convenience". This turns out to be terribly inconvenient if the author does not mention that fact.

Assume for the purposes of discussion that $F(J)$ is described as a rigid rotator alone. This will permit a simplified but generalized discussion of spectra. The wave numbers for each transition are then

$$\check{\nu} = \check{\nu}_o + [B_v' \, J'(J' + 1) - B_v'' \, J''(J'' + 1)] \qquad (9.104)$$

Clearly, various molecules will depart from this formalism depending upon centrifugal distortion, the symmetric top correction, the electronic coupling case, etc., but the following basic discussion remains useful. In absorption ($J'' \to J'$ transitions), if we start from a generic rotational level J, then

the R branch is: $\check{\nu} = \check{\nu}_o + F'(J + 1) - F''(J)$
the Q branch is: $\check{\nu} = \check{\nu}_o + F'(J) - F''(J)$
the P branch is: $\check{\nu} = \check{\nu}_o + F'(J - 1) - F''(J)$ (9.105)

where the parentheses indicate functional J dependence. Inserting the rigid rotator expression produces

the R branch is: $\check{\nu} = \check{\nu}_o + 2B_v' + (3B_v' - B_v'')J + (B_v' - B_v'')J^2$
the Q branch is: $\check{\nu} = \check{\nu}_o + (B_v' - B_v'')J + (B_v' - B_v'')J^2$
the P branch is: $\check{\nu} = \check{\nu}_o - (B_v' + B_v'')J + (B_v' - B_v'')J^2$

The R and P branches can be written with a common expression:

$$\breve{\nu} = \breve{\nu}_\circ + (B'_v + B''_v)m + (B'_v - B''_v)m^2 \qquad (9.106)$$

where $m = J + 1$ in the R branch and $m = -J$ in the P branch. Recall that because we are discussing two different electronic levels, $B'_v \neq B''_v$. At small J, the values of $\breve{\nu}$ are dominated by the linear term in equation (9.106), but the quadratic term takes over at large J. This explains the more complex structure in the electronic transition presented in Figure 9.4. The rotational band starts a progression but reverses after the quadratic term takes over. The point where the band reverses is called a "band head". If $B'_v > B''_v$, there will be a band head in the P branch that is "degraded to the blue" (reverses towards shorter wavelengths), and so forth. Q branches can also exhibit this behavior.

9.2.6 Selection Rules, Degeneracy, and Notation

We have already discussed some selection rules, but the rules for diatomic electronic transitions for Hund's cases a and b are compiled here for convenience. For rotation in general

$$\Delta J = 0, \pm 1, \text{ with } J = 0 \nleftrightarrow J = 0 \qquad (9.107)$$

We actually allow $\Delta J = \pm 2$, but the O and S transitions are very weak relative to the P, Q and R transitions.

There is no restriction on Δv, except that Franck-Condon principles apply (see Chapter 10).

Parity must change in a heteronuclear molecular electronic transition [same as equation (9.83)]:

$$+ \leftrightarrow - \quad + \nleftrightarrow + \quad - \nleftrightarrow -$$

The rules are similar for homonuclear diatomics, although the notation is different. In the case of electronic transitions between two Σ states, $\Sigma^+ \leftrightarrow \Sigma^+$, $\Sigma^- \leftrightarrow \Sigma^-$, but $\Sigma^+ \nleftrightarrow \Sigma^-$.

Electronic quantum numbers follow the rule

$$\Delta\Lambda = 0, \pm 1 \text{ and } \Delta S = 0 \qquad (9.108)$$

In Hund's case a

$$\Delta\Sigma \ = \ 0$$
$$\Delta\Omega \ = \ 0, \pm 1 \qquad\qquad (9.109)$$

In Hund's case b

$$\Delta N = 0, \pm 1, \quad \text{except } \Delta N \neq 0 \text{ in } \Sigma \leftrightarrow \Sigma \text{ transitions} \qquad (9.110)$$

Some details have been neglected here. It would be best to consult a source such as the book by Herzberg [17] for the finer details of specific cases.

Degeneracies are a matter of spectral resolution again. In many cases, two levels may be very slightly split, but the splitting cannot be resolved by the instrument and in fact a single line broadening function fits the measured spectral profile well. If that is the case, then the two slightly split levels can be treated as a degenerate source for a single line. If, on the other hand, the experimental data are best represented by two individual lines with separate broadening functions, one should treat the levels as two nondegenerate levels. Electronic degeneracy can be generally described by

$$g_e = (2S + 1)\phi \qquad\qquad (9.111)$$

where $\phi=1$ for $\Lambda = 0$ and $\phi=2$ for $\Lambda \neq 0$ (e.g. a lambda doubled level). In specific cases the energy levels can actually differ slightly. As one example, the spin split in Hund's case a molecules may be too large to consider them equal in energy.

As before, vibrational levels are not degenerate.

Rotational degeneracy in Hund's case a is

$$g_J = 2J + 1 \qquad\qquad (9.112)$$

Note that lambda doubling has already been accounted for in equation (9.111) and so it is not counted here. In Hund's case b, rotational degeneracy is given by

$$g_N = 2N + 1 \qquad\qquad (9.113)$$

For heteronuclear diatomic molecules, the nuclear spin degeneracy is given by

$$g_I = (2I_a + 1)(2I_b + 1) \tag{9.114}$$

Homonuclear diatomic molecule spin degeneracy is a function of J, as shown in Table 2.1. It is necessary to find such tabulations for homonuclear molecules.

Spectroscopic notation is fairly straightforward for diatomic molecules. The ground state is labeled the "X-state", the next level is labeled the "A-state", the next beyond A is the "B-state", and so forth. The full electronic state designation typically goes as

$$\bigcirc^{(2S+1)} \triangle_\Omega^\diamond \tag{9.115}$$

where

$\bigcirc$	represents the letter designating electronic levels (X, A, B,...)
$(2S+1)$	is the multiplicity
$\triangle$	represents the capital Greek letter designating the value of Λ (Σ, Π, Δ, ... for $\Lambda = 0$, 1, 2, ...)
$\diamond$	is a parity designator, "+" for a symmetric ψ and "$-$" for an asymmetric ψ
Ω	$= (\Lambda + \Sigma)$, when there is spin coupling to the internuclear axis

For example, $A^2\Sigma^+$ describes the first excited electronic state in OH. Often the succeeding subscript is left off. The subscript is irrelevant in a Hund's case b molecule anyway. Note that the ground state of OH is labeled $X^2\Pi$, with no succeeding superscript. This is done because Π states are lambda doubled, and in this case the two Λ components have different parity. It is thus not possible to label a universal parity on this Π state. Note also that the parity selection rules do not allow every $\Delta J = 0, \pm 1$ transition, because half the Π states have incorrect parity in this specific case.

Transitions are labeled with these letter designators, with the lower level on the right and an arrow indicating in which direction the transition occurs. As an example, the designation

$$A^2\Sigma^+ \leftarrow X^2\Pi$$

indicates an absorption from a ground electronic state to a first excited state (e.g. in OH). For very well-known transitions, this is often shortened to $A \leftarrow X$.

Vibrational band notation is written as (v', v'') (upper, lower), while rotational transitions are labeled by P, Q, R, and so forth. As already mentioned, there are rotational transitions with nonzero transition moments for $\Delta J = \pm 2$. These are labeled the O branch for $\Delta J = -2$ and S branch for $\Delta J = +2$. The rotational transition format is not quite as standardized, because transitions depend upon the angular momentum coupling case. The following general format is used:

$$^{\diamond}\triangle_{\alpha,\beta} \; (J'' \text{ or } N'') \tag{9.116}$$

where

$\diamond$	= S, R, Q, P, O for $\Delta N = +2, +1, 0, -1, -2$; but when $\Delta J = \Delta N$, or when ΔN makes no sense for the coupling case, the term $\diamond$ is left off
$\triangle$	= S, R, Q, P, O for $\Delta J = +2, +1, 0, -1, -2$
α, β	are related to Hund's case b. α is used for the upper level, $\alpha=1, 2, 3, ...$ for $J' = (N+S), (N+S-1), (N+S-2), ...$ and β is for the lower level, $\beta=1, 2, 3, ...$ for $J'' = (N+S), (N+S-1), (N+S-2), ...$; when $\alpha = \beta$, only one value is used

As one example, one might say that the "$P_1(8)$-line of the (0,0) band of the $A^2\Sigma^+ \leftarrow X^2\Pi$ electronic transition" was observed. Note that all of the necessary information for specifying the states is included in the notation.

9.3   Polyatomic Molecules

As more atoms are added to a molecule, the structure, energy states and spectra all become even more complex. Spectral features become specific to each molecule, and it becomes impossible to generalize results. As extra lines are added to a spectrum, and as these lines broaden somewhat, fine structure becomes masked by overlaps. With added complexity, spectral and structural features are not understood as well. This trend reaches a point where spectroscopic measurement of number density is simply calibrated.

Because this topic is complex and multifaceted, we will simply introduce a few concepts so that they will not come as a surprise when scanning the literature. Significantly more detail can be found in volumes such as the book by Herzberg [18].

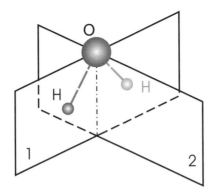

Figure 9.10: σ symmetry planes (e.g. H_2O).

9.3.1 Symmetry and Point Groups

Diatomic molecules are always linear (except for the need to apply symmetric top notions when the electronic structure requires it), while polyatomic molecules are often three-dimensional. Linear alignment is thus a simplifying case of the full three-dimensional problem. The symmetry of a three-dimensional molecule has a great deal to do with spectroscopic response. Highly symmetric molecules will have many rotational and vibrational modes that are equal. This concept is clearly associated with degeneracy. It becomes critical to understand molecular symmetry in order to understand molecular spectra. It is important to mention that, by "symmetry" in this context, we mean physical symmetry of a collection of various atoms.

σ symmetry

This category designates a plane of symmetry; by reflection (interchange of physical masses) at a plane, the molecule looks exactly as it did before. In the case of water (see Figure 9.10), for example, there is a symmetry plane containing the two H atoms and the O atom (plane number 1 in the figure). Clearly, if we reflect masses across that plane the molecule will look the same. Diatomic molecules have an infinite number of equivalent symmetry planes containing the internuclear axis. A second H_2O symmetry plane (plane number 2) can be found normal to the first. If we reflect the two hydrogen atoms across that plane it will look the same, because hydrogen atoms are identical.

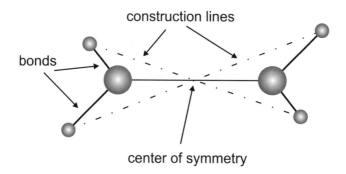

Figure 9.11: Locating a center of symmetry.

i symmetry

This form designates a center of symmetry. If one can draw lines from each atom through the center and find another atom at identical spacing from the center, then there is a center of symmetry (see Figure 9.11). A molecule can have only one center of symmetry. Homonuclear diatomic molecules always have i symmetry.

C_p symmetry

C_p designates a p-fold axis of rotational symmetry. A good example is ammonia (NH_3), as shown in Figure 9.12. The C_3 line passes through the center of the nitrogen atom and through the center of the triangle established by the three hydrogen atoms. Even though one hydrogen atom cannot be distinguished from another, the H atoms in Figure 9.12 have been labeled for the purposes of discussion. Imagine that the molecule is rotated about the C_3 axis so that H number 1 is now in the position formerly occupied by number 2, and so forth. The new configuration will be identical to the former configuration. This can be done 3 times with ammonia before returning to the original state. The p in C_p designates the angle of one rotation via $360°/p$. In an atom with C_p symmetry, for every atom not on the axis there must be $p-1$ other equal atoms, spaced the same distance from the C_p axis and in the same plane. Linear molecules will always have a C_∞ axis passing along the internuclear axis.

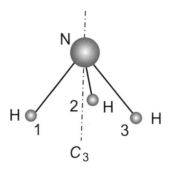

Figure 9.12: A three-fold axis of symmetry passing through the center of the large atom and through the center of the triangle established by the three small atoms (e.g. NH_3).

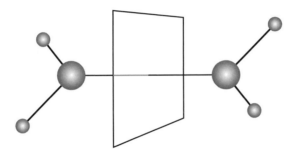

Figure 9.13: Schematic of S_p symmetry: one twofold rotation followed by reflection across the plane generates the same structure.

S_p symmetry

S_p symmetry requires two operations in sequence; a single p-fold rotation and then reflection across a plane normal to the C_p axis (see Figure 9.13). Note that a two-fold rotation followed by reflection is the same as i symmetry. Those two symmetry operations are part of the same point group (defined below). In fact, S_p symmetry is guaranteed for molecules that have C_p symmetry and a σ plane of symmetry normal to the C_p axis. This is not the only time that S_p symmetry occurs, but it is guaranteed for that case.

Point groups

It is possible to generalize notions about symmetry operations. For example, a molecule cannot have C_3 and C_4 symmetry in the same

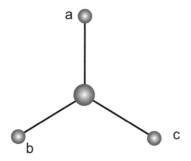

Figure 9.14: Schematic of an ammonia-like molecule with several symmetry properties.

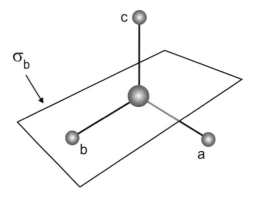

Figure 9.15: Reflection of a and c across σ_b.

direction. If two symmetry planes cross at right angles, the intersection will contain a C_2 axis. A C_2 axis and a plane of symmetry normal to the axis means there will also be a center of symmetry, and so forth.

If two symmetry operations are carried out in succession, the result can be the same as one other symmetry operation if they are all part of the same point group. Consider the ammonia-like molecule in Figure 9.14. There is a σ plane containing the bond between atom b and the center, call it the σ_b plane. Atoms a and c can be reflected across this plane, as shown in Figure 9.15. Now the system can be rotated about the C_3 axis, as shown in Figure 9.16, to produce the molecule depicted in Figure 9.17. This outcome is identical to the outcome from reflection of atoms a and b across σ_c in the original molecule, as shown in Figure 9.18. Hence, $\sigma_b \times C_3$ is equivalent to σ_c. They are members of a point group. There are a large number of point groups, and they

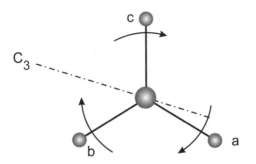

Figure 9.16: Rotation about C_3.

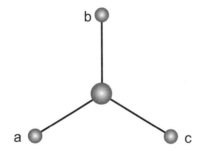

Figure 9.17: Outcome of reflection of a and c across σ_b, followed by rotation about C_3.

use a second letter designation system. We will not delve into them here because it is a very large topic. Interested readers should consult Herzberg [18, 19].

9.3.2 Rotation of Polyatomic Molecules

Interactions between rotation, vibration, and electronic states are stronger in polyatomic molecules. A good example is ozone. At low J, the molecule is bent at an angle of 116°, but at higher values of J, centrifugal distortion straightens the molecule. This has a great effect upon the electronic potential, which also controls vibration. Despite these complications, the Born-Oppenheimer approximation is applied to polyatomic molecules.

For rotation of a nonvibrating molecule, $F(J)$ will again depend upon $B \propto (1/I)$, where we now have a three-dimensional molecule and that affects I (the moment of inertia). If I is evaluated for every

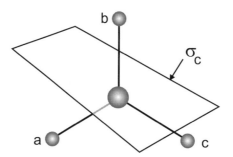

Figure 9.18: Reflection of a and b across σ_c.

direction in space, there will occur perpendicular axes along which I is either a maximum or a minimum. These are called the principal axes of the molecule. If a molecule has axes of symmetry, they will be principal axes. A plane of symmetry is always perpendicular to a principal axis.

Polyatomic molecules are classified by their three principal moments of inertia. An "asymmetric top" has unequal principal moments of inertia. A "symmetric top" has two equal principal moments of inertia, but the third is different. A "spherical top" has all three equal.

Linear polyatomic molecules

The behavior of linear molecules is fairly easy to generalize. For this reason they make a good case study for the interaction of symmetry and spectral response. As molecules become more complex, however, it is no longer possible to generalize. We therefore devote more space to simple systems, and assume readers who care about a particular complex system will find the necessary material.

Linear polyatomic molecules belong to two possible point groups. If there is a plane of symmetry perpendicular to the internuclear axis, this forms a $D_{\infty h}$ point group (see Figure 9.19). It has a ∞-fold axis because a linear molecule can be rotated about the internuclear axis into any angle and preserve symmetry. There are also an infinite number of C_2 axes perpendicular to the C_∞, an infinite number of planes containing C_∞, and one plane perpendicular to C_∞. Alternatively, linear molecules can be without a plane of symmetry (Figure 9.20). This is a $C_{\infty v}$ point group. It has a C_∞ axis along the internuclear axis, and an infinite number of planes containing C_∞.

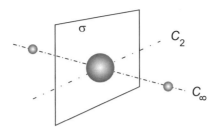

Figure 9.19: A linear molecule with a plane of symmetry (e.g. CO_2).

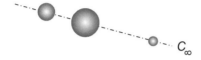

Figure 9.20: A linear molecule without a plane of symmetry (e.g. HCN).

These two point groups have rotational term energies that have the same form as a diatomic molecule:

$$F(J) = \frac{B}{c} J(J+1) - \frac{D^2}{c} J^2(J+1)^2 + \cdots \qquad (9.117)$$

where, as before, B is the rotational constant and D is a centrifugal distortion term.

Parity is much the same. In most molecules, ψ is symmetric in the ground state. Since vibrational wave functions are symmetric, parity then depends upon the rotational wave function. Rotational degeneracies depend upon the point group.

Spherical top molecules

If a molecule has two or more C_3 or higher axes, the moments of inertia going through the center of mass are all equal. A tetrahedral structure (T_d point group) like CH_4 fits this description (see Figure 9.21). In such a case, each bond between an outer atom and the central atom is a C_3 axis.

Here, the rotational term energy is given by

$$F(J) = \frac{B}{c} J(J+1) \qquad (9.118)$$

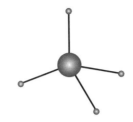

Figure 9.21: A tetrahedral molecule (e.g. CH_4).

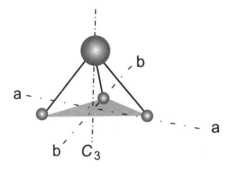

Figure 9.22: A symmetric top molecule.

In addition to the normal $2J+1$ degeneracy, every time the molecule is reoriented (in a C_3 way) it looks the same in a rotational sense. This generates a second $2J + 1$ degeneracy. The total degeneracy for a spherical top is thus $(2J + 1)^2$.

Symmetric top molecules

These are molecules with a C_3 or higher axis. Figure 9.12 contains one example. Figure 9.22 demonstrates the symmetric top nature of such a molecule. In a plane perpendicular to C_3, such as the one connecting the three smaller atoms, the moment about the line a-a is the same as the moment about b-b. In fact, the moment about any line in that plane is the same. There certainly are other symmetric top cases, but this one is easy to visualize.

The rotation of such a molecule becomes more complex. It is necessary to define a "figure axis" that will serve as the former z axis or internuclear axis for angular momentum coupling from electronic and diatomic spectroscopy. In Figure 9.22, the C_3 axis would serve well. Now the quantum number Λ is replaced by $\mathcal{K}$, representing the component of electronic orbital momentum along the figure axis. Then

$$J = \mathcal{K}, \; \mathcal{K}+1, \; \mathcal{K}+2, ... \tag{9.119}$$

and

$$F(J,\mathcal{K}) = \frac{B}{c} \, J(J+1) + \frac{(A-B)}{c}\mathcal{K}^2 \tag{9.120}$$

where $\mathcal{K}^2(A-B)/c$ is a polyatomic symmetric top term. A and B are given by

$$B = \frac{h}{8\pi^2 I_B}$$
$$A = \frac{h}{8\pi^2 I_A} \tag{9.121}$$

Here, I_B is the moment about the figure axis, and I_A is the moment about the many other axes that all produce equal moments.

Issues such as parity and degeneracy are much more complex. It is now necessary to investigate each molecule as a separate case.

Asymmetric top molecules

These molecules have no threefold or higher axes, and all the moments are different. Here, even energy levels cannot be specified by general formulas. Large collections of formulas with corrections are required. It is now imperative to investigate each molecule as a separate case.

9.3.3 Vibrations of Polyatomic Molecules

A molecule with N nuclei has 3N−6 vibrational degrees of freedom. This is because there are three degrees of positional freedom for each atom, producing a total of 3N for the molecule. The minus 6 occurs because we do not care to describe the position of the center of mass, neither do we care about the angular orientation of the molecule. For vibrational analysis we care simply about the positions of the nuclei relative to each other. For a linear molecule, it takes only two degrees of freedom to orient the molecule, so there are 3N−5 vibrational degrees of freedom.

These vibrations can have very complex behavior, but they can be decomposed into fairly straightforward "normal modes". One will often hear talk of a "CH stretch" for example (as it turns out, the CH stretch is not a true normal mode, but in many molecules it approximates one). Polyatomic molecular vibrational behavior is a superposition of these normal modes.

In quantum mechanics, by introducing normal coordinates for normal modes, it becomes possible to separate the Schrödinger equation into individual equations for each normal mode. The total energy is then

$$E = E_1 + E_2 + ...E_{3N} \qquad (9.122)$$

Here the six degrees of freedom associated with position and orientation are carried along for mathematical convenience. They will produce six values of zero for vibrational energy. Each energy level, not surprisingly, is

$$E_i = h\nu_{e,i}\left(v_i + \frac{1}{2}\right), \qquad v_i = 0, 1, 2, ... \qquad (9.123)$$

In addition, it is now possible to have degenerate vibrational modes, depending upon the structure of the molecule. These concepts can not be generalized further and it becomes important to consider each molecule individually.

9.3.4 Electronic Structure

Here the electronic structure is solved as before, but it is significantly more complex. It can be different in three dimensions, for example. Energy levels for spectroscopic application can be found in the literature. Again, it is necessary to consider each molecule individually. Indeed, for very complex molecules one simply calibrates the spectrometer with a chemical technique.

9.4 Conclusions

As with the last chapter, this chapter has presented the very basic ideas in molecular spectroscopy. No specific examples have been provided, but they can be found in several books, Eckbreth [1], Herzberg [16,

17, 18, 19], Banwell and McCash [14] and Levine [68], and in the open literature.

To this point we have discussed where in the electromagnetic spectrum one can find signatures from atoms and molecules. In the next chapter we discuss how to analyze the strength of each response.

Chapter 10

RESONANCE RESPONSE

One can find many different representations for resonance response throughout the literature. This is unfortunate, but it has arisen because different communities have different needs. Some communities report line strengths or absorption cross-sections, while others report oscillator strengths or Einstein A coefficients, and still others provide band strengths.

In this chapter, we first develop a quantum expression for the Einstein A coefficient. Following that, the ways in which common representations can be related to each other (and to the A coefficient) are discussed. It is tempting to present formulas for conversion. Hillborn [27] does that, and in fact he presents a table for conversion from one form to another. His approach is usually helpful; that paper is a good standard reference. Unfortunately, there are very many variants. Approaches may differ in the basis for the formalism (e.g. on a per wavelength-bandwidth basis vs a per wavenumber-bandwidth basis), units (MKSA vs Gaussian), the normalization convention for the rotational factors, and so forth. Organization of units alone may pose a problem. As an example, most publications consolidate units. In such a case, a s^{-1}, which represents ν in Hz, will cancel an s, which represents time. As another example, a cm^{-1}, which represents a wave number, will cancel a cm, which represents distance. For those new to this field, to search out what a number and set of units actually represents can seem almost like a puzzle game.

This chapter will not present a series of conversion tables, because this problem is larger than one table. Instead, we provide detail on the development of expressions. These details should make it possible to understand and adjust to subtle differences from one author to the next.

10.1   Einstein Coefficients

The simplest way to generate a quantum expression for the parameters that govern absorption and emission of light is to start with equation (5.91), reproduced here:

$$A_{nm} = \frac{\omega^3 |\vec{\mu}_{ed}|^2}{6h\epsilon_\circ c^3}$$

The following approach is also described both by Measures [12] and by Verdeyen [26]. We focus on the dipole moment of the atom or molecule, $\vec{\mu}_{ed}$. In the electromagnetic treatment, the subscript "ed" denoted an electric dipole. Here, in recognition that we are dealing with a transition between two quantum levels, we rename it $\vec{\mu}_{nm}$, where n denotes the upper level and m denotes the lower level, as before.

To begin the treatment, we recognize that atoms or molecules remaining in a thermodynamically stable, stationary state will not radiate. An expectation value for the quantum dipole occuring between level m and itself can be expressed as

$$\langle \vec{\mu} \rangle = \langle \psi_m | \hat{\mu} | \psi_m \rangle = 0 \tag{10.1}$$

This relation is proven in most quantum texts.

An atom or molecule in an excited state ψ_n, however, can radiate during a transition to a lower state ψ_m. The atom or molecule will make a transition over a period of time, during which it resides in a mixture of the two states:

$$|\psi\rangle = |c_n \psi_n + c_m \psi_m\rangle \tag{10.2}$$

The values of the c coefficients change over time, but we assume they change slowly relative to optical frequencies. The following limits apply to radiation from level n to level m:

$$\begin{aligned}
\text{at } t = 0, \ |c_n|^2 &= 1, \ |c_m|^2 = 0 \\
\text{at } t \to \infty, \ |c_n|^2 &= 0, \ |c_m|^2 = 1
\end{aligned}$$

$$\tag{10.3}$$

and for a two-level system

$$|c_n|^2 + |c_m|^2 = 1 \tag{10.4}$$

The expectation value for the quantum dipole moment during this process is then described by

$$\langle \vec{\mu}_{nm} \rangle = \langle c_n \psi_n + c_m \psi_m | \hat{\mu} | c_n \psi_n + c_m \psi_m \rangle \tag{10.5}$$

which becomes

$$\begin{aligned}
\langle \vec{\mu}_{nm} \rangle &= |c_n|^2 \langle \psi_n | \hat{\mu} | \psi_n \rangle + |c_m|^2 \langle \psi_m | \hat{\mu} | \psi_m \rangle \\
&\quad + c_n^* c_m \langle \psi_n | \hat{\mu} | \psi_m \rangle + c_m^* c_n \langle \psi_m | \hat{\mu} | \psi_n \rangle
\end{aligned} \tag{10.6}$$

As before, the first two terms in equation (10.6) are zero, and the last two terms can be combined using the definition of a complex conjugate. We also take the real part of the combined expression, because we wish to calculate the properties of an observable:

$$\langle \vec{\mu}_{nm} \rangle = 2 \operatorname{Re}[c_n^* c_m \langle \psi_n | \hat{\mu} | \psi_m \rangle] \tag{10.7}$$

Next, we assume that the term $c_n^* c_m$ is slowly varying and nearly equal to 1:

$$\langle \vec{\mu}_{nm} \rangle = 2 \vec{\mu}_{nm} \tag{10.8}$$

where $\vec{\mu}_{nm}$ is the matrix element of the dipole moment (also known as the transition moment) given by

$$\vec{\mu}_{nm} \equiv \langle \psi_n | \hat{\mu} | \psi_m \rangle \tag{10.9}$$

This term "matrix element" will reoccur. It is perhaps appropriate to discuss the name now. Even though we are working with continuous functions, they are all eigenfunctions specified by discrete quantum numbers. Whenever we deal with mixed quantum numbers, we enter a multidimensional space that is described naturally by matrix mathematics. Much more could be said about this point, and much more is said in the references.

Using equation (10.8) in equation (5.91) we find an expression for the Einstein A coefficient:

$$\boxed{A_{nm} = \frac{2 \, \omega_{nm}^3 |\vec{\mu}_{nm}|^2}{3 \, h\epsilon_\circ c^3 g_n}} \tag{10.10}$$

which assumes a random orientation of the dipole. The only real surprise here is the term g_n, the degeneracy of the level from which the process starts. As written in equation (10.10), the Einstein A coefficient is for the entire collection of states with matched energy and dipole moment (although it is not obvious from the notation), and hence an individual transition A coefficient must be reduced by this number count.

This development of equation (10.10) was a little ad hoc, especially the assumption that the term $c_n^* c_m$ is slowly varying and nearly equal to 1. Reader skepticism would be justified. Equation (10.10) can be found several ways, however. As an alternative, the time-dependent Schrödinger equation can be solved using perturbation techniques. To solve for time decay in this case is appropriate because A_{nm} is related to the excited state decay time, as shown in equation (3.60). The outcome of the perturbation solution is identical to equation (10.10). That development was not shown here because the development provided above went much faster; the perturbation solution can be found in both Levine [68] and Schwabl [58]. In addition, the density matrix equations of quantum mechanics will reproduce equation (10.10) in the steady-state limit. The density matrix equations are developed in Chapter 14, where they are taken to the steady state limit and equation (10.10) is reproduced.

The quantum expression [equation (10.10)] is not much different from the classical expression [equation (5.91)]. The quantum-mechanical details, however, have been simply bundled into the expression for the matrix element. Note that if the matrix element for the dipole moment is zero, there will be no resonance response of the atom or molecule. This can happen when the symmetry of the wave functions for states n and m are the same. In this case, the integrand in equation (10.9) is odd, and the integral is zero. This situation corresponds to a forbidden transition. Transition rules (allowed vs forbidden transitions) are discussed in Chapters 8 and 9. Even for a forbidden transition, there can be a weak interaction due to a magnetic dipole or an electric quadrupole interaction. A classic example is the case of a homonuclear molecule (e.g. O_2 or N_2). Because the vibrational wave functions are symmetric on inversion, there is no resonance response for rotational/vibrational transitions in the ground state (e.g. in the infrared).

10.1.1 Franck-Condon and Hönl-London factors

The simplicity of equation (10.10) belies some necessary detail, which we discuss in the following treatment. As discussed in Chapter 9, molecules have quantum levels representing electronic, vibrational, and rotational states. If we seek an Einstein coefficient simply for an electronic transition, for example, then equation (10.10) serves quite well, if we can find an expression for the electronic dipole moment inside the integration. Such a case would occur with an atomic transition, or an electronic transition in a molecule where we have included all of the vibrational and rotational transitions.

We begin by focusing on the full electronic/vibrational/rotational problem, following which we briefly discuss vibrational/rotational transitions in the lower electronic state. We rely upon the Born-Oppenheimer approximation, written here as

$$\Psi(\vec{r},t) = \Psi_e(\vec{r},t)\Psi_v(\vec{r},t)\Psi_r(\vec{r},t) \tag{10.11}$$

where $\Psi_e(\vec{r},t)$ is the electronic state wave function, $\Psi_v(\vec{r},t)$ is the vibrational state wave function, and $\Psi_r(\vec{r},t)$ is the rotational state wave function.

If we assume that a measurement is spectrally broad enough to cover an entire rotational manifold within a vibrational transition, we can focus just on the interaction between electronic and vibrational modes for the time being. We will relax this requirement and include rotational modes when the discussion on vibrational contributions is complete.

The dipole matrix element is given by equation (10.9). The dipole moment is expressed as the product of a charge and a distance, which can be separated into vector components. Since it is a linear sum, we can separate the total molecular dipole moment into

$$\hat{\mu} = \hat{\mu}_e + \hat{\mu}_n \tag{10.12}$$

where the subscript "n" denotes a nuclear dipole moment. The dipole matrix element can then be written in integral form (to avoid potential confusion) as

$$\vec{\mu}_{nm} = \int \psi_{n,e}^* \psi_{n,v}^* \, \hat{\mu}_e \, \psi_{m,e}\psi_{m,v} \, d\tau + \int \psi_{n,e}^* \psi_{n,v}^* \, \hat{\mu}_n \, \psi_{m,e}\psi_{m,v} \, d\tau \tag{10.13}$$

The vibrational eigenfunctions are self-adjoint (see Herzberg [15]). Since the electronic terms are separable from the nuclear terms, the second term in equation (10.13) can be written as

$$\int \psi_{n,v} \; \hat{\mu}_n \; \psi_{m,v} \; d\tau_n \int \psi^*_{n,e}\psi_{m,e} \; d\tau_e \qquad (10.14)$$

This term is equal to zero whenever $n \neq m$ (i.e. during an electronic transition) because $\psi^*_{n,e}$ and $\psi_{m,e}$ are orthogonal, and the integral over electronic states is an inner product. Equation (10.13) is then written in terms of the electronic dipole moment as

$$\vec{\mu}_{nm} = \int \psi_{n,v}\psi_{m,v} \; dR \int \psi^*_{n,e} \; \hat{\mu}_e \; \psi_{m,e} \; d\tau \qquad (10.15)$$

where the first integral is taken over the internuclear distance R because the vibrational wave functions are written as a function of internuclear distance alone. That integration does not vanish because the wave functions $\psi_{n,v}$ and $\psi_{m,v}$ are written for two different electronic levels. They therefore have different potential wells and different equilibrium internuclear spacing. If they had been written for the same electronic state, this term would also be zero.

When these multiple-mode configurations occur, there will be a sifting process associated with the highest level term, just as we sifted out the second term in equation (10.13) by orthogonality of the electronic wave functions. Because the Einstein coefficient goes as the modulus squared [see equation (10.10)], the term

$$\left| \int \psi_{n,v}\psi_{m,v} \; dR \right|^2 \qquad (10.16)$$

determines the relative strength of the vibrational response, and it is called the Franck-Condon factor. Equation (10.15) can now be explained in logical terms. We define the electronic term integral in equation (10.15) as an electronic dipole matrix element:

$$\vec{\mu}_e \equiv \int \psi^*_{n,e} \; \hat{\mu}_e \; \psi_{m,e} \; d\tau \qquad (10.17)$$

The dipole matrix element for a molecule is therefore the electronic dipole matrix element times the square root of the Franck-Condon factor (neglecting rotation for now):

$$\vec{\mu}_{nm} = \vec{\mu}_e \left[\int \psi_{n,v}\psi_{m,v} \; dR \right] \qquad (10.18)$$

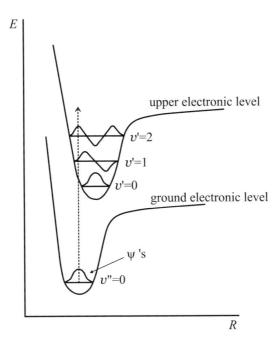

Figure 10.1: Schematic of the Franck-Condon overlap integral.

The Franck-Condon factor is a fractional multiplier for the relative strength of certain vibrational transitions within the electronic transition. The term:

$$\int \psi_{n,v} \psi_{m,v} dR \qquad (10.19)$$

is an overlap integral between the two wave functions and it serves as a transition probability amplitude. For electronic transitions, the overlap integral is often depicted as shown in the cartoon of Figure 10.1. As before, the two large curves in Figure 10.1 represent two electronic energy levels. Within each of these curves, several vibrational levels are depicted. The ground electronic state shows just one (for $v'' = 0$). The upper electronic energy level in the figure contains three vibrational levels. They are characterized by vibrational quantum numbers $v' = 0$, 1, or 2. The waveforms located at each vibrational level in Figure 10.1 are the wave functions for each. The overlap integral representing the Franck-Condon factor is thus represented by the arrow pointing upwards. Where the ground-state wave function is large (has high probability amplitude), the excited state wave function is not for $v' = 0$ or 1. At $v' = 2$, however, the overlap is strong. The Frank-Condon factor for absorption from $v'' = 0$ to $v' = 0$ or 1 is therefore very small in Figure 10.1, while for $v'' = 0$ to $v' = 2$ it is large. The resonance response is therefore strongest for $v'' = 0$ to $v' = 2$. For emission, the same process applies except that more emission lines are probable (see Herzberg [15]).

We can now write the Frank-Condon factor with more common notation:

$$S_{v'v''} \equiv | \int \psi_{v'} \, \psi_{v''} \mathrm{d}R|^2 \qquad (10.20)$$

Because the fractional response must add to one for all possibilities from one level, the most common sum rule for $S_{v'v''}$ is

$$\sum_{v'} S_{v'v''} = 1 \qquad (10.21)$$

for absorption and

$$\sum_{v''} S_{v'v''} = 1 \qquad (10.22)$$

for emission.

Accounting for rotational transitions within the same framework is now quite simple. We do it by defining the Hönl-London factors:

$$S_{J'J''} \equiv | \int \psi_r^{*'} \psi_r'' \, \mathrm{d}\tau|^2 \qquad (10.23)$$

where (as before) J is the rotational quantum number. The Hönl-London factors are sometimes called "line strengths". Herzberg provides a series of expressions for calculation of these factors, as does Kovacs [70]. Zare [66] shows in detail how these factors can be accessed through the Clebsch-Gordon coefficients, and provides similar results. The most common sum rule for the Hönl-London factors is

$$\sum_{J'} S_{J'J''} = 2J'' + 1 \qquad (10.24)$$

for absorption and

$$\sum_{J''} S_{J'J''} = 2J' + 1 \qquad (10.25)$$

for emission. They do not sum to one because there are several degenerate levels, and the degeneracy for a rotational state is

$$g_J = 2J + 1 \qquad (10.26)$$

The total dipole matrix element for the molecule is then

$$\vec{\mu} = \vec{\mu}_e \sqrt{S_{v'v''}} \sqrt{S_{J'J''}} \qquad (10.27)$$

The total molecular Einstein A coefficient is now generated by modifying equation (10.10) to account for all of these effects:

$$A_{nm} = \frac{2\,\omega_{nm}^3}{3\,h\epsilon_o c^2} \frac{|\vec{\mu}_e|^2 S_{v'v''} S_{J'J''}}{g_n} \qquad (10.28)$$

where g includes all of the degeneracies of state n. The units for A_{nm} are s^{-1}.

To this point we have discussed combined electronic, vibrational and rotational transitions. For purely vibrational/rotational transitions within an electronic ground state, we use an integral similar to equation (10.13), but the naturally occuring dipole moment provides the overall resonance response. In such a case, the ground-state electronic wave function is not separable from the vibrational, in the sense of the Born-Oppenheimer approximation. Because there is a natural dipole moment, the combined overlap integral does not go to zero. This integral determines the strength of each vibrational transition, and it is usually strongest for $\Delta v = 1$, the "fundamental band". It is smaller for $\Delta v = 2$ (the "first overtone") and even smaller for $\Delta v = 3$ (the "second overtone"). The vibrational transition strengths are usually reported in terms of a band oscillator strength (see below). The strength of each rotational transition within a vibrational band is then determined by the Hönl-London factors, or line strengths. Homonuclear molecules do not have strong naturally occurring dipole moments, the combined overlap integral is negligible, and as a result they are not infrared active.

One might reasonably ask where these data come from. The available data are based upon measurements or *ab initio* calculations. In many cases the values for A_{nm} are supplied for every level (see, for example, the LIFBASE compilation [13] for several important electronic transitions). Such an approach minimizes confusion and the possibility for mistakes, but it consumes large amounts of paper or memory. Alternatively, an A-coefficient based upon $\vec{\mu}_e$ is supplied, together with Frank-Condon factors. It is then assumed that a source such as Herzberg [17] or Kovacs [70] will be used for the Hönl-London factors. Much of the important infrared data is compiled in a database named HITRAN [69].

When dealing with *ab initio* calculations of the transition matrix, it is very important that the degeneracy g_n in equation (10.28) be defined consistently with the definition of the dipole moment operator and with the sum rule for the Hönl-London factor (rotational line strengths). Otherwise absolute calculations of composite line intensities may be in error by some constant multiplicative factor. One source of error results from ambiguities in the definition of the operator for the electronic transition moment for perpendicular transitions. As one example, Whiting *et al.* [71] define the parallel operator as $\hat{\mu}_z$ and the perpendicular operator as $\sqrt{1/2}(\hat{\mu}_x \pm \hat{\mu}_y)$. The $\sqrt{1/2}$ avoids a factor of 2 error in the calculation of composite line intensities. This is actually a straightforward notion, but such details must be investigated before data can be used reliably.

One should also be aware of the resolution of reported measurements. The spin multiplet or Λ doubling may be resolved in the measurement. If it is resolved, it is not counted as a degeneracy. A simple rule of thumb is to ensure that each of these levels has been accounted for, but not over-counted.

The Einstein B coefficients are used to calculate the rates for stimulated emission and absorption. The probability of a stimulated transition is given by the product of the appropriate B coefficient and the spectral energy density ρ_ν of the radiation field that stimulates the process. In Section 3.4, the B coefficients were written in terms of the Einstein A coefficient. Furthermore, it was shown that the ratio of the B coefficients is equal to the ratio of the degeneracy of the two levels participating in the transition. The results [equations (3.95) and (3.96)] are repeated here for convenience.

$$\frac{A_{nm}}{B_{nm}} = \frac{8\pi h\nu^3}{c^3}$$

and

$$g_n B_{nm} = g_m B_{mn}$$

The product of the B coefficient and the spectral energy density is a stimulated rate, and it has units of inverse time. It is very important to note that the definition of B depends on the units of the spectral energy density. Spectral energy density is given in terms of energy per unit volume per unit frequency interval. Therefore, the B coefficient has units of volume times frequency interval per energy per unit time. The key difference encountered in formulations is the definition of the frequency interval (ν in Hz, ω in rad/s, or wave number $\check{\nu}$ in cm^{-1}). We have defined the B coefficient in equation (3.95) by writing the spectral energy density as ρ_ν with units of J/m^{3}Hz. The units of the B coefficient are then m^{3}Hz /Js.

If, on the other hand, one formulates a problem in terms of angular frequency ω, the spectral energy density ρ_ω has units of J/m^3(rad/s). Using these units, the differential frequency interval is dω=2π dν [for conversion of equation (3.50), for example]. Therefore, the appropriate B coefficient is equal to the product of 2π and the coefficient given in equation (3.95):

$$B_{nm}^\omega = \frac{2\pi^3 c^3}{h\omega^3} A_{nm} \tag{10.29}$$

where the superscript indicates that the coefficient is on an angular frequency basis.

When the frequency is expressed in terms of wave numbers $\check{\nu} = (1/\lambda) = \nu/c$, the differential frequency interval in cm^{-1} is given by d$\check{\nu} = \frac{1}{c}d\nu$. The B coefficient in this case is related to the A coefficient as follows:

$$B_{nm}^{\check{\nu}} = \frac{1}{8\pi hc\,(1/\lambda)^3} A_{nm} \tag{10.30}$$

10.2 Oscillator Strengths

The oscillator strength was first presented in equation (5.34) of Section 5.2 as a way to bring the classical description of dipole oscillator into agreement with measurements or with quantum calculations. Now consider the spectral absorption coefficient obtained from the ERT [equation (3.72)] and that obtained by considering a classical dipole oscillator [equation (5.33)]. To set the two expressions on an equal footing, we assume that all population is in the lower state m so that stimulated emission is neglected in equation (3.72):

$$\kappa_{\text{actual}} = \frac{h\nu}{c} N_m B_{mn}\, Y_\nu \tag{10.31}$$

and from the discussion on the Lorentz atom we have equation 5.33:

$$\kappa_{\text{Lorentz}} = \frac{N_m e^2}{4 m_e \epsilon_o c} \left[\frac{1/\tau}{(\omega_o - \omega)^2 + 1/(2\tau)^2} \right]$$

where the terms in square braces are the Lorentzian lineshape function. If we multiply the Lorentzian coefficient by an absorption oscillator strength f_{mn}, and equate the two expressions, we obtain

$$\kappa_{\text{actual}} = f_{mn} \kappa_{\text{Lorentz}} \tag{10.32}$$

We then integrate over all frequencies ν, making it possible to relate the oscillator strength and the Einstein B coefficient:

$$\boxed{f_{mn} = \frac{4 \epsilon_o m_e \, h\nu_{nm}}{e^2} B_{mn}} \tag{10.33}$$

Note that in the integration of κ_{actual}, the lineshape function is sufficiently narrow for us to assume that $\nu = \nu_{nm}$.

Some authors write f_{mn} in terms of the classical electron radius r_e. In SI units, the classical electron radius is

$$r_e = \frac{1}{4\pi\epsilon_o} \frac{q_e^2}{m_e c^2} = 2.818 \times 10^{-15} \text{ m} \tag{10.34}$$

and

$$f_{mn} = \frac{h\nu_{nm}}{\pi r_e \, c^2} B_{mn} \tag{10.35}$$

If, instead, we were to consider a stimulated emission process, we would have obtained the following relationship between the emission oscillator strength and the Einstein B coefficient for stimulated emission:

$$\boxed{f_{nm} = \frac{4 \epsilon_o m_e \, h\nu_{nm}}{q_e^2} B_{nm} = \frac{h\nu_{nm}}{\pi r_e \, c^2} B_{nm}} \tag{10.36}$$

Clearly, by analogy to equation (3.96),

$$\boxed{g_n f_{nm} = g_m f_{mn}} \tag{10.37}$$

Using equation (3.95), we can relate the oscillator strength to the Einstein A coefficient:

$$\boxed{f_{nm} = \frac{\epsilon_o m_e \, c^3}{2\pi \, q_e^2 \, \nu_{nm}^2} A_{nm} = \frac{c}{8\pi^2 r_e \, \nu_{nm}^2} A_{nm}} \tag{10.38}$$

10.3   Absorption Cross-sections

A simple measurement of resonance response involves the absorption of a tunable or broad-band source. In that case, the experiment can be modeled with a Beer's law expression as in equation (3.82):

$$\frac{I(z)}{I(0)} = \exp(-\kappa_\nu z)$$

Recall that the spectral absorption coefficient contains the lineshape function as well as the dependence on the population in the upper and lower states, as shown in equation (3.72):

$$\kappa_\nu = \frac{h\nu}{c}(N_m B_{mn} - N_n B_{nm})Y_\nu$$

We can factor out of this expression the total species number density N_{tot} and write Beer's law in terms of a spectral absorption cross-section σ_ν, which has units of m^2:

$$\boxed{\frac{I(z)}{I(0)} = \exp(-\sigma_\nu N_{\text{tot}} z)} \tag{10.39}$$

where

$$\boxed{\sigma_\nu \equiv \frac{h\nu}{c}\left(\frac{N_m}{N_{\text{tot}}}B_{mn} - \frac{N_n}{N_{\text{tot}}}B_{nm}\right)Y_\nu} \tag{10.40}$$

Cross-sections are most useful when both scattering and absorption are used in the same ERT formalism (e.g in sooting flames or atmospheric research). Scattering is typically described in terms of a differential cross-section, and to introduce absorption into that system it is easiest to define it in a related fashion.

For a sample which is in equilibrium, both the population fractions N/N_{tot} and the lineshape function depend upon temperature. Furthermore, the lineshape function may depend on the partial pressures of all molecular and atomic species in the system. Therefore, when using reported spectral absorption cross-sections, it is very important either to match the conditions under which the measurement was made or properly to extrapolate the expression to apply to the new conditions. In the latter case, the molecular structure must be known in order to calculate the population distribution properly.

In some cases, most notably for IR studies, researchers choose to report their measurements in terms of spectrally integrated cross-sections. If the measurement is produced with a narrow-band laser and stimulated emission is negligible, it can be directly related to the Einstein coefficients. For this specific case, the following relationship applies:

$$\sigma_\circ = \int \sigma_\nu d\nu = \frac{h\nu_{nm}}{c}\frac{N_m}{N_{\text{tot}}}B_{mn} \qquad (10.41)$$

assuming stimulated emission is negligible. Beer's law can be written as

$$\frac{I(z)}{I(0)} = \exp(-\sigma_\circ\, Y_\nu\, N_{\text{tot}} z) \qquad (10.42)$$

The integrated cross-section presented here has units of $m^2 Hz$. In some cases, cross-sections may be reported in other units. As one example, the unit bandwidth may be reported in wave numbers, or the number density may be reported in partial pressure. In any case, it is a simple exercise in dimensional analysis to convert units.

10.4   Band Oscillator Strengths

Thus far we have written the Einstein coefficients, oscillator strengths, and cross-sections for either atomic transitions or single rotational transitions of a molecule. A table of all of the Einstein coefficients for a molecular band can become quite extensive. In these cases, a band strength for either a vibrational band or an entire electronic system may be introduced. A useful discussion of band strengths is presented by Lucht *et al.* [28], and we summarize the results here.

We have shown in equation (10.28) that the Einstein A coefficient for a single rotation line is proportional to the product of an electronic transition probability, the Franck-Condon factor, and the Hönl-London factor. Using equations (10.28), (10.37), and (10.38), we can write the absorption oscillator strength as follows:

$$f_{J''J'} = \frac{8\pi^2\, m_e\, \nu_{nm}}{3h\, q_e^2\, g_m}|\vec{\mu}_e|^2\, S_{v'v''}\, \frac{S_{J'J''}}{2J''+1} \qquad (10.43)$$

where we have included rotational quantum numbers as a subscript on the oscillator strength to emphasize that it applies to a single rotational line within a vibrational band of the nm electronic transition. We also explicitly wrote the rotational degeneracy as $2J'' + 1$ in the denominator which implies that the degeneracy term g_m includes all other degeneracy factors for the band. Vibrational levels are not degenerate, so these additional degeneracies originate in the electronic configurations (e.g. spin splitting, lambda doubling, and so forth).

Owing to the summation rules for the Franck-Condon and Hönl-London factors [equations (10.21) and (10.24)], we can define an electronic oscillator strength that is independent of vibrational and rotational quantum numbers:

$$f^{el}_{mn} = \frac{8\pi^2 \, m_e \, \nu_{nm}}{3h \, q_e^2 \, g_m} |\vec{\mu}_e|^2 \tag{10.44}$$

Similarly, a vibrational oscillator strength (commonly called a "band strength") which describes a particular vibrational band within a particular electronic system, can be written as follows:

$$\boxed{f_{v''v'} = f^{el}_{mn} S_{v'v''}} \tag{10.45}$$

These are typically reported as direct measurements, especially when it is inappropriate to separate the electronic and vibrational terms, as the Born-Oppenheimer approximation does. Using this definition, the absorption oscillator strength for an individual transition can be written in terms of the Hönl-London factor:

$$\boxed{f_{J''J'} = f_{v''v'} \frac{S_{J'J''}}{2J'' + 1}} \tag{10.46}$$

The introduction of electronic and vibrational oscillator strengths considerably simplifies the amount of information one needs to model a molecular band. However, this formalism assumes that the electronic, vibrational, and rotational degrees of freedom are independent. In fact, this is only an approximation, and we find tabulations of A coefficients for all of the rotational lines such as those contained in LIFBASE [13] more useful.

10.5   Conclusions

Here we have developed relationships among the most common representations for resonance response, all in MKSA units. It is necessary to point out, however, that we have not yet discussed the effect of polarization orientation on resonance reponse, which is discussed in Chapter 12.

Unfortunately, there is no system by which all of these response formalisms can be rationalized. Instead, it is necessary for individual researchers to delve completely into the sources of information in order to ensure that the outcome they seek has been rationalized and put into the system that the individual researcher prefers.

Chapter 11

LINE BROADENING

11.1 Introduction

Broadening of spectral lines is discussed in many sources on spectro-scopic diagnostics (e.g. Eckbreth [1], Demtröder [2], Measures [12], and others listed in the Bibliography). Broadening can strongly affect resonance response. At the center of the spectral line, for example, spectroscopic response decreases as the line broadens. The integrated response across the entire line remains fixed, however, because the line-shape function Y_ν is subject to a normalization condition:

$$\int_{-\infty}^{+\infty} Y_\nu \, d\nu = 1 \tag{11.1}$$

Clearly, broadening must be understood and described properly in or-der to understand a spectroscopic signal. There are three important regimes. When the width of an optical source spectrum [described by $L(\nu)$] is much narrower than the spectral line (described by Y_ν), the source spectrum can be treated as a delta function relative to the spec-tral line. Here, the optical source could be a narrow bandwidth ("single frequency") laser used for a Beer's law absorption measurement, for ex-ample. In such a case, the spectral lineshape function Y_ν is sampled at specific discrete values of ν (or ω, or λ). The optical response is thus determined by the lineshape function Y_ν. Conversely, when the source spectrum is broad relative to the spectral line (e.g. the source is a short pulse laser or a lamp, and no other spectral filter is used), the spectral line can be treated like a delta function relative to the source. Here, the entire spectral line has been integrated [$L(\nu)$ is assumed constant over Y_ν, and so it falls out of spectral integrals used to describe signal generation, producing a spectral integral equal to equation (11.1) which is evaluated to 1]. The integrated optical response then does not have a spectral dependence. Finally, if the widths are comparable, then a spectral overlap integral may be necessary (i.e. an integral over all frequencies that includes the product $L(\nu)Y_\nu$, together with other spectral response functions that may affect the measurement).

Broadening effects are generally classified as either homogeneous or inhomogeneous. Homogeneous broadening refers to those cases for which all molecules interact with the electromagnetic field in the same way. Inhomogeneous broadening, on the other hand, refers to broadening mechanisms for which the interaction differs for different classes of molecules. Some authors provide more detail than others, but all of the sources mentioned describe homogeneous broadening via collisions, inhomogeneous broadening introduced by the Doppler effect, and a combination of the two by convolution.

This text contains a classical mechanical model for homogeneous broadening, presented in Chapter 5. In the Lorentz treatment, broadening was shown to be caused by a characteristic decay time. In many sources, that same Lorentzian expression [as found in equation (5.33)] is simply modified by insertion of phenomenological decay times (e.g. natural decay time via the Einstein A-coefficient, characteristic collisional times caused by dephasing collision rates, and so forth). Homogeneous broadening is also discussed in Chapter 14, where coherence dephasing times are included in the density matrix equations (DME). Those times genuinely represent decay of atomic or molecular coherences, and that formalism can describe phenomena such as power broadening, but it cannot be used to generate an analytical expression for Y_ν in a straightforward way. In fact, only by comparing the steady-state DME to an accepted expression for Y_ν does it become possible to compare the DME with the ERT. Both Heitler [72] and Levine [68] use a perturbation solution to the time-dependent Schrödinger equation to develop the Lorentzian with $1/A_{nm}$ as the decay time, but that treatment does not model dephasing. Steinfeld [21] uses a quantum-based approach that is similar to the classical approach presented here.

Inhomogeneous broadening in gases is caused by the Doppler effect, because each atomic or molecular velocity class has a different Doppler shift relative to the optical signal. In many books, the Doppler profile is simply presented. In others (e.g. Demtröder [2]), a straightforward ERT analysis of the Doppler shift makes use of the Maxwellian velocity distribution to develop an expression for broadening. In the density matrix equations (Chapter 14), the Doppler shift is included as a frequency detuning, and the resulting velocity-dependent material polarization is integrated over velocity classes using the Maxwellian distribution.

In this chapter, a more unified spectral theory is developed based in large measure upon the presentation by Shore [73]. A similar treatment for homogeneous broadening is presented by Demtröder [2]. This approach has been adopted here for several reasons. It parallels spectral techniques used in electronic communications, which are closely related to ideas in control theory and in turbulence research. It also allows various phenomena to be included into the broadening expressions without having to change from one formalism to another. In what follows, therefore, we present the basic spectral notions within which this theory is framed. First, we review a general spectral theory for random processes. This theory, based upon the Wiener-Khinchine relation, generates a power spectral density via autocorrelation of random processes. We then show how the phenomena that produce line broadening can be described in terms of random processes, thus generating a lineshape function (based upon the power spectral density). Collisional dephasing and Doppler shifts can be incorporated in a straightforward way. Following that, the Voigt profile is presented.

11.2 A Spectral Formalism

Engineers typically encounter spectral behavior in the fields of communications, signals and systems, and control theory. It is common to describe an entire system in terms of transfer functions for each device forming the system. These transfer functions are spectral representations for the impulse response of the device, generated using Fourier transforms. The output signal spectrum of a system can be described as the input signal spectrum multiplied by the transfer functions. The Fourier transform of the output spectrum then describes the time dependence of the output signal. According to the convolution theorem, this analysis is the same as taking the convolution of the input with the various device impulse response functions. Large portions of undergraduate control theory courses are devoted to this technique.

The signals described by these transform techniques are continuous and deterministic; they can be fully described in terms of the signal generation system. The phenomena that control spectral lineshapes are not deterministic, however, they are random. Collisional broadening, for example, is controlled by random collisions that interrupt the coherences of the atom or molecule. Coherences are states of the atom or molecule that couple radiant energy between stationary energy states. These coherences oscillate at the same frequency as the light that will be absorbed or emitted (they are described in much more detail, and the role of collisional dephasing is described analytically, in Chapter 14). As a second example, Doppler shifts are associated with random velocity alignments in the gas, as described by the Maxwellian velocity distribution. Because these are random processes, it is necessary to recast transform notions that are based upon deterministic expressions into a formalism based upon statistics. Once recast, these transform techniques can be used with descriptions of dephasing and Doppler shifts to generate expressions for spectral lineshapes.

Techniques for describing random processes in terms of a power spectral density can be found in many texts on communication systems (see, for example, Lathi [74]). In what follows, we first discuss statistics of random variables, building towards a description of random processes. Following presentation of the Wiener-Khinchine relation, expressions for the collisional and Doppler lineshapes are developed.

Statistics of random variables

For our purposes, we define random variables as discrete numbers that are independent of time. For a random variable ξ, the probability of observing ξ in the range $(\xi, \; \xi + \Delta\xi)$ is $p(\xi)\Delta\xi$, where $p(\xi)$ is the probability density function (PDF). This statement can also be written as

$$\mathcal{P}(\xi_1 < \xi \leq \xi_2) = \int_{\xi_1}^{\xi_2} p(\xi) \, \mathrm{d}\xi \tag{11.2}$$

where $\mathcal{P}$ is the probability of this event. Because ξ must have some finite value:

$$\int_{-\infty}^{+\infty} p(\xi)\,\mathrm{d}\xi = 1 \tag{11.3}$$

Various moments of the PDF can be calculated. The nth moment is defined as

$$n^{\text{th}} \text{ moment} \equiv \int_{-\infty}^{+\infty} \xi^n p(\xi)\,\mathrm{d}\xi \tag{11.4}$$

The average value of ξ is the first moment:

$$\overline{\xi} = \int_{-\infty}^{+\infty} \xi p(\xi)\,\mathrm{d}\xi \tag{11.5}$$

where the overbar denotes an average.

Statistics of random processes

Random processes (also called stochastic processes), such as those leading to spectral broadening, are continuous functions of time [e.g. $\xi(t)$] within some time window. A measurement of the random *variable* ξ produces a number, and ξ can be specified statistically by measuring it a large number of times. In contrast, measurement of the random *process* $\xi(t)$ produces a time-waveform that must be measured repeatedly at each value of time in order to specify the process fully. A collection of these waveforms $[\xi(t)]$ is often called an ensemble. Each waveform in an ensemble is actually deterministic. We can, for example, construct a specific, detailed description of one molecular trajectory that produces a particular Doppler shift relative to an input laser beam. We have no idea, however, when that particular trajectory will occur. Which waveform one will measure in an individual observation is thus a random process.

It is not possible to consider $\xi(t)$ simply a random variable that changes with time. If that were true, $\xi(t)$ could be completely specified by the PDF $p(\xi)$ as it changes with time. Lathi provides a good example of the error associated with such a simple notion. Imagine we have a particular ensemble of waveforms $\xi(t)$ leading to a determination of $p(\xi)$. Now imagine that same set of waveforms is time compressed by some constant value. If a new PDF were measured, it would yield $p(\xi)$ again. The change in frequency represented by the time compression is not included in the PDF. This occurs because $p(\xi)$ is first order in

ξ. The frequency content of a random process can be measured by correlating amplitudes at two times.

Consider a measurement of ξ at two times t_1 and t_2. The real autocorrelation $G_\xi^R(t_1, t_2)$ is defined in terms of these two observations:

$$G_\xi^R(t_1, t_2) \equiv \overline{\xi(t_1)\xi(t_2)} = \overline{\xi_1 \xi_2} \qquad (11.6)$$

Extending the definition of an average given by equation (11.5) produces an expression for evaluation of $G_\xi^R(t_1, t_2)$:

$$G_\xi^R(t_1, t_2) = \int_{-\infty}^{+\infty} \int_{-\infty}^{+\infty} \xi_1 \xi_2\, p(\xi_1, \xi_2)\, d\xi_1\, d\xi_2 \qquad (11.7)$$

where $p(\xi_1, \xi_2)$ is the joint PDF, which is second order.

Thus far we have defined averages of sample points. In contrast, the time average can be defined as

$$\widetilde{\xi(t)} \equiv \lim_{T \to \infty} \frac{1}{2T} \int_{-T}^{T} \xi(t)\, dt \qquad (11.8)$$

where $1/2T$ has been used so that the limits of integration can span from $-T$ to $+T$, and the rippled overbar on $\widetilde{\xi(t)}$ denotes a time average. A time-based autocorrelation function can be defined:

$$\mathcal{G}_\xi^R(\tau) \equiv \lim_{T \to \infty} \frac{1}{2T} \int_{-T}^{T} \xi(t)\xi(t + \tau)\, dt \qquad (11.9)$$

A "stationary process" does not depend upon absolute time; it is possible to define the initial time t_o however one likes, because the statistics do not change with t_o. In such a case

$$G_\xi^R(t_1, t_2) = \mathcal{G}_\xi^R(t_2 - t_1) = \mathcal{G}_\xi^R(\tau) \qquad (11.10)$$

where $\tau \equiv (t_2 - t_1)$. This is an idealization; truly stationary processes do not exist. All the same, one can assume stationarity over the times of concern to broadening.

Power spectral densities

The power spectral density (PSD) for a deterministic signal is found by taking the Fourier transform of the signal. For a random process, however, the exercise is somewhat more complicated. We define the

PSD $S_\xi(\omega)$ for the random process $\xi(t)$ as the ensemble average of the PSDs for the sample functions:

$$S_\xi(\omega) \equiv \lim_{T \to \infty} \frac{|X_T(\omega)|^2}{T} \qquad (11.11)$$

where $X_T(\omega)$ is the Fourier transform of the truncated random process $\xi(t) \prod(t/T)$. In this notation, $\prod$ denotes a top-hat sample function, not a product; it indicates that $\xi(t)$ has been sampled with unit gain over a time period T. This power spectral density is actually a lineshape function that has not yet been formally normalized.

The Wiener-Khinchine relation dictates that $S_\xi(\omega)$ is also the Fourier transform of the autocorrelation function $\mathcal{G}_\xi^R(\tau)$ [equation (11.9)]. This relation is proven in most signal processing texts (see, for example, reference [74]). To find an expression for $S_\xi(\omega)$ using autocorrelation proves to be much more straightforward than to find $X_T(\omega)$ for every random process. In what follows, we show how this can be done for spectral lines.

11.3   General Description of Optical Spectra

Even steady radiation consists of "trains of waves that vary periodically with time" (Shore [73]). Even when these waves have the same carrier frequency, the phase variations between waves will cause spectral broadening. In order to describe this broadening, one must describe the phase variations, find the autocorrelation function for these phase variations, and then solve for $S(\omega)$ by taking the Fourier transform of the autocorrelation.

Imagine we have an electric field located at some fixed position (so that position dependence can be neglected while spectral features are analyzed). It has the following Fourier decomposition:

$$E(t) = \int_{-\infty}^{+\infty} e^{-i\omega t} E(\omega) \, d\omega \qquad (11.12)$$

For the case in question, the electric field is not really continuous and it becomes necessary to use a time-truncated (to time T) field denoted by $E_T(t)$. In this case the transform of $E_T(t)$ provides the spectral form:

$$E_T(\omega) = \frac{1}{2\pi} \int_{-\infty}^{+\infty} e^{i\omega t} E_T(t) \, dt \tag{11.13}$$

and

$$E(\omega) = \lim_{T \to \infty} E_T(\omega) \tag{11.14}$$

Our goal is to find the spectral distribution around the carrier frequency for this field.

Spectral density

Sensors detect irradiance, as shown in equation (4.78):

$$I = \frac{\epsilon_o c}{2} |E_o|^2$$

where $|E_o|$ is the absolute value of the amplitude of the electric field. The expression has already been averaged over a large time (many cycles). We therefore care about a time-averaged value of $E^2(t)$, defined here in the same form used throughout this chapter. If we assume a stationary process, then

$$\overline{E^2(t)} = \overline{E^2(0)} = \lim_{T \to \infty} \frac{1}{2T} \int_{-T}^{+T} E_T^2(t) \, dt \tag{11.15}$$

In this case, $T \to \infty$ simply means times much larger than other relevant timescales.

Rayleigh's theorem, which is a specialized version of Parseval's theorem, states that

$$\int_{-\infty}^{+\infty} E_T^2(t) \, dt = 2\pi \int_{-\infty}^{+\infty} |E_T(\omega)|^2 \, d\omega \tag{11.16}$$

If we divide by $2T$ and take the limit as $T \to \infty$, we find

$$\lim_{T \to \infty} \frac{1}{2T} \int_{-\infty}^{+\infty} E_T^2(t) \, dt = \lim_{T \to \infty} \frac{2\pi}{2T} \int_{-\infty}^{+\infty} |E_T(\omega)|^2 \, d\omega \tag{11.17}$$

The left-hand side of equation (11.17) produces $\overline{E^2(t)}$. In addition, we define

$$S(\omega) \equiv \lim_{T \to \infty} \frac{2\pi}{2T} \mid E_T(\omega) \mid^2 \tag{11.18}$$

where $S(\omega)$ is the Wiener-Khinchine spectral density. Then

$$\overline{E^2(t)} = \int_{-\infty}^{+\infty} S(\omega)\, d\omega \tag{11.19}$$

Multiplying by $\epsilon_\circ c/2$ yields

$$\frac{\epsilon_\circ c}{2}\, \overline{E^2(t)} = I = \frac{\epsilon_\circ c}{2} \int_{-\infty}^{+\infty} S(\omega)\, d\omega = \int_{-\infty}^{+\infty} I(\omega)\, d\omega \tag{11.20}$$

which is equivalent to equation (3.14).

Wiener-Khinchine relation

The term $\mid E_T(\omega) \mid^2$ in equation (11.18) can be found through a double integration:

$$\mid E_T(\omega) \mid^2 = \left(\frac{1}{2\pi}\right)^2 \int_{-T}^{+T} dt \int_{-T}^{+T} e^{i\omega(t-t')} E(t)E(t')\, dt' \tag{11.21}$$

which uses a time-truncated Fourier transform as defined by Shore [73], and takes advantage of the fact that $E(t)$ is real. Now define a new τ by $\tau \equiv (t - t')$ and multiply by 2π:

$$2\pi \mid E_T(\omega) \mid^2 = \frac{1}{2\pi} \int_{-T}^{+T} e^{i\omega\tau}\, d\tau \int_{-T}^{+T} E(t' + \tau)E(t')\, dt' \tag{11.22}$$

Now divide by $2T$ and take the limit as $T \to \infty$:

$$S(\omega) = \frac{1}{2\pi} \mathrm{Re} \int_{-\infty}^{+\infty} e^{i\omega\tau} \mathcal{G}(\tau)\, d\tau = \mathcal{G}(\omega) \tag{11.23}$$

where the possibility of a complex integral is recognized by selection of the real component. The autocorrelation $\mathcal{G}^R(\tau)$ is defined by

$$\mathcal{G}^R(\tau) = \lim_{T \to \infty} \frac{1}{2T} \int_{t-T}^{t+T} E(t + \tau)E(t)\, dt = \overline{E(t + \tau)E(t)} \tag{11.24}$$

Equation (11.23) is a statement of the Wiener-Khinchine relation; $S(\omega)$ is the Fourier transform of the autocorrelation function $\mathcal{G}(\tau)$. Because E is stationary, $\mathcal{G}^R$ does not depend upon t, it depends only on τ:

$$\mathcal{G}^R(\tau) = \lim_{T \to \infty} \frac{1}{2T} \int_{-T}^{+T} E(\tau)E(0)\, dt \qquad (11.25)$$

Because it is an autocorrelation between $t = 0$ and τ, $\mathcal{G}^R(\tau)$ is thus a measure of the field coherence.

Because $\mathcal{G}^R$ is real and symmetric about $\tau = 0$, we can write equation (11.23) as

$$S(\omega) = \frac{1}{\pi} \mathrm{Re} \int_0^{+\infty} e^{i\omega\tau} \mathcal{G}^R(\tau)\, d\tau \qquad (11.26)$$

Line profiles

The electric fields produced by an atom or molecule in emission, or an electric field interacting with an atom or molecule in absorption, will experience random field dephasing events. The spectral line profile can be developed by taking the Fourier transform of the autocorrelation of these fields. We begin by normalizing (at the peak) the autocorrelation, because the Fourier transform of a normalized autocorrelation will produce an area-normalized spectral density, which is a property of the PSD that we seek. We therefore define a normalized autocorrelation $g(\tau)$ using equation (11.24):

$$g(\tau) \equiv \frac{\mathcal{G}^R(\tau)}{\mathcal{G}^R(0)} = \frac{\overline{E(t+\tau)E(t)}}{\overline{E(t)E(t)}} \qquad (11.27)$$

where $g(0)$ is indeed 1. Now define a spectral profile $Y_\omega(\omega - \omega_o)$ by

$$Y_\omega(\omega - \omega_o) \equiv \frac{S(\omega)}{\int_{-\infty}^{+\infty} S(\omega)\, d\omega} \qquad (11.28)$$

Equation (11.19) provides an expression for the denominator:

$$\int_{-\infty}^{+\infty} S(\omega)\, d\omega \;=\; \overline{E(t)E(t)}$$
$$=\; \mathcal{G}^R(0) \qquad (11.29)$$

Therefore

$$Y_\omega(\omega - \omega_\circ) = \frac{\frac{1}{2\pi} \text{Re} \int_{-\infty}^{+\infty} e^{i\omega\tau} \mathcal{G}^R(\tau) \, d\tau}{\mathcal{G}^R(0)} \qquad (11.30)$$

which simplifies to

$$\boxed{Y_\omega(\omega - \omega_\circ) = \frac{1}{\pi} \text{Re} \int_0^{+\infty} e^{i\omega\tau} g(\tau) \, d\tau} \qquad (11.31)$$

This development has been based entirely upon classical arguments. A similar quantum-mechanical treatment produces the same result (see, for example, Steinfeld [21]), where $g(\tau)$, the normalized correlation function, is replaced by an ensemble average of atomic or molecular dipole correlation functions.

11.4 Homogeneous Broadening

In homogeneous broadening, some event in time causes the correlation between the atomic or molecular coherence and the light to break down. This can be caused by decay of population (natural broadening), or by a collision that interrupts the coherent interaction between the atom or molecule and the light (pressure, or collision broadening). These phenomena can be described in a general way by changes in phase. The electric field can be described by

$$E = E_\circ e^{-i\omega_\circ t + i\phi(t)} \qquad (11.32)$$

where positional dependence has been removed for simplicity.

Here, the phase $\phi(t)$ is what shifts randomly with time, and it is the phase terms that must be correlated. The normalized autocorrelation is defined in equation (11.27). One can then use the definitions for $\overline{E(t+\tau)E(t)}$ and $\overline{E(t)E(t)}$ given in equation (11.24) to evaluate $g(\tau)$. This is done by folding the phase shift terms into the correlations, but by excluding the carrier terms $\exp(-i\omega_\circ t)$ from the correlation integrals. The outcome is

$$g(\tau) = e^{-i\omega_\circ \tau} \, \overline{e^{i\phi(t+\tau) - i\phi(t)}} \qquad (11.33)$$

The correlation term $\overline{e^{i\phi(t+\tau) - i\phi(t)}}$ is equal to 1 during the time between interruptions, because ϕ is constant in between interruptions.

During an interruption, however, the correlation vanishes. The correlation term must therefore be the number 1, multiplied by a weighting function that describes the probability that time τ can pass without interruption. Following Shore, we adopt an exponential probability to give

$$g(\tau) = e^{-i\omega_o\tau - \gamma\tau} \tag{11.34}$$

where γ is a phase interruption rate, or $1/\gamma$ is the mean time to interruption.

Now we can find the lineshape function via equation (11.31):

$$
\begin{aligned}
Y_H(\omega - \omega_o) &= \frac{1}{\pi}\mathrm{Re}\int_0^{+\infty} e^{i\omega\tau}\left[e^{-i\omega_o\tau - \gamma\tau}\right]\,d\tau \\
&= \frac{1}{\pi}\mathrm{Re}\int_0^{+\infty} e^{[i(\omega-\omega_o)-\gamma]\tau}\,d\tau
\end{aligned}
\tag{11.35}
$$

where the subscript ω on Y in equation (11.31) has been replaced by subscript H here to designate a general class of homogeneous broadening. Evaluation of the integral produces

$$Y_H(\omega - \omega_o) = \frac{1}{\pi}\,\mathrm{Re}\left[\frac{1}{\gamma - i(\omega - \omega_o)}\right] \tag{11.36}$$

To separate this expression into real and imaginary components, multiply the numerator and denominator by $\gamma + i(\omega - \omega_o)$. The real part is

$$\boxed{Y_H(\omega - \omega_{nm}) = \frac{1}{\pi}\left[\frac{\gamma}{(\omega - \omega_{nm})^2 + \gamma^2}\right]} \tag{11.37}$$

where we have changed the notation of ω_o to ω_{nm} in recognition that the line is centered on the frequency of a spectroscopic transition, making equation (11.37) match other forms in this book. Note that the original Lorentzian formalism [see, for example, equation (5.33)] had $(\omega_o - \omega)^2$ in the denominator. Here the two frequencies are switched. The time-dependent perturbation solution to natural broadening generates the same sequence. This difference does not matter, as the term is squared and the curve is symmetric.

Because $g(\tau)$ is normalized at the peak, equation (11.37) is normalized across the area of the curve, as expected. Indeed, whenever a Lorentzian is in the form

$$Y_H = \frac{a}{\pi} \frac{1}{(x - x_o)^2 + a^2} \tag{11.38}$$

the integral will be 1:

$$\frac{a}{\pi} \int_{-\infty}^{+\infty} \frac{1}{(x - x_o)^2 + a^2} \, dx = 1 \tag{11.39}$$

The "full width at half-maximum" (FWHM) of the spectral profile generated by equation (11.37) is 2γ. Sometimes γ is replaced by $\Gamma = 2\gamma$, and equation (11.37) is then written as

$$Y_H(\omega - \omega_{nm}) = \frac{1}{2\pi} \left[\frac{\Gamma}{(\omega - \omega_{nm})^2 + \left(\frac{\Gamma}{2}\right)^2} \right] \tag{11.40}$$

which makes the FWHM equal to Γ.

The Lorentzian profile has been developed using the angular frequency ω $(= 2\pi\nu)$ because harmonic behavior is required (the 2π ensures harmonic behavior). In spectroscopy, however, it is more common to use line shape functions that are based upon ν, because ν is closely related to the spectral feature one actually measures (wavelength λ or wave number $\breve{\nu}$).

To convert equation (11.37) to a form that is written in terms of ν, one must convert both ω and the widths that are written in terms of ω ($d\omega$ and γ in this case) via $\omega = 2\pi\nu$. This conversion is performed by enforcing the normalization criterion via

$$
\begin{aligned}
Y_H(\nu - \nu_{nm}) \, d\nu &= Y_H(\omega - \omega_{nm}) \, d\omega \\
Y_H(\nu - \nu_{nm}) &= 2\pi Y_H(\omega - \omega_{nm})
\end{aligned}
\tag{11.41}
$$

because $d\omega = 2\pi \, d\nu$. Following the normalization criteria this ν-based Lorentzian function assumes the following form:

$$Y_H(\nu - \nu_{nm}) = \frac{\Delta\nu_H}{2\pi} \frac{1}{(\nu - \nu_{nm})^2 + \left(\frac{\Delta\nu_H}{2}\right)^2} \tag{11.42}$$

which is normalized according to equation (11.38), and has an FWHM equal to $\Delta\nu_H$.

Lifetime broadening, as the name suggests, is due to the finite lifetimes of the excited and ground states (see the discussion on T_n and T_m combining to give T_1, which then contributes to T_2 in Chapter 14). For a ground-state transition, lifetime broadening is due to the excited-state lifetime given by the Einstein coefficient for spontaneous emission:

$$\Delta\nu_H = \frac{1}{2\pi} A_{nm} \tag{11.43}$$

The $1/2\pi$ converts from an ω-based formalism. For strong atomic transitions, A_{nm} is of the order of 10^8 s^{-1}, corresponding to lifetime broadening of the order of 10 MHz. Molecules, on the other hand, typically have much weaker transition moments, and therefore smaller lifetime broadening. This mechanism is usually a very minor effect compared with other broadening mechanisms with the exception occuring in low-pressure, low-temperature environments such as encountered in molecular or atomic beam studies.

Pressure broadening is a collisional effect; it is often termed collisional broadening. In this case, elastic, or dephasing, collisions give rise to broadening. These collisions randomly interrupt the phase of the coherent oscillations of a dipole moment. Reorientation and inelastic collisional terms can also contribute, in different amounts, depending upon the atom or molecule. The FWHM is related to the rate of the dephasing collisions γ_{nm}:

$$\Delta\nu_H = \frac{1}{\pi} \gamma_{nm} \tag{11.44}$$

where $\gamma_{nm} = \gamma/2$. We can describe the dephasing collision rate for a particular collision partner as follows:

$$\gamma_{nm} = N_w \overline{\sigma_{nm,w} v} \tag{11.45}$$

where N_w is the number density of the collision partner, $\sigma_{nm,w}$ is the collisional dephasing cross-section with collision partner w, and v is the relative velocity between collision species. Assuming a Maxwellian distribution of velocity and a velocity-independent cross-section, we can write

$$\gamma_{nm} = N_w \sigma_{nm,w} \sqrt{\frac{8k_B T}{\pi \mu}} \tag{11.46}$$

where μ is the reduced mass.

Typical collisional broadening cross-sections result in a pressure broadening FWHM which is of the order of 10 GHz at atmospheric pressure. However, this parameter can be very sensitive to the species and collision partner, and in most cases one must resort to empirical data for the specific case being studied.

Calculation of γ_{nm} can actually become more complicated than simply finding a value for $\sigma_{nm,w}$ and plugging into equation (11.46). Often γ_{nm} is written twice, in terms of dephasing for the two levels n and m, and the two are summed to form a total dephasing rate. In a few cases, one will find $\Delta\nu_{\mathrm{H}} = \gamma_{nm}/(2\pi)$, which requires a different definition for γ_{nm}. As one often finds in spectroscopy, it is critical to understand fully the formalism used by an author reporting collisional broadening data, in order to use the data correctly. For many optical diagnostics applications, the values for the broadening parameters are actually unknown. In combustion, as an example, the collisional dephasing cross-section will depend strongly upon the collider mix, which changes rapidly in a flame. To overcome this problem, experimentalists often acquire the spectral profile and fit a model for a profile to it. This approach is discussed in more detail below.

11.5   Inhomogeneous Broadening

There are many forms of inhomogeneous broadening, including isotopic effects and hyperfine splitting, but we will simply focus on Doppler broadening here.

Doppler broadening results from the distribution of velocities in a thermal ensemble. For atoms or molecules traveling with nonzero velocity components along the optical propagation direction, the optical frequency experiences a random shift (a velocity-dependent detuning) within the rest frame of the absorber. In order to model this, we write an electric field that is a function of position:

$$E = E_o e^{i\mathbf{k}\cdot\mathbf{r} - i\omega_o t} \tag{11.47}$$

where

$$\mathbf{r} = \mathbf{r}_o + \mathbf{v}t \qquad (11.48)$$

and $\mathbf{v}$ is the atomic or molecular velocity. It is the random orientation and magnitude of $\mathbf{v}$ that causes random relative frequency shifts. The autocorrelation is developed in a way similar to the homogeneous case, by folding the random wavelength shift into the correlation integral:

$$g(\tau) = e^{-i\omega_o\tau}\overline{e^{i\mathbf{k}\cdot\mathbf{v}\tau}} \qquad (11.49)$$

In order to evaluate the ensemble average represented by $\overline{\exp(i\mathbf{k}\cdot\mathbf{v}\tau)}$, we assume a Maxwellian velocity distribution [e.g. we apply equation (2.45)]:

$$g(\tau) = e^{-i\omega_o\tau}\left(\frac{m}{2\pi k_B T}\right)^{3/2}\int_{-\infty}^{+\infty} e^{\frac{-mv^2}{2k_B T}}\, e^{i\mathbf{k}\cdot\mathbf{v}\tau} \qquad (11.50)$$

This integral can be evaluated by splitting $\mathbf{v}$ into Cartesian components and evaluating separate integrals for each. As an example, the x-directed integral (without the preceding exponential and fraction) is

$$\int_{-\infty}^{+\infty} e^{\left(\frac{-mv_x^2}{2k_B T}+ik_x v_x\tau\right)}\, dv_x \qquad (11.51)$$

which is equal to

$$\left(\frac{2\pi k_B T}{m}\right)^{1/2} e^{\left(\frac{(k_x\tau)^2}{4}\frac{2k_B T}{m}\right)} \qquad (11.52)$$

Equation (11.50) is then solved to give

$$g(\tau) = e^{-i\omega_o\tau}e^{-\frac{1}{4}\left[k^2\frac{2k_B T}{m}\tau^2\right]} \qquad (11.53)$$

This expression for the autocorrelation is then inserted into equation (11.31):

$$Y_D(\omega - \omega_o) = \frac{1}{\pi}\mathrm{Re}\int_0^{+\infty} e^{i\omega\tau}e^{-i\omega_o\tau}e^{-\frac{1}{4}\left[k^2\frac{2k_B T}{m}\tau^2\right]}\, d\tau \qquad (11.54)$$

The integral evaluates to

$$Y_D(\omega - \omega_{nm}) = \frac{1}{\sqrt{\pi}\Delta\omega_D} \exp\left[-\left(\frac{\omega - \omega_{nm}}{\Delta\omega_D}\right)^2\right] \tag{11.55}$$

where the Doppler width $\Delta\omega_D$ is given by

$$\Delta\omega_D \equiv \omega_{nm}\sqrt{\frac{2k_BT}{mc^2}} \tag{11.56}$$

The equalities $k = 2\pi/\lambda$ and $\nu = c/\lambda$ have been used to generate this result. Equation (11.55) is normalized according to equation (11.1).

To convert equation (11.55) to a ν-based expression is straightforward, and it generates

$$Y_D(\nu - \nu_{nm}) = \frac{1}{\sqrt{\pi}\Delta^e\nu_D} \exp\left[-\left(\frac{\nu - \nu_{nm}}{\Delta^e\nu_D}\right)^2\right] \tag{11.57}$$

where the Doppler $1/e$ width $\Delta^e\nu_D$ is given by

$$\Delta^e\nu_D \equiv \nu_{nm}\sqrt{\frac{2k_BT}{mc^2}} \tag{11.58}$$

Equation (11.57) is more commonly written as

$$Y_D(\nu - \nu_{nm}) = \sqrt{\frac{4\ln 2}{\pi}}\frac{1}{\Delta\nu_D} \exp\left[-4\ln 2\left(\frac{\nu - \nu_{nm}}{\Delta\nu_D}\right)^2\right] \tag{11.59}$$

where the Doppler FWHM is given by

$$\Delta\nu_D = \nu_{nm}\sqrt{\frac{8\ln 2\, k_BT}{m\, c^2}} \tag{11.60}$$

Equations (11.57) and (11.59) are equivalent, and they are normalized. Plugging constants into the Doppler width expression provides

$$\Delta\nu_D = 7.16 \times 10^{-7}\, \nu_{nm}\sqrt{\frac{T}{M}} \tag{11.61}$$

where the temperature is given in kelvins, and the mass is given in atomic mass units. Taking the potassium D_1 line as an example, the Doppler FWHM is 770 MHz at room temperature, and 2 GHz at 2000 K.

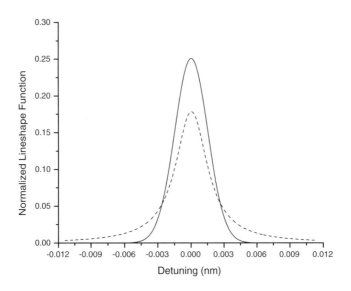

Figure 11.1: Gaussian (solid) and Lorentzian (dashed) lineshape functions (the broadening parameters used are specific to the potassium D_1 line in a 2000 K flame).

11.6 Combined Mechanisms: the Voigt Profile

For reference, both a homogeneous (Lorentzian) and an inhomogeneous (Gaussian) lineshape function are plotted in Figure 11.1. Note that the Gaussian function is more peaked and drops quickly in the wings, whereas the Lorentzian function has very large wings.

Unfortunately, most systems are not described exclusively by either homogeneous or inhomogeneous broadening. Instead, it is often a combination of both types of mechanisms that produces the broadening. In this case, we typically describe the lineshape with the Voigt profile.

The Voigt profile is defined as the convolution of a Lorentzian and a Gaussian function. This makes sense, because every atom or molecule within a specific velocity group will be homogeneously broadened. The total spectral response therefore scans the Lorentzian across the Doppler profile (assuming that the Lorentzian is independent of velocity), just like a mathematical convolution. The Voigt profile has the long wings characteristic of a Lorentzian profile, but in a relative sense it will be more strongly peaked at line center than a Lorentzian, reflecting the influence of the Gaussian function. The Voigt profile is written as

$$Y_V(\nu) = \int_{-\infty}^{+\infty} Y_D(\nu' - \nu_{nm}) \, Y_H(\nu - \nu') \, d\nu' \qquad (11.62)$$

Unlike the Gaussian and Lorentzian functions, the Voigt profile cannot be expressed in a simple analytical form. Therefore, it is necessary to compute the convolution integral for particular broadening parameters numerically. For computation, it is desirable to nondimensionalize the integral. To this end, we introduce a nondimensional detuning parameter x:

$$\boxed{x \equiv \frac{\sqrt{4 \ln 2}}{\Delta \nu_D} (\nu - \nu_{nm})} \qquad (11.63)$$

Next, we define a nondimensional broadening parameter a:

$$\boxed{a \equiv \sqrt{\ln 2} \, \frac{\Delta \nu_H}{\Delta \nu_D}} \qquad (11.64)$$

Obviously, for low-pressure combustion systems where Doppler broadening dominates, $a \ll 1$, while for high-pressure combustors $a \gg 1$. Finally, for atmospheric combustion systems, $a \sim 1$.

Using these definitions, the Voigt profile can be written in terms of a nondimensional Voigt function $V(x; a)$:

$$\boxed{Y_V(x; a) = \frac{\sqrt{4 \ln 2}}{\sqrt{\pi} \, \Delta \nu_D} V(x; a)} \qquad (11.65)$$

where

$$\boxed{V(x; a) \equiv \frac{a}{\pi} \int_{-\infty}^{+\infty} \frac{e^{-x'^2}}{(x - x')^2 + a^2} \, dx'} \qquad (11.66)$$

$V(x; a)$ is actually the real part of the error function for complex argument, and as such it can be found in most mathematical databases.

In Section 11.4 it was mentioned that one often does not know the relevant homogeneous broadening parameters. In such a case, it is common to measure a spectrally resolved lineshape. That line can then be fitted with a Voigt function wherein the value of a is adjusted within reasonable bounds to optimize the fit. Such fits are often achieved by use of the convolution theorem. One takes the Fourier transforms of equations (11.42) and (11.59) and compares those with an FFT of the spectral data. A nonlinear fitting routine can be used to minimize the error of the fit.

11.7   Conclusions

This chapter has developed the most common, important relations that describe broadening of spectral lines. It has by no means exhausted the topic. One should be aware that there are other, less common broadening mechanisms that can change the spectral profile. As one example, high power lasers can saturate the center of a profile but not the wings. This causes power broadening. As a second example, the collisional dephasing cross-sections are velocity dependent. Because collisions change velocity, there is a collisional narrowing phenomenon. It will be observable only in highly resolved spectral data. Many molecules have "predissociated" upper electronic states. When light is absorbed, placing the molecule into one of these states, the molecule dissociates after a characteristic time that is fairly short. This causes another, relatively broad homogeneous profile. One should be aware that broadening mechanisms other than those described by the Voigt profile exist. If fits to data are poor, it could be due to a real, physical phenomenon.

For absolute determination of concentration, it is often best to perform a spectrally resolved measurement and then to integrate the data spectrally. Because the profile is normalized, the outcome will be relatively independent of errors in the spectral model used. If a light source is used (in absorption for example) that has a spectral width near that of the absorption linewidth, it will be necessary to deconvolve the source profile from the data. The error involved may overcome advantages to this approach.

Chapter 12

POLARIZATION

12.1 Introduction

Polarization as a subject does not receive as much attention as broadening does, for example. The reason for this tendency to bypass polarization effects is not clear, but the subject can be somewhat confusing. As an example, consider the versions of Beer's law presented in equations (3.81) and (3.82). There is no mention of optical polarization in those equations. In fact, the formalism was originally developed to describe randomly polarized light interacting with randomly oriented absorbers. Note, however, that the absorbers are represented by the Einstein B coefficients, which are directly related to the Einstein A coefficients. The development of the A coefficient [see, for example, equation (10.28)] was couched in terms of a linear dipole oscillator. Alternatively, the B coefficient can be derived from the imaginary index found in the Lorentz atom treatment; again, a linear dipole. Thus far the dipole has been linearly polarized. Is there then an inconsistency with the random orientation in Beer's law? As a second example, relativistic quantum mechanics dictates that the value of photon spin is $m_p = \pm 1$, meaning that photons are allowed only right- and left-hand circular polarization. Some atomic or molecular transitions, however, absorb only linearly polarized light. The interaction occurs between a photon and an atom. How does this happen? Finally, imagine that a polarized laser beam has been used to excite a collection of atoms or molecules, which will then fluoresce. Will the fluorescence be polarized as well? Our goal in the following section is not to address every issue related to polarization, but we do hope to resolve questions that can cause misunderstanding.

Polarization is discussed by several authors. Feofilov [75] discusses polarization of transitions, absorption, and emission. Siegman [25] uses a classical approach to discuss polarization orientation and resonance response in the context of laser gain. Zare and co-workers [66, 76, 77] discuss polarization of the transition in terms of the angular momenta of molecules. Their primary application is the study of molecular fragment orientation, and the use of these observations to describe the parent molecular states. Doherty and Crosley [78] have used the results published by Feofilov and Zare to infer the degree of polarization in LIF, and they reported observations that confirm their predictions. Recently, Lucht *et al.* [79] have analyzed orientational dependence in absorption and emission using the density matrix equations (see Chapter 14 for a description of this formalism). Their results are in agreement with these other sources. These are the materials upon which we base the discussions in this chapter. The discussion below is abbreviated because these issues are either straightforward, or they are complex but ancillary to the subject of this text. In the latter case, we make reference to available materials and only briefly describe the outcome.

12.2   Polarization of the Resonance Response

The elementary dipole problem has been presented above, for the classical case, culminating in expressions for the susceptibility and for the Einstein A coefficient. These expressions are derived for a dipole in terms of a linear polarization. Atoms and molecules can also absorb circularly polarized light. In the classical formalism, this is described by the "elementary electric rotor", which is represented by two linear dipoles oscillating in the same plane and out of phase with each other by $\pi/2$. The two oscillators add coherently, resulting in a description of circular polarization (see, for example, Figure 4.8 a). Indeed, it is even possible to describe Zeeman splitting using classical arguments [75]. Rather than enter into these developments, however, more general results can be derived by applying quantum mechanics to the one-electron atom.

Recall that the Einstein A coefficient is given in terms of the matrix element of the dipole moment [see equation (10.10)]:

$$A_{nm} = \frac{2 \, \omega_{nm}^3 |\vec{\mu}_{nm}|^2}{3 \, h\epsilon_\circ c^2 g_n}$$

and the matrix element of the dipole moment is defined by equation (10.9):

$$\vec{\mu}_{nm} \equiv \langle \psi_n | \hat{\mu} | \psi_m \rangle$$

or in integral form

$$\vec{\mu}_{nm} = \int \psi_n^* \, \hat{\mu} \, \psi_m \, d\tau \tag{12.1}$$

In Cartesian space, this can be written as

$$\vec{\mu}_{nm}^x = q_e \int \psi_n^* \, x \, \psi_m \, d\tau$$

$$\vec{\mu}_{nm}^y = q_e \int \psi_n^* \, y \, \psi_m \, d\tau$$

$$\vec{\mu}_{nm}^z = q_e \int \psi_n^* \, z \, \psi_m \, d\tau \tag{12.2}$$

Assume the electron is in a central force field. The wave function is then written in the well-known spherical formalism for the one-electron atom [see equation (8.63)]:

$$Y_{\ell,m}(\theta, \varphi) = \sin^{|m|}\theta \left[\sum_{j=0}^{\ell-|m|} a_j \cos^j \theta \right] \frac{e^{im\varphi}}{\sqrt{2\pi}} = P_{\ell m}(\theta) e^{im\varphi} \tag{12.3}$$

where $P_{\ell m}(\theta)$ are Legendre polynomials. The angles used in this and the following expressions are defined in Figure 12.1.

At this point it is important to remember that the quantum numbers n (for energy) and m (the magnetic quantum number, often written m_ℓ) are different from the subscripts we are using on the wave functions to represent the upper and lower states (n and m) of a transition. In equation (8.63), the subscript ℓ was left off the m because it was obvious what m was. Here, we will write expressions with the subscript ℓ, to remove one source of confusion.

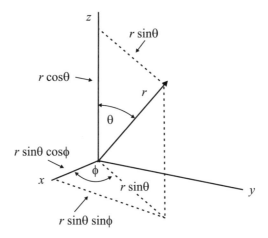

Figure 12.1: Coordinate system used for the one-electron atom.

We designate the lower state using both the subscript m and the double-primed quantum numbers. In Dirac notation

$$\psi_m = |n''\ell''m_\ell''\rangle \qquad (12.4)$$

Similarly, the upper state is designated by the subscript n and the single-primed quantum numbers:

$$\psi_n = |n'\ell'm_\ell'\rangle \qquad (12.5)$$

Equation (12.3) can be substituted into equation (12.2) (assuming unit r). In the x direction, as an example, we have

$$
\begin{aligned}
\vec{\mu}_{nm}^{\,x} &= q_e \int \psi_n^* \, x \, \psi_m \, \mathrm{d}\tau \\
&= q_e \int e^{-im_\ell'\varphi} P_{\ell'm_\ell'}(\theta) \, \sin\theta \, \cos\varphi \, e^{im_\ell''\varphi} \\
&\quad \times P_{\ell''m_\ell''}(\theta) \sin\theta \, \mathrm{d}\theta \, \mathrm{d}\varphi \\
&= q_e \int_0^{2\pi} e^{-im_\ell'\varphi} \, \cos\varphi \, e^{im_\ell''\varphi} \, \mathrm{d}\varphi \int_0^\pi P_{\ell'm_\ell'}(\theta) \\
&\quad \times P_{\ell''m_\ell''}(\theta) \, \sin^2\theta \, \mathrm{d}\theta
\end{aligned}
\qquad (12.6)
$$

Again, we have assumed unit r which generates the differential volumetric terms used in the integration; see Figure 12.1 for details. We can recast $\cos\varphi$ using the Euler representation:

$$\cos\varphi = \frac{1}{2}\left(e^{i\varphi} + e^{-i\varphi}\right) \qquad (12.7)$$

Using this result in equation (12.6), we find

$$
\vec{\mu}_{nm}^{\,x} \;=\; \frac{q_e}{2}\int_0^{2\pi} e^{i(m_\ell''-m_\ell'+1)\varphi} + e^{i(m_\ell''-m_\ell'-1)\varphi}\; d\varphi
$$

$$
\times \int_0^{\pi} P_{\ell'm_\ell'}(\theta)P_{\ell''m_\ell''}(\theta)\; \sin^2\theta\; d\theta \tag{12.8}
$$

Similarly, using the Euler representation for $\sin\varphi$, we obtain

$$
\vec{\mu}_{nm}^{\,y} \;=\; \frac{q_e}{2i}\int_0^{2\pi} e^{i(m_\ell''-m_\ell'+1)\varphi} - e^{i(m_\ell''-m_\ell'-1)\varphi}\; d\varphi
$$

$$
\times \int_0^{\pi} P_{\ell'm_\ell'}(\theta)P_{\ell''m_\ell''}(\theta)\; \sin^2\theta\; d\theta \tag{12.9}
$$

and

$$
\vec{\mu}_{nm}^{\,z} \;=\; q_e\int_0^{2\pi} e^{i(m_\ell''-m_\ell')\varphi}\; d\varphi
$$

$$
\times \int_0^{\pi} P_{\ell'm_\ell'}(\theta)P_{\ell''m_\ell''}(\theta)\; \sin\theta\cos\theta\; d\theta \tag{12.10}
$$

When $\Delta m_\ell = 0$ ($m_\ell'' = m_\ell'$), the first integrals in equations (12.8) and (12.9) go to zero, but the first integral in equation (12.10) does not. Moreover, the second integral in equation (12.10) does not vanish (consistent with the selection rule $\Delta\ell = \pm 1$). In such a case, $\vec{\mu}_{nm} = \vec{\mu}_{nm}^{\,z}$, while $\vec{\mu}_{nm}^{\,x} = \vec{\mu}_{nm}^{\,y} = 0$; the dipole is aligned along z. This is termed a "π transition", and it is linearly polarized along z.

Conversely, when $\Delta m_\ell = \pm 1$, $\vec{\mu}_{nm}^{\,x} = \mp i\vec{\mu}_{nm}^{\,y}$ and $\vec{\mu}_{nm}^{\,z} = 0$. In such a case, there is a phase shift of $\pm\pi/2$ between $\vec{\mu}_{nm}^{\,x}$ and $\vec{\mu}_{nm}^{\,y}$ (note that $\pm e^{i\pi/2} = \pm\,i$). In this case, $\vec{\mu}_{nm}$ is the coherent sum of the components in $\vec{\mu}_{nm}^{\,x}$ and $\vec{\mu}_{nm}^{\,y}$. They add to form a circularly polarized transition, termed a "σ transition". Again the selection rule $\Delta\ell = \pm 1$ ensures that the θ-integral in equations (12.8) and (12.9) does not vanish.

The same results apply to multi-electron atoms and to molecules. The transition will be linearly or circularly polarized depending entirely upon the change in magnetic quantum number. Note that this fairly restrictive finding applies to a single resonance response. In the next chapter on scattering (Chapter 13), polarization features of light scattered from molecules can potentially be more complex. It will therefore be necessary to describe scattering in terms of a polarizability tensor, providing a more complex polarization response. Because that is a much larger subject, we present the full details in Chapter 13.

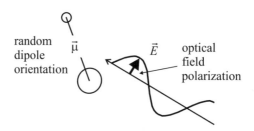

Figure 12.2: Randomly oriented linear dipole interacting with a linearly polarized beam of light.

12.3   Absorption and Polarization

Both the classical and the quantum approaches can be used to describe the orientational effects on the absorption interaction, with the same outcome. Emission, however, is not adequately described by the classical approach. Here, we will briefly present the classical approach to orientation effects in absorption. In the following section, we will summarize the quantum-based findings for emission.

Siegman [25] presents a classical treatment of orientation and optical interactions. In general, lasers can create all of the possible interactions, depending upon the gain medium and the polarization imposed upon the intracavity beam. Siegman defines a specialized susceptibility and oscillator strength in order to deal with these issues. Here, we focus on a subset of the problem because this text discusses gas-phase species, i.e. randomly oriented absorbers. For that reason we will present a simplified treatment that can be applied to the gas-phase case.

The necessary relationship between the radiation field and the dipole can be found using one case, that of a linearly polarized beam interacting with randomly oriented linear dipoles, as depicted in Figure 12.2. The probability that radiation will be absorbed is proportional to $\vec{\mu} \cdot \vec{E}$. In order to deal with the possible orientations of the dipole, it is necessary to describe two coordinate systems: a lab frame (in which the optical beam resides) and a dipole-based frame. One then writes relationships between the two coordinate systems in terms of Euler angles, and these then lead to a direction cosine matrix. For linearly polarized light, one integrates the direction cosines over all possible orientations of the dipole (see Chapter 13).

This exercise actually depends upon the polarization orientation embedded in the formalism one is using to describe the atomic or molecular response. Moreover, this response factor could be in the form of an Einstein coefficient, an oscillator strength, an absorption coefficient, an absorption cross section, a susceptibility, a polarizability, or an index. In all of these cases, the published resonance response factor may result from an experiment or a calculation. In both cases, it is necessary for the author to describe the work in sufficient detail for the reader to understand what the relative polarization states were when the response factor was determined. If nothing is said about polarization, one usually assumes the result is for random optical polarization and randomly oriented absorbers. It is really incumbent upon the author to provide orientation information, but this is often overlooked.

Chapter 13 presents a detailed treatment of Euler angles, rotation matrices, and integration over the Euler angles. Equation 13.20, reproduced just below, provides an expression for a space averaged, mean polarizability for a scattering molecule:

$$a = \frac{1}{3} \left(\alpha_{xx} + \alpha_{yy} + \alpha_{zz} \right)$$

The variables α_i are the polarizability in directions $i = x, y, z$ (where $x, y,$ and z are in the lab frame). The overall response of a randomly oriented dipole to linear input polarization, therefore, has a factor of $1/3$ in the space average. In this case, not all of the dipoles are aligned for optimum interaction with the radiation, as they would be in a crystal for example. This reduces the probability of absorption because the absorber can be randomly oriented within three-dimensional space, but the light polarization is fixed. In the extreme case of orthogonality between the optical and absorber polarization, for example, the light is simply not absorbed. While we don't repeat the entire treatment for resonance response in this chapter, the notion of multiplying the average dipole response function by $1/3$ when irradiating a randomly oriented dipole with a polarized beam is repeated time and again. This is an unsurprising and straightforward result. In this book, therefore, we simply present an algorithm for the adjustment of response factors for differences in polarization states. For those who require more detailed treatment, Lucht *et al.* [79] discuss more complex polarization interactions.

If the input light has a pure polarization state (linear or circular), if the absorber response factor (describing the interaction between the optical beam and the absorber) is for a pure polarization state, and if the absorbers are randomly oriented, then it is necessary to multiply the absorber response factor by 1/3.

Alternatively, if the input light has a pure polarization state, if the absorber response factor is for random orientation of a dipole (e.g. the source of the number has already been divided by 3), and if the absorbers are randomly oriented, then it is necessary to leave the response factor as it is.

If the input light is randomly polarized, if the absorber response factor is for a pure polarization state, and if the absorbers are randomly oriented, then it is necessary to leave the absorber response factor as-is. The image to use here is a random distribution of linear polarization vectors interacting with a random distribution of linear dipoles. This is equivalent to the case of a linearly polarized beam interacting with an array of properly-oriented linear dipoles (e.g. along a crystalline axis).

The multiplier will typically be either 1 or 1/3 for gas-phase species. The multiplier is 1 if the absorber response factor is appropriate for the case in use, and it is 1/3 when the polarization states are mismatched.

There is one case where the multiplier will be 3. In this situation, the response factor has already been divided by 3, but it was unnecessary. Such a case would occur if the light were randomly polarized, the absorbers randomly oriented, but the response factor had been divided by 3 on the assumption that the input light would be linearly polarized.

Again, if polarization is not mentioned in the source for the response factor, one should assume it is for the case of random polarization and random orientation (a case that also applies to pure polarization and aligned absorbers).

12.4 Polarized Radiant Emission

Optical radiation emission from flowfields typically occurs from chemically excited states (chemiluminescence, often in the visible and/or ultraviolet); in order to establish thermal equilibrium (thermal emission, typically in the infrared); or because absorption has created a non-Boltzmann distribution which will relax spectroscopically within the excited-state lifetime (laser-induced fluorescence). Another possibility is optical scattering from the molecules or particles in the flow. That phenomenon is discussed in Chapter 13. In the case of chemiluminescence or thermal emission, the radiators are randomly oriented and the light is therefore unpolarized.

In contrast to thermally or chemically induced sources, laser-induced fluorescence (LIF) relies upon absorption of a laser beam (usually polarized) to create the excited state, which then emits. As discussed above, absorbers that are properly aligned to the polarization state of the beam are preferentially excited. If the excited state remains undisturbed, the polarization state of the emitted light will preserve the polarized nature of the absorption transition, and hence the laser polarization. This can affect the measured signal irradiance even if there are no polarization-selective optical components in the collection system. LIF measurements are made by relating the detected optical signal to the absorber/emitters using a rate equation model [38]. That model assumes that the emission pattern is isotropic. In the extreme case of stationary linear dipoles, the spatial emission pattern would follow the image presented in Figure 5.4. Orientation of the molecule with respect to the signal detection system, and with the rate equation model, would be critical if absolute measurements were claimed. A circularly polarized transition will emit in a different pattern, reflecting the fact that it can be described as a coherent addition of two linear dipoles oscillating in the same plane but out of phase by $\pi/2$. Feofilov [75] discusses this radiance pattern in further detail. This extreme example of experimental anisotropy is not observed often in combustion systems because the excited state exists for a period of time during which the atom or molecule rotates, collides, transfers energy to other rotational or vibrational states, and so forth. All the same, polarization-based anisotropy has been observed.

Starting from the work of Zare [77], Doherty and Crosley [78] have investigated polarization in LIF emission signals. Lucht *et al.* have found results that agree with Doherty and Crosley, using a DME approach [79]. Researchers who must concern themselves with this issue in detail are encouraged to consult these references. Polarization is important in cases where absolute measurements are claimed but no calibration has been performed. The brief discussion that follows is based upon the results presented by Doherty and Crosley.

Doherty and Crosley calculated the "degree of polarization" of the LIF emission that one would expect from a molecule in a collision-free environment, because collisional effects on polarization are not sufficiently characterized to permit accurate modeling. In addition, they performed LIF experiments, observing emission from the first electronic transition of OH, recording the signal levels from various lines, and sampling various polarization states. They found that at atmospheric pressure there was a significant degree of polarization that was preserved in the flame environment. More recent short-pulse experiments [80], [81] demonstrate preservation of polarization during the initial portion of the overall emission process.

Polarization can be modified before emission via several processes [78]: (1) Rotation of the molecule precesses about $\vec{J}$, and this could lead to some rotation of m within the lab frame (where m is the total magnetic quantum number for the molecule). Under typical LIF measurement conditions, a molecule can rotate roughly 1000 times before emission occurs. (2) Elastic depolarizing collisions can change m with no effect on J. (3) Collisions can change both J and m, although this is most probable for small ΔJ. With some colliders, however, m can be preserved, and with it the polarization orientation of the excitation beam. (4) Collisional quenching can remove population from the excited state nonradiatively. When the rates for processes 1 and 2 are significantly less than for 3 and 4, the emission that occurs is strongly polarized. This is evidently the case for the experiments described by Doherty and Crosley.

In their analysis, Doherty and Crosley evaluate the polarization states by assuming that the LIF signals will be proportional to

$$I_{\parallel} \propto \frac{1}{2J_i + 1} \sum |\langle n_f J_f m_f | \mu_z | n_e J_e m_e \rangle|^2 |\langle n_e J_e m_e | \mu_z | n_i J_i m_i \rangle|^2 \quad (12.11)$$

$$I_{\perp} \propto \frac{1}{2J_i + 1} \sum |\langle n_f J_f m_f | \mu_x | n_e J_e m_e \rangle|^2 |\langle n_e J_e m_e | \mu_z | n_i J_i m_i \rangle|^2 \quad (12.12)$$

The term $\langle n_f J_f m_f |$ represents the full state vector, including quantum numbers n, J, and m for the final state (subscript f). Subscript e denotes the excited state, and subscript i denotes the initial state. It is assumed that the polarization will be observed along lab frame coordinate z, and that x is normal to it. The summation is taken over all m_e and m_f for every m_i, and then summed over all m_i and divided by $2J_i + 1$ to obtain an average.

Doherty and Crosley express the dipole moment of the molecule in the lab frame using rotation matrices. The relative amplitudes given in equations (12.11) and (12.12) can then be expressed in terms of Clebsch-Gordan coefficients, which are the projection probabilities of one rotation state onto another [66]. Here, Doherty and Crosley state that "The polarization characteristic of an individual line depends not on line strength or coupling scheme, but only on the photon polarization and branch type". By branch type, they refer to ΔJ, which makes sense in terms of the allowed Δm and the resulting polarization. Doherty and Crosley then provide a detailed table of transitions and expected degree of polarization. They then describe experiments that demonstrate significant preservation of polarization at atmospheric pressure. They also make suggestions that should mitigate these effects, the simplest being to sample several nearby transitions from different branches, thus mixing polarization and more nearly imitating an isotropic emitter.

12.5 Photons and Polarization

Relativistic quantum mechanics dictates that the spin quantum number for photons must be $m_p = \pm 1$, with spin angular momentum given by $\pm\hbar$. The only possible conclusion is that individual photons have left- and right-circular polarization exclusively. We associate $m_p = +1$ with left-circularly polarized light and $m_p = -1$ with right-circularly polarized light. In contrast, we have discussed the fact that atomic or molecular π transitions ($\Delta m = 0$) absorb linearly polarized light. The absorption process involves a single absorber and a single photon. How can this be? An even simpler question is: what happens to a circularly polarized single photon as it encounters a linear polarizer? These are typical of the problems one can come across when describing quantum outcomes and comparing them with experience.

In the discussion on electromagnetic plane waves, it was shown that two basis polarization states can be used to describe any polarization state (see, for example, Figures 4.7 and 4.8). That is the key to the photon polarization conundrum just posed. In what follows, we present a condensed version of a discussion offered by Baggott [57]. The act of measurement is often called a "collapse of the wave function". A specific observation occurs only after the wave function collapses into

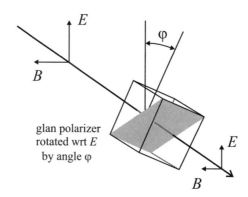

Figure 12.3: Vertically polarized electromagnetic wave passing through a glan polarizer that has been rotated around the axis by angle φ relative to the vertical.

the observed state, having encountered the "measurement operator". The photon interaction cannot be considered without also considering the measurement operator, because the photon no longer exists in isolation. The observation states are projections of the wave function, each with an assigned probability. The experiments Baggott describes are all related to the polarization properties of photons, the outcomes of which are consistent with the discussion presented here.

We start with a classical electromagnetic wave passing through a polarizer, as shown in Figure 12.3. The polarization of the input wave, denoted by the orientation of the electric and magnetic vectors, can be expressed as a coherent sum of two orthogonal states, as shown in Figures 4.7 and 4.8. After passing through the polarizer, the relative phases of the basis vectors have not changed, so the polarization is not rotated. Instead, the relative contributions of the basis vectors have changed, reducing the magnitude of the resultant, i.e. the vertically polarized electric vector depicted at the polarizer output. Just by rotating the polarizer a little, we do not suddenly stop the beam altogether. Only when the polarizer is rotated to $\varphi = 90°$ is the beam fully attenuated. Simple geometry shows that the amplitude of E at the polarizer output is proportional to $\cos \varphi$, and therefore the irradiance is proportional to $\cos^2 \varphi$.

Now consider the notion of a single, linearly polarized photon as it passes through the same polarizer (now a measurement operator). Imagine that the photon enters the polarizer with a state vector $|\psi_v\rangle$, where the subscript v denotes vertical polarization. As it passes through the polarizer, there is a possibility that the photon will acquire a new state vector called $|\psi_v'\rangle$. The probability of this occurrence is

$$|\langle \psi_v' | \psi_v \rangle|^2 = \cos^2 \varphi \qquad (12.13)$$

In short, $|\langle \psi_v' | \psi_v \rangle|^2$ is the probability of projection from $|\psi_v\rangle$ to $|\psi_v'\rangle$. $\langle \psi_v' | \psi_v \rangle$ is then the projection amplitude. Here, we are already describing expectation values (e.g. statistical quantities), and these quantum ideas match the polarization decompositions depicted in Figures 4.7 and 4.8. Note also that

$$|\langle \psi_h' | \psi_h \rangle|^2 = \cos^2 \varphi$$
$$|\langle \psi_v' | \psi_h \rangle|^2 = \sin^2 \varphi$$
$$|\langle \psi_h' | \psi_v \rangle|^2 = \sin^2 \varphi \qquad (12.14)$$

where subscript h denotes horizontal polarization. Circular polarization (subscript R and subscript L) can be used to describe linear polarization:

$$|\langle \psi_v | \psi_R \rangle|^2 = 1/2$$
$$|\langle \psi_h | \psi_R \rangle|^2 = 1/2$$
$$|\langle \psi_v | \psi_L \rangle|^2 = 1/2$$
$$|\langle \psi_h | \psi_L \rangle|^2 = 1/2 \qquad (12.15)$$

Left-circularly polarized light hitting the polarizer will be projected into vertical polarization (defined now by the polarizer itself) with half the original irradiance. The other half is rejected by the polarizer.

If one must consider single photons, it is necessary to think in terms of the collapse of the wave function upon measurement. A photon has a 50% chance of collapse into a linearly polarized representation upon overlap with the polarizer (or π transition).

Now imagine we have a linearly polarized laser beam tuned to a π transition. The linearly polarized beam consists of equal parts of left- and right-circularly polarized photons, but the coherent sum of the two is a full-irradiance, vertically polarized beam. When such a beam is decomposed at the absorber (similar to a polarizer), the left-circularly polarized photons can be decomposed into half-vertical and half-horizontal. The same can be said for the right-circularly polarized photons, except that the rejected horizontal components are π out of phase with each other (see Baggott [57] for phase expressions in the probability amplitudes), and the two rejected components cancel. That being the case, the two half-components that remain must necessarily add to give the full irradiance of the beam, polarized linearly. Otherwise the quantum system would not match the macroscopic world as the photon population grows. Does that then mean that our formalism for the Einstein coefficients is off by 1/2? No, those are written for large, statistical samples that match the macroscopic world in which we make measurements.

12.6   Conclusions

This chapter has described what one must do in order to deal with questions regarding polarization in resonance diagnostics using a polarized laser beam. It also addresses one theoretical question regarding the behavior of photons. While that question is not critical for the proper performance of a diagnostic, to address such questions helps develop depth of understanding.

Chapter 13

RAYLEIGH AND RAMAN SCATTERING

13.1 Introduction

The term "scattering" is applied to Rayleigh and spontaneous Raman (called simply "Raman" in this book) signal generation because Compton-style arguments describing a collision between a photon and a molecule are often used to explain these phenomena. Rayleigh signal generation is termed "elastic scattering" because the photon leaves with the same energy (wavelength) that it had before the interaction. A green laser beam, for example, is visible to the eye as it traverses through air owing to elastic scattering. If one looks towards the laser (not looking directly into the beam, of course), one can see point scattering from dust (often called "Mie" scattering) and a much more uniform but weaker beam which is generated by molecular Rayleigh scattering. If one then goes to the laser and views the beam as it traverses away from the source, most of the dust appears to have vanished but the Rayleigh signal remains. This is because scattering from large particles favors the forward direction while Rayleigh scattering is more isotropic. If the laser is vertically polarized (in the lab frame) and one looks down at the beam, it nearly disappears. If one views it from the side, however, Rayleigh scattering is strong, for reasons to be discussed. Raman scattering, on the other hand, is "inelastic" in the sense that the signal is shifted in wavelength from the incident light; Raman photons leave with energy that is different from that of the incoming

317

photons. Unfortunately, Raman scattering from a gas is far too weak to be observed by the naked eye.

Because this book focuses upon spectroscopic measurement, scattering from particulate will not be discussed (references [34], [35] and [36] provide background on particulate scattering). Rayleigh scattering (which does not involve spectroscopy) will be discussed because it is closely related to Raman scattering (a spectroscopic phenomenon). Moreover, spectroscopists often use Rayleigh scattering to measure temperature (via density) or to calibrate other measurement systems (the optical collection and electronic signal measurement components of an LIF instrument, for example). As such, it is a general-purpose technique for researchers who use optical diagnostics.

It is important to point out that the fundamental expressions for Rayleigh scattering can be developed in more than one way. Kerker [35], for example, develops detailed expressions for particulate scattering and then simplifies them in the small-particle limit to achieve the Rayleigh result. The other approach, taken here, describes Rayleigh scattering via molecular polarizability, a formalism that can also explain Raman scattering. Both approaches generate the same expression for the Rayleigh scattering cross-section. Because this is a book on molecular behavior, we will take the molecular approach.

Rayleigh scattering originates from the aggregate of all the scatterers in a measurement volume. It therefore indicates a species-averaged density. Raman scattering, on the other hand, separates molecular signatures by spectroscopic wavelength shifts, making it possible to delineate number densities for the various species that scatter with sufficient optical strength to register in a photomultiplier tube or multichannel plate. It is a popular diagnostic for multispecies detection because it is linearly dependent upon number density, it is independent of the collisional environment and absolute calibration is readily accomplished. Unfortunately, Raman scattering is very weak. It can therefore be used only to detect species that exist in relatively large concentrations. The combination of Raman scattering for major species and LIF for minor species works well and it is a common approach (see, for example, Barlow *et al.* in [3] and the references therein).

Rayleigh scattering is caused by an interaction between the optical electromagnetic field and the equilibrium polarizability of a molecule. It is similar to the Lorentz atom discussed in Chapter 5 (excitation

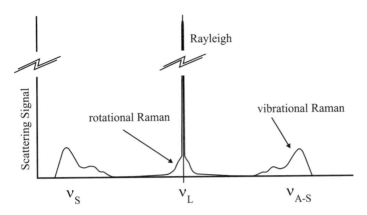

Figure 13.1: Schematic of Rayleigh, and rotational and vibrational Raman spectra. Rayleigh scattering occurs at the laser frequency ν_L while Raman scattering is shifted away by the rotational or vibrational Raman shift (vibrational shift = $\nu_{v-Raman}$). The vibrational anti-Stokes Raman shown is shifted to higher frequency ($\nu_{A-S} = \nu_L + \nu_{v-Raman}$), while the Stokes Raman is shifted to lower frequency ($\nu_S = \nu_L - \nu_{v-Raman}$). The rotational Raman also has Stokes and anti-Stokes shifts.

of a dipole followed by emission in this case), but the interaction is nonresonant. Raman scattering is caused by an interaction between the optical electromagnetic field and a dipole that is in motion (a dynamic polarizability due to rotation and vibration). The electrical analog to Raman scattering would be an RF mixer, wherein the carrier (the frequency of a laser beam) mixes with the ro/vibrational modes of the molecule (dipole), producing sum- and difference-frequency side-bands, as depicted in Figure 13.1. The input radiation is not resonant with a vibrational or ro-vibrational transition; any laser beam can be used, but green to UV lasers are favored because the signal scales with frequency taken to the fourth power. The laser radiation does induce a transition, however, either upwards (inducing a Stokes shift in the scattered light) or downwards (inducing an anti-Stokes shift in the scattered light); energy conservation is thus assured. Some treatments, in fact, describe a transition to a virtual state in the molecule that is induced by the incoming photon. The molecule then rapidly (within 1 ps) relaxes to the final state, emitting a photon. Considering the speed of this process, the author prefers explanations that utilize a time-dependent polarizability.

The magnitude of the Raman vibrational anti-Stokes component is less than the magnitude of the vibrational Stokes component, especially at room temperature, because the anti-Stokes component must originate from an excited level. The number of molecules in that level, which would be prepared to participate in the scattering process, will be smaller. Furthermore, as the vibrational Raman shift increases, the energy level spacing increases and the difference between Stokes and anti-Stokes amplitudes grows.

The vibrational Raman spectra depicted in Figure 13.1 have some unresolved structure. That is because each Raman-active vibrational transition has rotational structure. Rotational Raman shifts (in wavelength or frequency) are much smaller than vibrational shifts. Indeed, if the laser used is sufficiently narrowband, and if a good-quality blocking filter is used, one can observe purely rotational Raman at the base of the Rayleigh signal (as shown in Figure 13.1). In reality, Raman scattering is typically 1000 times weaker than Rayleigh scattering (and Rayleigh scattering is usually 1000 times weaker than the input irradiance). Because Raman is so weak, intense lasers are used to generate the signal. It is thus imperative to separate spectrally the Rayleigh and Raman signals. If a narrow-band laser is used, then the Rayleigh spectrum is narrow and easy to block with a laser line blocking filter before the collected signal enters the spectrometer used for spectral separation of the Raman peaks.

Rayleigh and Raman scattering are utilized somewhat differently to the resonance techniques discussed elsewhere in this book. They are both self-calibrating in room air, or in reacting flows that have reached chemical equilibrium. Because scattering is relatively isotropic in gases, it also becomes unnecessary to concern oneself with most of the orientational issues that are faced in solids and in some liquids. Depolarization ratios for Raman scattering can provide important information regarding molecular structure, but, if one simply needs a measurement of number density these details can be avoided. One then calibrates the instrument under well-known conditions and then corrects the measurement for changes in temperature and laser irradiance. For flowfield diagnostics, it is necessary to know how the signal scales, but not to worry overmuch about absolute values for scattering cross-sections (for example).

This chapter is based largely upon the work of Long [32], with some reliance upon McCartney [82], and Bhagavantam [83]. At the point in time when this book went to press, a new text by Long [84] had just become available. The new book presents a comprehensive and unified treatment of Raman scattering, based primarily upon time-dependent perturbation treatments of quantum mechanics. The new book should be used as a primary reference by researchers who plan to specialize in Raman scattering. In such a case, the semi-classical treatment presented here can be considered an introduction to the material in reference [84]. In this book, we review the older treatment provided by the earlier book (reference [32]) because we have not discussed time-dependent perturbation approaches in this text.

Rayleigh and Raman scattering are also discussed by Eckbreth [1] and by Demtröder [2]. Reviews on Rayleigh scattering and its application to flowfield measurements have been provided by Carter [85], Pitts and Kashiwagi [86], and Zhao and Hiroyasu [37]. Raman scattering is discussed in the relatively early workshop volume edited by Lapp and Penney [87] and the review article by Lederman [88]. The application of Raman scattering to multispecies detection is discussed by Barlow *et al.* in reference [3].

13.2 Polarizability

The Lorentz atom development in Chapter 5 serves as the starting point for the description of Rayleigh and Raman scattering. Equation (5.1) defines the polarizability of a molecule:

$$\vec{\mu} = \alpha \vec{E}(r)$$

where the units of α are (Cm^2/V) or (C^2m^2/J). The discussion in Chapter 5 neglected dipole orientational effects, but they cannot be ignored here. The molecular polarizability is actually a second-rank tensor, requiring that we rewrite equation (5.1) as

$$\vec{\mu} = \vec{\alpha} \cdot \vec{E}(r) \tag{13.1}$$

which can be expanded to

$$\begin{aligned}
\mu_x &= \alpha_{xx}E_x + \alpha_{xy}E_y + \alpha_{xz}E_z \\
\mu_y &= \alpha_{yx}E_x + \alpha_{yy}E_y + \alpha_{yz}E_z \\
\mu_z &= \alpha_{zx}E_x + \alpha_{zy}E_y + \alpha_{zz}E_z
\end{aligned} \qquad (13.2)$$

The subscripts on E indicate the direction of field polarization. This can also be represented in matrix form by

$$\begin{bmatrix} \mu_x \\ \mu_y \\ \mu_z \end{bmatrix} = \begin{bmatrix} \alpha_{xx} & \alpha_{xy} & \alpha_{xz} \\ \alpha_{yx} & \alpha_{yy} & \alpha_{yz} \\ \alpha_{zx} & \alpha_{zy} & \alpha_{zz} \end{bmatrix} \begin{bmatrix} E_x \\ E_y \\ E_z \end{bmatrix}$$

which is often written in shorthand as

$$\{\mu\} = \{\alpha\}\{E\} \qquad (13.3)$$

The absolute values for the α_{ij} depend upon the choice of coordinates, but there are polarizability invariants that will prove useful. Because polarizability is a tensor, a linearly polarized input beam can induce a three-dimensional response, as depicted in Figure 13.2.

The polarizability tensor that would describe Figure 13.2a. is

$$\begin{bmatrix} \alpha_{xx} & 0 & 0 \\ 0 & \alpha_{yy} & 0 \\ 0 & 0 & \alpha_{zz} \end{bmatrix}$$

For Figure 13.2b. it is

$$\begin{bmatrix} \alpha_{xx} & 0 & 0 \\ 0 & \alpha_{yy} & \alpha_{yz} \\ 0 & \alpha_{zy} & \alpha_{zz} \end{bmatrix}$$

and for Figure 13.2c. it is

$$\begin{bmatrix} \alpha_{xx} & \alpha_{xy} & \alpha_{xz} \\ \alpha_{yx} & \alpha_{yy} & \alpha_{yz} \\ \alpha_{zx} & \alpha_{zy} & \alpha_{zz} \end{bmatrix}$$

Another way to represent the polarizability is in the form of an ellipsoid, defined by

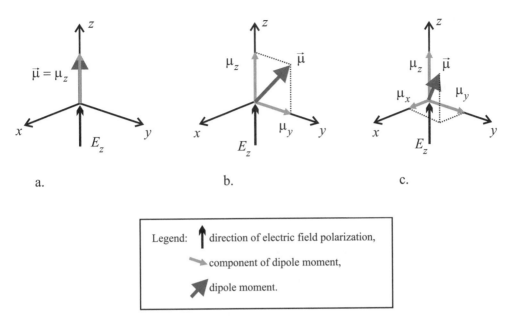

a.

b.

c.

Legend: direction of electric field polarization,

component of dipole moment,

dipole moment.

Figure 13.2: Induced dipole components and the resultant vector for a single, linearly polarized input beam, but with various polarizability tensors (see discussion in text).

$$\alpha_{xx}x^2 + \alpha_{yy}y^2 + \alpha_{zz}z^2 + 2\alpha_{xy}xy + 2\alpha_{yz}yz + 2\alpha_{zx}zx = 1 \qquad (13.4)$$

When represented graphically, this ellipsoid is centered at $(0,0,0)$ but its principal axes (x', y', z') are not necessarily aligned along the x, y, z axes of the lab frame (see Figure 13.3) because of the cross-terms $(2\alpha_{xy}xy + 2\alpha_{yz}yz + 2\alpha_{zx}zx)$.

The distance from the center to the surface of the ellipsoid in any direction represents $1/\sqrt{\sigma_R}$, where σ_R is the component of molecular polarizability along this direction. If an applied electric field is polarized along that same direction, the dipole moment will be given by $\mu = \alpha_R \mid E \mid$.

When the field polarization is aligned along the principal axes of the ellipsoid (x', y', z'), we can write

$$\mu_{x'} = \alpha_{x'x'}E_{x'}$$
$$\mu_{y'} = \alpha_{y'y'}E_{y'}$$

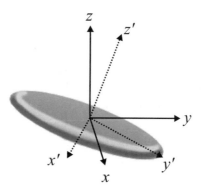

Figure 13.3: Schematic of the polarization ellipsoid for a randomly oriented molecule. The principal axes of the ellipsoid are aligned with coordinates x', y', z', not the lab frame x, y, z. The components of the ellipsoid along x, y, z can be found via direction cosines.

$$\mu_{z'} = \alpha_{z'z'} E_{z'} \tag{13.5}$$

and $\alpha_{x'y'} = \alpha_{x'z'} = \cdots = \alpha_{z'x'} = 0$. In the coordinate frame x', y', z', the ellipsoid has the equation

$$\alpha_{x'x'} x'^2 + \alpha_{y'y'} y'^2 + \alpha_{z'z'} z'^2 = 1 \tag{13.6}$$

with semi-minor axes of $1/\sqrt{\alpha_{x'x'}}, 1/\sqrt{\alpha_{y'y'}}, 1/\sqrt{\alpha_{z'z'}}$. The largest polarizability is associated with the shortest axis.

The polarizabilities along the principal axes provide a simplified formalism. The input laser beam will be polarized in the lab frame, however, and molecules in the gas-phase will rotate freely. It is not possible to align the field polarization with the principal axes of the ellipsoid. It is therefore necessary to find the lab frame (x, y, z) components of a randomly oriented ellipsoid. This can be done via direction cosines:

$$\alpha_{xy} = \sum_{x'y'} \alpha_{x'y'} \cos(xx') \cos(yy') \tag{13.7}$$

where (xx') is the angle between x and x', and the summation is taken over all possible pairs of Cartesian axes in the x', y', z' system.

Direction cosines and Euler angles

A general formalism for rotation of the polarization ellipsoid axes is described by Bhagavantam [83], and it is done using Euler angles. Zare [66] provides a concise description of Euler angles and their use in molecular systems. Both Marion and Thornton [56] and Goldstein [89] also provide generalized discussions on Euler angles and rotation matrices. As it turns out, there are several variants on the Euler angle formalism. For example, left-handed coordinate systems are sometimes used (as in Bhagavantam [83]). These differences will generate rotation matrices that look different, but if one remains consistent within the chosen system, spatial averages should always produce the same outcome. Here, Zare's right-handed presentation is emulated because that book describes molecular orientations. It seems that this book and Zare's book should at least be consistent with each other.

Euler angles are used simply to map coordinates from one frame (e.g. the polarizability ellipsoid principal axes, called "principal axes frame" here) into another frame (e.g. the lab frame, which will contain the laser polarization). This mapping involves three rotations of the coordinate axes, through the Euler angles θ, ϕ, and χ. The Euler angle mapping then provides expressions for the direction cosines used in equation (13.7).

The two Euler angles θ and ϕ are the standard spherical polar coordinates used in the past. θ measures the rotation from the vertical axis (in the lab frame), while ϕ (an azimuthal angle) measures rotation within the x/y plane. χ is a second azimuthal angle. It is defined as the angle of rotation away from the intersection fold between the x', y' plane and the x, y plane. This intersection fold is perpendicular to both z' and z.

To map the coordinates from the principal axes frame into the lab frame, we perform the following three operations (see Figure 13.4):

i. $R_Z(\phi)$ rotation - rotation through angle ϕ about z'. This also carries the y' axis to y_i, onto the intersection fold.

ii. $R_{IF}(\theta)$ rotation - rotation through angle θ about the intersection fold, which carries the z' into z_{ii} (which is the same as z).

iii. $R_Z(\chi)$ rotation - counterclockwise rotation through angle χ about z.

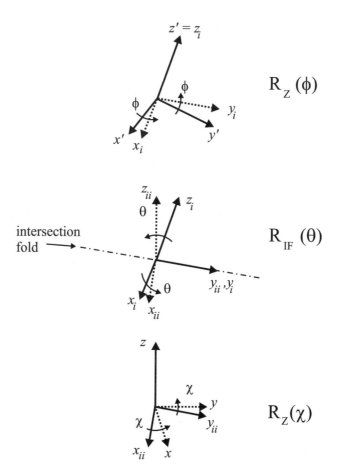

Figure 13.4: Schematic of Euler angle rotations from the principal axes frame into the lab frame.

The total rotation from x', y', z' to x, y, z can be described in terms of a rotation matrix:

$$\begin{bmatrix} x \\ y \\ z \end{bmatrix} = \vec{R} \begin{bmatrix} x' \\ y' \\ z' \end{bmatrix}$$

where $\vec{R}$ is the total rotation matrix given by $\vec{R} \equiv \vec{R}_Z(\chi)\vec{R}_{IF}(\theta)\vec{R}_Z(\phi)$. Each individual matrix is developed via direction cosines. As one example, a rotation by $R_Z(\phi)$ alone is written as

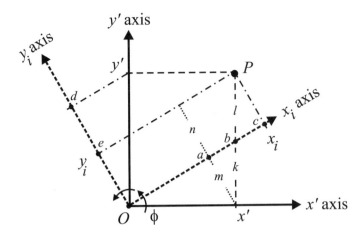

Figure 13.5: Schematic of $\vec{R}_Z(\phi)$ rotation.

$$\begin{bmatrix} x_i \\ y_i \\ z_i \end{bmatrix} = \vec{R}_Z(\phi) \begin{bmatrix} x' \\ y' \\ z' \end{bmatrix}$$

Because there are several variants on Euler angles and it can therefore be difficult to compare formalisms, we will demonstrate in detail how one of these rotation matrices can be found. The $\vec{R}_Z(\phi)$ rotation shown in Figure 13.4 is really a two-dimensional process since z does not change during this rotation. The x, y frame is shown in Figure 13.5, but it has been realigned for easier viewing. The z axis is located at O and it comes out of the page. In order to convert from the x'/y' frame to the x_i/y_i frame, we simply locate point P in both frames. To locate point x_i relative to x', one must find

$$x_i = \overline{Oa} + \overline{ab} + \overline{bc} \tag{13.8}$$

where O, a and b are points along the x_i axis in Figure 13.5. These distances are given by

$$\begin{aligned} \overline{Oa} &= x' \cos \phi \\ \overline{ab} &= k \sin \phi \\ \overline{bc} &= l \sin \phi \\ \overline{ab} + \overline{bc} &= (k+l) \sin \phi = y' \sin \phi \end{aligned} \tag{13.9}$$

Therefore

$$x_i = x' \cos \phi + y' \sin \phi \qquad (13.10)$$

Similarly

$$y_i = \overline{Od} - \overline{de} \qquad (13.11)$$

These distances are given by

$$
\begin{aligned}
\overline{Od} &= y' \cos \phi \, (= m + n) \\
\overline{de} &= m = x' \sin \phi
\end{aligned}
\qquad (13.12)
$$

Therefore

$$y_i = -x' \sin \phi + y' \cos \phi \qquad (13.13)$$

The direction cosine for z is simply multiplication by 1, since $z' = z_i$ in this first step. Equations (13.10) and (13.13) can then be written in matrix form as

$$
\begin{bmatrix} x_i \\ y_i \\ z_i \end{bmatrix} = \begin{bmatrix} \cos \phi & \sin \phi & 0 \\ -\sin \phi & \cos \phi & 0 \\ 0 & 0 & 1 \end{bmatrix} \begin{bmatrix} x' \\ y' \\ z' \end{bmatrix}
$$

or

$$
\vec{R}_Z(\phi) = \begin{bmatrix} \cos \phi & \sin \phi & 0 \\ -\sin \phi & \cos \phi & 0 \\ 0 & 0 & 1 \end{bmatrix}
$$

Using similar geometric arguments

$$
\vec{R}_{IF}(\theta) = \begin{bmatrix} \cos \theta & 0 & -\sin \theta \\ 0 & 1 & 0 \\ \sin \theta & 0 & \cos \theta \end{bmatrix}
$$

and

$$
\vec{R}_Z(\chi) = \begin{bmatrix} \cos \chi & \sin \chi & 0 \\ -\sin \chi & \cos \chi & 0 \\ 0 & 0 & 1 \end{bmatrix}
$$

It is necessary to ensure positive rotation in a right-handed system to achieve these results.

Table 13.1: Rotation matrix values

	x	y	z
x'	$c\phi\, c\theta\, c\chi - s\phi\, s\chi$	$s\phi\, c\theta\, c\chi + c\phi\, s\chi$	$-s\theta\, c\chi$
y'	$-c\phi\, c\theta\, s\chi - s\phi\, c\chi$	$-s\phi\, c\theta\, s\chi + c\phi\, c\chi$	$s\theta\, s\chi$
z'	$c\phi\, s\theta$	$s\phi\, s\theta$	$c\theta$

Both Zare and Bhagavantam provide a table of combined rotation matrices that generate expressions for the total direction cosine as required by equation (13.7). Table 13.1 contains a similar presentation using the definitions provided here (it agrees with the table presented by Zare). In order to fit Table 13.1 on one page width, "cos" has been abbreviated to "c" and "sin" has been abbreviated to "s".

To use Table 13.1, the definition of the direction cosines in equation 13.7 must be recalled, e.g. "(xx')" is the angle between x and x'". To find the full direction cosine for angle (xx'), therefore, one finds x along the horizontal and x' along the vertical. Then

$$\cos(xx') = [\cos\phi\cos\theta\cos\chi - \sin\phi\sin\chi] \qquad (13.14)$$

Space-fixed polarizability ellipsoid

Imagine we have a polarizability ellipsoid fixed in space, thus defining the principal axes frame x', y', z'. One simplifying aspect of the principal axes frame is that $\alpha_{x'y'} = \alpha_{x'z'} = \cdots = \alpha_{z'x'} = 0$; all of the off-diagonal elements of the polarizability matrix are zero in this frame.

To use equation (13.7) (repeated here for convenience)

$$\alpha_{xy} = \sum_{x'y'} \alpha_{x'y'} \cos(xx')\cos(yy')$$

one must find all of the individual terms in the summation and then add them. Such a task is uncomplicated but messy. Here we provide just one example; we will find α_{zz} in the lab frame based upon known α values in the principal axes frame. It is perhaps easiest to organize the direction cosines first, as shown below:

x', y', z' axis pairs (summation index)	1st cos term	2nd cos term	product of direction cosines from Table 13.1
$x'x'$	$\cos(zx')$	$\cos(zx')$	$[-\sin\theta\cos\chi]^2$
$y'y'$	$\cos(zy')$	$\cos(zy')$	$[\sin\theta\sin\chi]^2$
$z'z'$	$\cos(zz')$	$\cos(zz')$	$\cos^2\theta$
$x'y'$	$\cos(zx')$	$\cos(zy')$	$[-\sin\theta\cos\chi][\sin\theta\sin\chi]$
$x'z'$	$\cos(zx')$	$\cos(zz')$	$[-\sin\theta\cos\chi][\cos\theta]$
$y'x'$	...	...	...
$z'x'$	...	...	...
$y'z'$	...	...	...
$z'y'$	...	...	...

The portion of this table containing off-diagonal elements (below the space) was not completed because none of those terms will contribute to the summation. This occurs because the off-diagonal elements of the polarizability matrix in the principal axes frame (which multiply these direction cosines within the summation) are all zero. Equation (13.7) then produces

$$\alpha_{zz} = \alpha_{x'x'}\sin^2\theta\cos^2\chi + \alpha_{y'y'}\sin^2\theta\sin^2\chi + \alpha_{z'z'}\cos^2\theta \qquad (13.15)$$

Similar calculations are used for the other components of the polarizability in the lab frame. Note that the off-diagonal terms are zero only in the principal axes frame, so it remains necessary to solve for the off-diagonal terms when going to the lab frame.

Space averages

The result just provided is for one molecule at one specific position (one set of Euler angles). Because gas-phase molecules are randomly oriented, laboratory measurements on an ensemble of scatterers will produce a space-averaged result. This problem can be approached by taking a space average of the square of each single-scatterer polarizability [like equation (13.15)]. We take the square because scattered irradiance will scale with μ^2 $(= \alpha^2 E^2)$. Then the contribution of the ensemble will be the number of scatterers times this result for one, space-averaged response.

To continue the example, the space average for α_{zz}^2 is given by

$$\overline{\alpha_{zz}^2} \equiv \frac{\int_0^\pi \int_0^{2\pi} \int_0^{2\pi} \alpha_{zz}^2 \sin\theta \, d\theta \, d\chi \, d\phi}{\int_0^\pi \int_0^{2\pi} \int_0^{2\pi} \sin\theta \, d\theta \, d\chi \, d\phi}$$

$$= \frac{1}{15} \Big[3\alpha_{x'x'}^2 + 2\alpha_{x'x'}\alpha_{y'y'} + 3\alpha_{y'y'}^2$$

$$+ 2\alpha_{x'x'}\alpha_{z'z'} + 2\alpha_{y'y'}\alpha_{z'z'} + 3\alpha_{z'z'}^2 \Big] \qquad (13.16)$$

If we make the definitions:

$$a \equiv (1/3)\Big(\alpha_{x'x'} + \alpha_{y'y'} + \alpha_{z'z'} \Big) \qquad (13.17)$$

$$\gamma^2 \equiv (1/2)\Big[\Big(\alpha_{x'x'} - \alpha_{y'y'} \Big)^2 + \Big(\alpha_{y'y'} - \alpha_{z'z'} \Big)^2$$

$$+ \Big(\alpha_{z'z'} - \alpha_{x'x'} \Big)^2 \Big] \qquad (13.18)$$

equation (13.16) can be more simply written as

$$\overline{\alpha_{zz}^2} = \frac{45a^2 + 4\gamma^2}{45} \qquad (13.19)$$

This result is more general than the treatment might imply. There are invariant values for the polarizability ellipsoid, no matter what the orientation or Cartesian frame. Using equation (13.7) and properties of the direction cosines, Long [32] states that, *in general*, the invariants are

$$a \equiv \frac{1}{3} (\alpha_{xx} + \alpha_{yy} + \alpha_{zz}) \qquad (13.20)$$

$$\gamma^2 \equiv \frac{1}{2} \Big[(\alpha_{xx} - \alpha_{yy})^2 + (\alpha_{yy} - \alpha_{zz})^2 + (\alpha_{zz} - \alpha_{xx})^2$$

$$+ 6 \Big(\alpha_{xy}^2 + \alpha_{yz}^2 + \alpha_{zx}^2 \Big) \Big] \qquad (13.21)$$

where a is the mean polarizability and γ is the anisotropy of the ellipsoid. Equations (13.20) and (13.21) are written for a coordinate system that is not necessarily aligned with the principal axes of the polarizability ellipsoid. When the coordinates are aligned with the principal axes, equations (13.17) and (13.18) apply. Note that when $\alpha_{x'x'} = \alpha_{y'y'} = \alpha_{z'z'}$, then $\gamma = 0$ and the ellipsoid is isotropic.

Long then provides expressions for the space averages of the polarizabilities. He does this using the same space averaging we have used, and the outcome is written in terms of the invariants:

$$\overline{\alpha_{xx}^2} = \overline{\alpha_{yy}^2} = \overline{\alpha_{zz}^2} = \frac{45a^2 + 4\gamma^2}{45} \qquad (13.22)$$

$$\overline{\alpha_{xx}\alpha_{yy}} = \overline{\alpha_{yy}\alpha_{zz}} = \overline{\alpha_{zz}\alpha_{xx}} = \frac{45a^2 - 2\gamma^2}{45} \qquad (13.23)$$

and

$$\overline{\alpha_{yx}^2} = \overline{\alpha_{yz}^2} = \overline{\alpha_{zx}^2} = \frac{\gamma^2}{15} \qquad (13.24)$$

We have not developed each of these relationships here because the presentation takes up a great deal of space. One can see, however, that they are consistent with the developments that have been provided. The early developers of these theories were required to perform large amounts of algebra, and it was necessary for them to keep many variables from becoming disorganized. As a modern alternative, one can use a mathematics software package to verify the relationships just provided. The author has chosen the software package path and can vouch for the relationships presented by Long.

13.3   Classical Molecular Scattering

The theory developed in this section introduces formalisms for both Rayleigh and Raman scattering. It will fail, however, to describe the amplitudes of Raman spectra accurately. At a minimum, this requires the semi-classical (electromagnetism together with quantum mechanics) approach outlined in Section 13.5.

We begin with a vibrating molecule that is not rotating, and we assume that the molecular polarizability is a function of the nuclear coordinates. The electric field actually couples to the electrons, but the electrons will rearrange themselves in response to nuclear motion.

The coordinate for a normal mode of vibration is written as Q. The polarizability tensor is then written as a Taylor expansion with respect to Q, about the equilibrium nuclear spacing and neglecting powers of Q higher than 1:

$$\alpha_{ij} = (\alpha_{ij})_\circ + \sum_k \left(\frac{\partial \alpha_{ij}}{\partial Q_k} \right)_\circ Q_k \qquad (13.25)$$

where the subscript $\circ$ denotes the equilibrium configuration, and the subscript k denotes a kth normal mode at frequency ω_k. The term $(\alpha_{ij})_\circ$ is called the "equilibrium polarizability tensor", while the partial derivative in equation (13.25) is defined as the "derived polarizability tensor" α_k':

$$\alpha_k' \equiv \left(\frac{\partial \alpha_{ij}}{\partial Q_k} \right)_\circ \qquad (13.26)$$

For one normal mode, therefore:

$$\vec{\alpha}_k = \vec{\alpha}_\circ + \vec{\alpha}_k' \, Q_k \qquad (13.27)$$

We assume that the electric field is harmonic: $\vec{E} = \vec{E}_\circ \cos \omega_\circ t$ (where $\omega_\circ$ is the frequency of the input light). Moreover, because the molecule vibrates, we assume that Q follows simple harmonic motion:

$$Q_k = Q_{k,\circ} \cos(\omega_k t + \delta_k) \qquad (13.28)$$

where δ_k is a phase shift. The polarizability is therefore time dependent, which generates time dependence in the dipole moment because

$$\vec{\mu} = \vec{\alpha}_k \cdot \vec{E} \qquad (13.29)$$

The dipole moment would then be given by

$$\vec{\mu} = \vec{\alpha}_\circ \cdot \vec{E}_\circ \cos \omega_\circ t + Q_{k,\circ} \vec{\alpha}_k' \cdot \vec{E}_\circ [\cos \omega_\circ t][\cos(\omega_k t + \delta_k)] \qquad (13.30)$$

The product of the two cosine terms generates sum and difference frequencies via trigonometric identity:

$$\vec{\mu} = \vec{\mu}(\omega_\circ) + \vec{\mu}(\omega_\circ - \omega_k) + \vec{\mu}(\omega_\circ + \omega_k) \qquad (13.31)$$

The terms on the right-hand side are defined by

$$\vec{\mu}(\omega_o) \equiv \vec{\mu}_o(\omega_o)\cos(\omega_o t) \tag{13.32}$$

$$\vec{\mu}(\omega_o - \omega_k) \equiv \vec{\mu}_o(\omega_o - \omega_k)\cos(\omega_o - \omega_k - \delta_k) \tag{13.33}$$

$$\vec{\mu}(\omega_o + \omega_k) \equiv \vec{\mu}_o(\omega_o + \omega_k)\cos(\omega_o + \omega_k + \delta_k) \tag{13.34}$$

and the amplitude terms are given by

$$\vec{\mu}_o(\omega_o) \equiv \vec{\alpha}_o \cdot \vec{E}_o \tag{13.35}$$

$$\vec{\mu}_o(\omega_o \pm \omega_k) \equiv \frac{1}{2}Q_{k,o}\vec{\alpha}'_k \cdot \vec{E}_o \tag{13.36}$$

There are thus three distinct components of the dipole moment. The first, $\vec{\mu}(\omega_o)$ at the optical frequency ω_o, describes Rayleigh scattering, while $\vec{\mu}(\omega_o - \omega_k)$ describes Stokes Raman scattering and $\vec{\mu}(\omega_o + \omega_k)$ describes anti-Stokes Raman scattering.

The polarizabilities $\vec{\alpha}_o$ and $\vec{\alpha}'_k$ retain their tensor characteristics. The principal axes of $\vec{\alpha}'_k$, however, do not necessarily coincide with the axes of $\vec{\alpha}_o$. It is possible, however, to develop and use invariants for $\vec{\alpha}'_k$, in addition to the invariants already developed for $\vec{\alpha}_o$. The tensor components of $\vec{\alpha}'_k$ can be negative or positive, while the components of $\vec{\alpha}_o$ will always be positive. This also means that a'_k (the mean value for α'_k) can be zero, whereas a_o cannot. The outcome is that Rayleigh and Raman scattering can have different directional properties and polarization states (or signal depolarization).

Equations (13.32) and (13.35) accurately describe the dipole moment used to predict Rayleigh scattering. Equations (13.33), (13.34) and (13.36) do a reasonable job of predicting Raman frequencies, vibrational selection rules, and so forth. They fail, however, to describe the vibrational scattering amplitudes correctly and they cannot accurately describe rotational amplitudes and selection rules. We will develop the expressions for Rayleigh scattering first, because this can be accomplished with the ideas in hand and the path taken can then be repeated quickly for Raman scattering. Following the presentation of Rayleigh theory, we describe the application of quantum mechanics to the areas of Raman theory where classical theory fails, and then the Raman formalism necessary for gas-phase diagnostics is described.

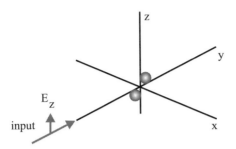

Figure 13.6: Space-fixed scatterer at the origin of Cartesian axes, illuminated by an electric field traveling along y with polarization in the z direction.

13.4 Rayleigh Scattering

Space-fixed molecule

We begin with a description for the scattering power from a space-fixed molecule, following which we deal with an ensemble of molecules with random orientation. Imagine the molecule is at the center of Cartesian axes as shown in Figure 13.6. Imagine as well that the scattering signal is observed along the x axis. The plane containing the optical path (y in this case) and the observation path (x axis here) is called the "scattering plane".

In order to find the power scattered into solid angle $d\Omega$ along x, it will be necessary to find the y and z components of $\vec{\mu}$, at the correct optical frequency. We do not concern ourselves with the component of $\vec{\mu}$ along x because there is no appreciable scattering signal from a component along x (see, for example, Figure 5.4). It will thus be necessary to find $\mu_{y,o}(\omega_o)$ and $\mu_{z,o}(\omega_o)$ to describe Rayleigh scattering. In Figure 13.6 the input wave is polarized perpendicular to the scattering plane ($E_x = E_y = 0$, but $E_y = 0$ in any case because a plane electromagnetic wave is transverse). For Rayleigh scattering

$$
\begin{aligned}
\mu_{y,o}(\omega_o) &= \alpha_{yz,o} E_{z,o} \\
\mu_{z,o}(\omega_o) &= \alpha_{zz,o} E_{z,o}
\end{aligned}
\qquad (13.37)
$$

These values for the dipole moment can then be used in equation (5.88) to describe the radiation emitted by such dipoles:

$$I(\theta) = \frac{\omega^4 |\vec{\mu}\,|^2}{32\pi^2 \epsilon_o c^3} \frac{\sin^2 \theta \hat{r}}{r^2}$$

In short, we have used the molecular polarizability to describe the formation of a dipole by an incoming electromagnetic wave, and the dipole then emits as described in Section 5.3 of Chapter 5. The irradiance in equation (5.88) is the power per unit area emitted, which we will designate dP/dA. Equation 5.88 can then be rearranged to give

$$dP = \frac{\omega^4 |\vec{\mu}\,|^2}{32\pi^2 \epsilon_o c^3} \left(\frac{dA}{r^2}\right) \sin^2 \theta \hat{r} = \frac{\omega^4 |\vec{\mu}\,|^2}{32\pi^2 \epsilon_o c^3} \, d\Omega \sin^2 \theta \hat{r} \qquad (13.38)$$

Then

$$\frac{dP}{d\Omega} = \frac{\omega^4 |\vec{\mu}\,|^2}{32\pi^2 \epsilon_o c^3} \sin^2 \theta \hat{r} \qquad (13.39)$$

This is not the same as the radiance J introduced in Chapter 3, because J is power per unit steradian per unit area, whereas this term is power per unit steradian. In the case under discussion, we observe $dP/d\Omega$ along x (at $\theta = \pi/2$), making $\sin^2 \theta = 1$. This is a very common (one might say it is the standard) geometry for molecular scattering measurements. Scattering patterns can be fairly complex, but at $\theta = \pi/2$ the pattern is simpler and for Rayleigh scattering the signal is also cleanly polarized. For this reason we will emphasize that geometry throughout this chapter. We can now rewrite $dP/d\Omega$ with the understanding that the magnitudes are of interest, and we recognize that the excitation frequency is ω_o:

$$\frac{dP}{d\Omega} = \frac{\omega_o^4 \mu^2}{32\pi^2 \epsilon_o c^3} \qquad (13.40)$$

This expression can be rewritten in terms of an input electromagnetic wave and an as yet unspecified polarizability:

$$\frac{dP}{d\Omega} = k'_\omega \omega_o^4 \alpha^2 E^2 \qquad (13.41)$$

where

$$k'_\omega \equiv \frac{1}{32\pi^2\epsilon_\circ c^3} \tag{13.42}$$

Written in terms of the optical wave numbers $\check{\nu}(\text{cm}^{-1})$, the same expression is

$$\frac{dP}{d\Omega} = k'_{\check{\nu}} \, \check{\nu}_\circ^4 \, \alpha^2 E^2 \tag{13.43}$$

where

$$k'_{\check{\nu}} \equiv \frac{\pi^2 c}{2\epsilon_\circ} \tag{13.44}$$

Finally, it would be better to write this expression in terms of the input irradiance, $I = (\epsilon_\circ c/2)E^2$. This produces

$$\frac{dP}{d\Omega} = k_{\check{\nu}} \, \check{\nu}_\circ^4 \, \alpha^2 I \tag{13.45}$$

where

$$k_{\check{\nu}} \equiv \frac{\pi^2}{\epsilon_\circ^2} \tag{13.46}$$

Now we invoke one of the standard scattering nomenclatures. The variable $^\perp dP/d\Omega_\perp$ indicates the measured power per unit steradian for input light that is polarized perpendicular to the scattering plane (polarized along z in Figure 13.6, indicated by the preceding superscript) and the scattered light is polarized perpendicular to the scattering plane (indicated by the succeeding subscript). One should be aware that some authors define "perpendicular polarization" as that which is normal to the polarization of the input beam. Once again, it is necessary to be careful with published results. In the particular example under discussion, we can write

$$^\perp\frac{dP}{d\Omega}_\perp = k_{\check{\nu}} \, \check{\nu}_\circ^4 \, \alpha_{zz,\circ}^2 I \tag{13.47}$$

$$^\perp\frac{dP}{d\Omega}_\| = k_{\check{\nu}} \, \check{\nu}_\circ^4 \, \alpha_{yz,\circ}^2 I \tag{13.48}$$

Both of these are written with the understanding that I is polarized perpendicular to the scattering plane (identified by the preceding superscript on $dP/d\Omega$). The total scattering power per sr in this example is then

$$\perp \frac{dP}{d\Omega} = k_{\check{\nu}}\; \check{\nu}_o^4 \left(\alpha_{zz,o}^2 + \alpha_{yz,o}^2 \right) I \tag{13.49}$$

If the input were polarized along x (parallel to the scattering plane), the total scattering power per sr would be

$$\| \frac{dP}{d\Omega} = k_{\check{\nu}}\; \check{\nu}_o^4 \left(\alpha_{zx,o}^2 + \alpha_{yx,o}^2 \right) I \tag{13.50}$$

Truly unpolarized light (randomly polarized), often called "natural polarization", would induce the following response:

$$^n \frac{dP}{d\Omega}_\perp = k_{\check{\nu}}\; \check{\nu}_o^4\; [\alpha_{zz,o}^2 + \alpha_{zx,o}^2] \frac{I}{2} \tag{13.51}$$

$$^n \frac{dP}{d\Omega}_\| = k_{\check{\nu}}\; \check{\nu}_o^4\; [\alpha_{yz,o}^2 + \alpha_{yx,o}^2] \frac{I}{2} \tag{13.52}$$

and the total signal is the sum of these two.

Space averages

Rayleigh scattering has no phase shift, which requires that the individual scattered fields add coherently. In a gas, however, random motion averages out any coherence effects and the Rayleigh signals add incoherently.

Gas phase molecules are free to reorient themselves. In order to model Rayleigh scattering from such a system, therefore, one calculates the induced dipole moment for each orientation and then averages over all possible orientations. That provides a space-averaged dipole moment that can then be used to find $dP/d\Omega$. The total scattering power is then found by multiplying this $dP/d\Omega$ by the total number of scatterers.

Samson [90] has performed the necessary space averaging, using the same formalism that was described and used in Section 13.2. He provides an effective polarizability term (polarizability squared because $dP/d\Omega \propto \alpha^2$) given by

$$\alpha_{\text{eff}}^2 \equiv a^2 \left(\frac{2 + \rho_n}{6 - 7\rho_n} \right) \tag{13.53}$$

where a is the average polarizability defined in equation (13.20). The term ρ_n is the "normal depolarization factor" defined by

$$\rho_n \equiv \frac{6\gamma^2}{45a^2 + 7\gamma^2} \tag{13.54}$$

where γ is the anisotropy of the ellipsoid defined in equation (13.21). Recall that γ is an invariant, and in the principal axes frame (x', y', z') it represents the differences between the polarizabilities along the three principal axes (the diagonal elements of the diagonalized polarizability matrix). It will always represent this difference between the diagonal elements of the diagonalized matrix, since it is an invariant. If the ellipsoid is reasonably isotropic, then $\gamma \cong 0$. The depolarization term in parentheses in equation (13.53) is very nearly $1/3$ for major species in the atmosphere and flames, because γ is very small for those compounds. McCartney [82] provides values for ρ_n for various atmospheric species. It is largest for carbon dioxide, reaching a value of $\rho_n \sim 0.08$. In all cases, these adjustments are very small and are often not even mentioned [which is why equation (13.53) was not developed in detail here].

The particle scattering formalism that is sometimes used to describe Rayleigh scattering also uses a depolarization term. It looks different because the depolarization terms are defined differently. They generate the same numerical adjustments to $dP/d\Omega$ (see, for example, Pitts and Kashiwagi [86]).

We can then describe Rayleigh scattering with a simple formalism:

$$\frac{dP}{d\Omega} = k_{\breve{\nu}} \, \breve{\nu}_o^4 \, \alpha_{\text{eff}}^2 \sin^2 \theta \, I \tag{13.55}$$

where we have replaced the dependence upon θ (the angle with respect to z). This expression is written for the input shown in Figure 13.6, and the scattering pattern is then the same as Figure 5.4 (with the dipole axis aligned along z). In such a case, the scattering signal is not dependent upon the azimuthal angle ϕ.

The Rayleigh formalism is not usually written in this way. It is typically presented in terms of the index of refraction. The conversion can be accomplished using expressions from Chapters 4 and 5. We begin with equation (5.12) for the dielectric constant:

$$\kappa_e = 1 + \frac{N\tilde{\alpha}}{\epsilon_\circ}$$

Rearranging gives

$$\kappa_e - 1 = \frac{N\tilde{\alpha}}{\epsilon_\circ} \tag{13.56}$$

In addition, equation (4.68) provides

$$\tilde{n} = n^R + in^I = \sqrt{\kappa_e}$$

Rayleigh scattering occurs far from resonance. We can therefore eliminate the imaginary parts of $\tilde{\alpha}$ and $\tilde{n}$. Setting $\kappa_e = (n^R)^2 = n^2$, and replacing α with α_{eff}, we obtain

$$\alpha_{\text{eff}} = \frac{(n^2 - 1)\epsilon_\circ}{N} \tag{13.57}$$

This can then be used in equation (13.55):

$$\frac{dP}{d\Omega} = \frac{\pi^2 \breve{\nu}_\circ^4}{N^2}(n^2 - 1)^2 \sin^2\theta \; I \tag{13.58}$$

where we have applied the definition of $k_{\breve{\nu}}$. The term $(n^2 - 1)^2$ can be approximated by $(n^2 - 1)^2 \cong 4(n - 1)^2$:

$$\boxed{\frac{dP}{d\Omega} = \frac{4\pi^2 \breve{\nu}_\circ^4 (n - 1)^2}{N^2} \sin^2\theta \; I} \tag{13.59}$$

A scattering cross-section can be defined by

$$\sigma \equiv \frac{dP/d\Omega}{I} \tag{13.60}$$

which generates an expression for the Rayleigh scattering cross-section:

$$\boxed{\sigma_{\text{Rayleigh}} = \frac{4\pi^2 \breve{\nu}_\circ^4 (n - 1)^2}{N^2} \sin^2\theta} \tag{13.61}$$

which, at $\theta = \pi/2$, is

$$\sigma_{\text{Rayleigh}}\left(\pi/2\right) = \frac{4\pi^2 \breve{\nu}_\circ^4 (n-1)^2}{N^2} \tag{13.62}$$

There are many other ways to define this scattering variable. Volume and mass scattering cross-sections are often used. Alternatively, the cross-section is sometimes integrated over all angles to generate a total scattering cross-section (e.g. for light extinction). These formalisms are dependent upon exactly what the experimentalist intends to measure, and how they intend to implement the experiment. Specific references should be consulted for the approach to specific problems.

Rayleigh line shapes

Despite the fact that Rayleigh scattering does not utilize a molecular resonance, there is homogeneous broadening to the Rayleigh line. It is caused by a collisional phenomenon called Brillouin scattering. At high pressures (not far above atmospheric) the line shape includes additional humps in the wings that grow progressively more prominent as the pressure rises, while the central peak decays. At sufficiently high pressure the central peak vanishes and only the two side lobes remain. These line shapes have been described by Tenti and co workers [91, 92] and by Young [93]. Rayleigh scattering is also Doppler broadened, as described in Chapter 11.

13.5   Raman Scattering

In this section we present what is primarily a classical electromagnetic treatment of Raman scattering (following reference [32]), with several required quantum mechanical adjustments. The quantum approach is required for development of accurate scattering cross-sections and for development of selection rules. In practical gas-phase measurements we simply use measured differential cross-sections, or calibrate the measurement. It remains necessary, however, to understand how to adjust these measurements (for change in temperature, for example), and to have a clear understanding of selection rules. The following treatment is therefore abbreviated. The goal is to develop familiarity with amplitude coefficients and selection rules, which will allow researchers to adjust published cross-sections or calibration coefficients appropriately.

Up to a point, we can adopt the same approach to Raman scattering as was taken for Rayleigh scattering. We begin, therefore, by adopting equations (13.33), (13.34), and (13.36). These define the dipole moment for Raman scattering, which can be combined with equation (5.88) as before. This approach will directly generate the following expression for a space-fixed molecule:

$$\perp \frac{dP}{d\Omega_\perp}\left(\theta = \frac{\pi}{2}\right) = k_{\check{\nu}}\,(\check{\nu}_\circ \pm \check{\nu}_k)^4\,(\alpha'_{zz,\circ})_k^2\,Q_{k\circ}^2\,I \qquad (13.63)$$

$$\perp \frac{dP}{d\Omega_\|}\left(\theta = \frac{\pi}{2}\right) = k_{\check{\nu}}\,(\check{\nu}_\circ \pm \check{\nu}_k)^4\,(\alpha'_{yz,\circ})_k^2\,Q_{k\circ}^2\,I \qquad (13.64)$$

and so forth. For freely rotating molecules, we can develop space averages that are quite similar to equations (13.22) and (13.24). For randomly oriented molecules, this will produce

$$\perp \frac{dP}{d\Omega_\perp}\left(\frac{\pi}{2}\right) = k_{\check{\nu}}\,(\check{\nu}_\circ \pm \check{\nu}_k)^4\,(\overline{\alpha'_{zz,\circ}})_k^2\,Q_{k\circ}^2\,I$$

$$= k_{\check{\nu}}\,(\check{\nu}_\circ \pm \check{\nu}_k)^4\,\left(\frac{45(a')_k^2 + 4(\gamma')_k^2}{45}\right)Q_{k\circ}^2\,I \qquad (13.65)$$

$$\perp \frac{dP}{d\Omega_\|}\left(\frac{\pi}{2}\right) = k_{\check{\nu}}\,(\check{\nu}_\circ \pm \check{\nu}_k)^4\,(\overline{\alpha'_{yz,\circ}})_k^2\,Q_{k\circ}^2\,I$$

$$= k_{\check{\nu}}\,(\check{\nu}_\circ \pm \check{\nu}_k)^4\,\left(\frac{(\gamma')_k^2}{15}\right)Q_{k\circ}^2\,I \qquad (13.66)$$

One can then describe differential Raman cross-sections $d\sigma/d\Omega$ by dividing $dP/d\Omega$ by NI, where N is the total number of scatterers.

There are several important differences between Rayleigh and Raman scattering, despite the similarities. In addition to the obvious wavelength dependence, Rayleigh scattering amplitudes are in phase with the input light while Raman amplitudes are not. The Rayleigh case has already been discussed; for gas-phase diagnostics we do not need to be concerned about it. Raman scattering has a random phase shift; it will always add incoherently.

Unfortunately, an accurate description of Raman scattering requires additional complexity. The amplitude of the scattering cross-section cannot be correctly described using classical mechanics. One can replace Q_{ko} with a more realistic term derived using quantum mechanics. This semi-classical approach is correct unless the laser wavelength is very near resonance, at which point a quantum-mechanical, time-dependent perturbation solution is required. It will not be required for our purposes.

Quantum polarizability

One can find a more accurate amplitude term simply by finding a quantum-mechanically correct dipole moment. We can define a Raman transition moment much like equation (10.9):

$$\vec{\mu} = \langle \psi_f \mid \vec{\alpha} \mid \psi_i \rangle \cdot \vec{E}_o \qquad (13.67)$$

where the subscript f denotes final and the subscript i denotes initial (if $f = i$ the description is for Rayleigh scattering, and if $f \neq i$ the description is for Raman scattering). This expression assumes that $\vec{E}_o$ is constant over the molecule (which assumes a small molecule). Because $\vec{E}_o$ is a vector and $\vec{\alpha}$ is a tensor, by Placzek polarizability theory [1] we can write

$$\begin{aligned}
\vec{\mu}_{xo} &= \{[\alpha_{xx}]E_{xo} + [\alpha_{xy}]E_{yo} + [\alpha_{xz}]E_{zo}\} \\
\vec{\mu}_{yo} &= \{[\alpha_{yx}]E_{xo} + [\alpha_{yy}]E_{yo} + [\alpha_{yz}]E_{zo}\} \\
\vec{\mu}_{zo} &= \{[\alpha_{zx}]E_{xo} + [\alpha_{zy}]E_{yo} + [\alpha_{zz}]E_{zo}\}
\end{aligned} \qquad (13.68)$$

where

$$\begin{aligned}
[\alpha_{xx}] &\equiv \langle \psi_f \mid \alpha_{xx} \mid \psi_i \rangle \\
[\alpha_{xy}] &\equiv \langle \psi_f \mid \alpha_{xy} \mid \psi_i \rangle \\
&\text{and so forth}
\end{aligned} \qquad (13.69)$$

We can now separate vibration and rotation. For example, we can write

$$[\alpha_{xy}] \equiv \langle \psi_{vf}\psi_{rf} \mid \alpha_{xy} \mid \psi_{vi}\psi_{ri} \rangle \qquad (13.70)$$

This separation is actually valid only if $\nu_o \gg \nu_k$ (where ν_k represents either rotation or vibration), if $\nu_o \ll \nu_{\text{electric}}$ (the electric energy divided by h), and if the ground electronic state is not degenerate. In some cases the expression can be readjusted for a degenerate ground state [32]. Here we assume the expression can be used for gas-phase diagnostics.

The polarizability ellipsoid principal axes coordinate frame (x', y', z') rotates with the molecule. Thus, the components of $\alpha_{x',y'}$ (for example) are functions of vibration only. Rotation is relative to the lab frame, not the principal axes frame. We therefore adapt equation (13.7) to produce

$$[\alpha_{xy}] = \sum_{x'y'} \langle \psi_{vf} \mid \alpha_{xy} \mid \psi_{vi} \rangle \langle \psi_{rf} \mid \cos(xx') \cos(yy') \mid \psi_{ri} \rangle \qquad (13.71)$$

The first inner product is defined as a vibrational polarizability:

$$[\alpha_{x'y'}]_{vf,vi} = \langle \psi_{vf} \mid \alpha_{xy} \mid \psi_{vi} \rangle \qquad (13.72)$$

To find the average vibrational term for rotating molecules, we can use the invariants as before [e.g. equation (13.22)].

Vibrational amplitude coefficients

To find the amplitude coefficients, we expand $\alpha_{x'y'}$ as before:

$$\alpha_{x'y'} = (\alpha_{x'y'})_\circ + \sum_k \left(\frac{\partial \alpha_{x'y'}}{\partial Q_k} \right)_\circ Q_k \qquad (13.73)$$

and this can be used in equation (13.72):

$$\begin{aligned}
[\alpha_{x'y'}]_{vf,vi} &= (\alpha_{x'y'})_\circ \langle \psi_{vf} \mid \psi_{vi} \rangle \\
&+ \sum_k \left(\frac{\partial \alpha_{x'y'}}{\partial Q_k} \right)_\circ \langle \psi_{vf} \mid Q_k \mid \psi_{vi} \rangle \qquad (13.74)
\end{aligned}$$

Because the wave functions are orthonormal, the first inner product on the right-hand side is

$$\langle \psi_{vf} \mid \psi_{vi} \rangle = \begin{cases} 0 & \text{for } v_{kf} \neq v_{ki} \\ 1 & \text{for } v_{kf} = v_{ki} \end{cases} \qquad (13.75)$$

where v_{kf} is the vibrational quantum number of the final state for normal mode k, and v_{ki} is the vibrational quantum number of the initial state for normal mode k. When $v_{kf} = v_{ki}$, there is no associated wavelength shift and this term describes Rayleigh scattering. If we were to continue to pursue this line of development, we would arrive at the same result for Rayleigh scattering as was developed with a classical argument just above.

The second inner product on the right-hand side of equation (13.74) is evaluated by Long, assuming a harmonic oscillator:

$$
\langle \psi_{vf} \mid Q_k \mid \psi_{vi} \rangle = \begin{cases} 0 \ \text{ for } v_{kf} = v_{ki} \\ \sqrt{v_{ki} + 1} \ b_{vk} \ \text{ for } v_{kf} = v_{ki} + 1 \\ \sqrt{v_{ki}} \ b_{vk} \ \text{ for } v_{kf} = v_{ki} - 1 \end{cases} \tag{13.76}
$$

where

$$
b_{vk} \equiv \frac{h}{8\pi^2 \nu_k} = \frac{h}{8\pi^2 c \breve{\nu}_k} \tag{13.77}
$$

The term b_{vk} is a quantum-mechanical replacement for Q_k; it is an amplitude coefficient. In this case, however, it is preceded by a term that is dependent upon v_{ki}. The Raman amplitudes thus depend upon the initial vibrational state.

Vibrational selection rules

We begin by focusing on the entire amplitude term for Raman scattering, taken from equation (13.74):

$$
\sum_k \left(\frac{\partial \alpha_{x'y'}}{\partial Q_k} \right)_\circ \langle \psi_{vf} \mid Q_k \mid \psi_{vi} \rangle \tag{13.78}
$$

In reality, the total vibrational wave function is the product of all the wave functions for each vibrational mode:

$$
\psi_{vi} = \prod_k \psi_{vik}(Q_k)
$$
$$
\psi_{vf} = \prod_k \psi_{vfk}(Q_k) \tag{13.79}
$$

The individual terms in summation 13.78 will be zero unless every term in the product $\prod_k$ is nonzero. For this to be true, each term in j (except when $j = k$) must have $v_{jf} = v_{ji}$ [but then the inner product is zero, see equation (13.76)]. When $j = k$, $v_{kf} = v_{ki} \pm 1$.

If $v_{kf} = v_{ki} + 1$, the kth term in (13.78) is then

$$\sqrt{v_{ki} + 1}\, b_{vk} \left(\frac{\partial \alpha_{x'y'}}{\partial Q_k} \right)_{\circ} \qquad (13.80)$$

or

$$
\begin{aligned}
[\alpha_{x'y'}]_{(vki+1,vki)} &= \sqrt{v_{ki} + 1}\, b_{vk} \left(\frac{\partial \alpha_{x'y'}}{\partial Q_k} \right)_{\circ} \\
&= \sqrt{v_{ki} + 1}\, b_{vk} (\alpha'_{x'y'})_k \qquad (13.81)
\end{aligned}
$$

If $v_{kf} = v_{ki} - 1$, the kth term in (13.78) is

$$[\alpha_{x'y'}]_{(vki-1,vki)} = \sqrt{v_{ki}}\, b_{vk} (\alpha'_{x'y'})_k \qquad (13.82)$$

In the harmonic oscillator approximation, only $\Delta v = \pm 1$ is allowed. In the anharmonic case, both overtones and combinations are allowed. In addition, hot bands will be shifted spectrally by a small amount. This actually offers the possibility of temperature measurement in a flame [94].

Note that $\Delta v = \pm 1$ is a necessary but not sufficient requirement for Raman activity. There must be a nonzero $[\alpha_{x'y'}]_{(vkf,vki)}$ for at least one normal mode. Long describes an algorithm based upon the symmetry and point-group classification of the molecule that can determine whether or not $[\alpha_{x'y'}]_{(vkf,vki)}$ will be nonzero for each normal mode.

Vibrational Raman scattering from a space-fixed molecule

If a space-fixed molecule is oriented so that the principal axes frame (x', y', z') aligns with the lab frame (x, y, z), then

$$[\alpha_{xy}]_{(vf,vi)} = [\alpha_{x'y'}]_{(vf,vi)} = \langle \psi_{vf} \mid \alpha_{xy} \mid \psi_{vi} \rangle \qquad (13.83)$$

When $v_f = v_i$, the polarizability describes Rayleigh scattering. For Stokes Raman scattering

$$
\begin{aligned}
[\mu_{xo}]_{(vki+1,vki)} &= \sqrt{v_{ki}+1}\; b_{vk}\{(\alpha'_{xx})_k E_{xo} + (\alpha'_{xy})_k E_{yo} \\
&\quad + (\alpha'_{xz})_k E_{zo}\} \\
[\mu_{yo}]_{(vki+1,vki)} &= \sqrt{v_{ki}+1}\; b_{vk}\{(\alpha'_{yx})_k E_{xo} + (\alpha'_{yy})_k E_{yo} \\
&\quad + (\alpha'_{yz})_k E_{zo}\} \\
[\mu_{zo}]_{(vki+1,vki)} &= \sqrt{v_{ki}+1}\; b_{vk}\{(\alpha'_{zx})_k E_{xo} + (\alpha'_{zy})_k E_{yo} \\
&\quad + (\alpha'_{zz})_k E_{zo}\} \quad\quad (13.84)
\end{aligned}
$$

To describe anti-Stokes scattering, we replace $\sqrt{v_{ki}+1}$ with $\sqrt{v_{ki}}$ in equation (13.84).

Vibrational Raman scattering from a space-averaged molecule

We consider Stokes Raman scattering at $\theta = \pi/2$ using a space-averaged, derived polarizability [32]:

$$
\begin{aligned}
\overline{\mu^2}_{vki+1,vki} &= (v_{ki}+1)\; b_{vk}^2\; \overline{(\alpha')}_k^2\; E_o^2 \\
&= (v_{ki}+1)\; b_{vk}^2 \left[\frac{45(a')_k^2 + 7(\gamma')_k^2}{45}\right] E_o^2 \quad (13.85)
\end{aligned}
$$

Here the 7 in front of $(\gamma')_k^2$ results because both directions of polarization are considered. Now the power scattered per unit solid angle is

$$
\frac{dP}{d\Omega} = \frac{(\omega_o - \omega_k)^4\; \overline{\mu^2}}{32\pi^2 \epsilon_o c^3} \quad (13.86)
$$

where $(\omega_o - \omega_k)$ is the frequency of the Stokes-shifted light. This can be rewritten in terms of wave numbers, and equation (13.85) can be substituted for the dipole moment to produce

$$
\frac{dP}{d\Omega} = \frac{\pi^2 c}{2\epsilon_o}(\breve{\nu}_o - \breve{\nu}_k)^4 (v_{ki}+1)\; b_{vk}^2 \left[\frac{45(a')_k^2 + 7(\gamma')_k^2}{45}\right] E_o^2 \quad (13.87)
$$

In terms of the input irradiance, this can be written as

$$\frac{\mathrm{d}P}{\mathrm{d}\Omega} = k_{\check{\nu}}(\check{\nu}_\circ - \check{\nu}_k)^4 (v_{ki} + 1)\, b_{vk}^2 \left[\frac{45(a')_k^2 + 7(\gamma')_k^2}{45} \right] I \qquad (13.88)$$

Alternatively, the definition of b_{vk}^2 can be used to write

$$\frac{\mathrm{d}P}{\mathrm{d}\Omega} = k_{\check{\nu}} \frac{h}{8\pi^2 c \check{\nu}_k} (\check{\nu}_\circ - \check{\nu}_k)^4 (v_{ki} + 1) \left[\frac{45(a')_k^2 + 7(\gamma')_k^2}{45} \right] I \qquad (13.89)$$

These expressions are actually for one allowed vibrational transition from one molecule. We could multiply by N (the total number of scatterers N=NV, where N is the number density and V is the sample volume) times the fraction in the vibrational level probed to obtain the total signal, but that would be incorrect in the harmonic oscillator limit. In this limit, the scattered signal could occur from a number of bands, all of them with $(v_{ki} + 1 \leftarrow v_{ki})$ (e.g. $1 \leftarrow 0$, $2 \leftarrow 1$, $3 \leftarrow 2$, etc.). All of these transitions have the same Raman shift in the harmonic oscillator limit. It is therefore necessary to account for these populations. In order to find the total scattering signal from all of the molecules in the sample volume, therefore, we replace the term $(v_{ki}+1)$ in equation (13.88) with

$$N \sum_i (v_{ki} + 1) f_{vki} \qquad (13.90)$$

where f_{vki} is the Boltzmann fraction for vibrational level i, given by equation (2.81). The vibrational energy used in equation (2.81) is given by the harmonic oscillator solution $E_v = (v_{ki} + 1)h\nu_e$ [equation (9.54)]. Long shows that

$$N \sum_i (v_{ki} + 1) f_{vki} = \frac{N}{1 - \exp\left(-\frac{h\nu_e}{k_B T}\right)} \qquad (13.91)$$

The total power scattered per unit solid angle is then

$$\frac{\mathrm{d}P}{\mathrm{d}\Omega} = \frac{k_{\check{\nu}} h}{8\pi^2 c \check{\nu}_k} \frac{(\check{\nu}_\circ - \check{\nu}_k)^4 \, N}{1 - \exp\left(-\frac{h\nu_e}{k_B T}\right)} \left[\frac{45(a')_k^2 + 7(\gamma')_k^2}{45} \right] I \qquad (13.92)$$

Note that this expression (which can be found in many references) applies to the harmonic oscillator limit, where we have included every band. If the bands are spectrally separated by anharmonicity, and one intends to investigate just one band, it is necessary to revert to equation (13.89), multiplied by the number in the originating vibrational level (as given by the Boltzmann expression).

An important outcome of this development is that the Raman signal $dP/d\Omega$ varies linearly with number of scatterers (usually represented by the product of number density and sample volume). It also depends upon quantum polarizabilities, which are often extracted from measurements of Raman cross-sections (defined below).

The differential Raman cross-section is usually defined as [1]

$$\frac{d\sigma}{d\Omega} \equiv \frac{\frac{dP}{d\Omega}}{NI} \qquad (13.93)$$

For Stokes Raman scattering from harmonic oscillators, collected at $\theta = \pi/2$, we find

$$\left[\frac{d\sigma}{d\Omega}\right]_{Stokes} = \frac{k_{\breve{\nu}}h}{8\pi^2 c \breve{\nu}_k} \frac{(\breve{\nu}_o - \breve{\nu}_k)^4}{1 - \exp\left(-\frac{h\nu_e}{k_B T}\right)} \left[\frac{45(a')_k^2 + 7(\gamma')_k^2}{45}\right] \qquad (13.94)$$

For anti-Stokes scattering under the same conditions, the cross-section is

$$\left[\frac{d\sigma}{d\Omega}\right]_{anti-Stokes} = \frac{k_{\breve{\nu}}h}{8\pi^2 c \breve{\nu}_k} \frac{(\breve{\nu}_o + \breve{\nu}_k)^4}{\exp\left(\frac{h\nu_e}{k_B T}\right) - 1} \left[\frac{45(a')_k^2 + 7(\gamma')_k^2}{45}\right] \qquad (13.95)$$

The ratio between the Stokes and anti-Stokes signals at $\theta = \pi/2$ will be

$$\frac{(dP/d\Omega)_{Stokes}}{(dP/d\Omega)_{anti-Stokes}} = \frac{(\breve{\nu}_o - \breve{\nu}_k)^4}{(\breve{\nu}_o + \breve{\nu}_k)^4} \exp\left(\frac{h\nu_e}{k_B T}\right) \qquad (13.96)$$

At low temperatures, the Stokes lines are much stronger than the anti-Stokes lines because the anti-Stokes signal must be generated by an excited vibrational state. At low temperatures, most of the population is found in the ground vibrational state, but this is not true of high temperatures. In fact, one form of Raman thermometry is to compare the amplitudes of the Stokes and anti-Stokes bands.

Vibrational Raman spectra

Vibrational Raman shifts $\Delta\breve{\nu}$ away from the carrier $\breve{\nu}_o$ are determined by the purely vibrational molecular spectra (the IR spectra) described in Chapter 9. For a harmonic oscillator, the wave number locations can be adapted from equation (9.54):

$$G_v = \frac{E_v}{hc} = \frac{\nu_e}{c}\left(v + \frac{1}{2}\right) = \omega_e\left(v + \frac{1}{2}\right) \tag{13.97}$$

where ν_e is the vibrational constant of the molecule. For an anharmonic oscillator, equation (9.77) (with just one anharmonic correction) applies:

$$G_v = \frac{E_v}{hc} = \omega_e\left(v + \frac{1}{2}\right) - \omega_e x_e\left(v + \frac{1}{2}\right)^2$$

The Raman shift is given by simply inserting the vibrational selection rule. For example, we can take the difference between G_{v+1} and G_v:

$$\Delta\breve{\nu} = \omega_e - 2(v + 1)\omega_e x_e \tag{13.98}$$

This expression applies to both the $v + 1 \leftarrow v$ and the $v + 1 \rightarrow v$ transitions. The outcome means that the wavelength shift actually depends upon the vibrational quantum number in the anharmonic case.

Rotational and ro/vibrational Raman amplitudes

To include rotation, we return to equation (13.71):

$$[\alpha_{xy}] = \sum_{x'y'}[\alpha_{x'y'}]_{vf,vi}\langle\psi_{rf} \mid \cos(xx')\cos(yy') \mid \psi_{ri}\rangle \tag{13.99}$$

If there is no change in the vibrational quantum number, then we observe rotational structure on the carrier (Rayleigh) signal:

$$[\alpha_{xy}] = \sum_{x'y'}[\alpha_{x'y'}]_o\langle\psi_{rf} \mid \cos(xx')\cos(yy') \mid \psi_{ri}\rangle \tag{13.100}$$

If the vibrational quantum number does change, rotational fine structure will accompany each vibrational feature. For Stokes scattering we write

$$[\alpha_{xy}] = \sqrt{v_{ki}+1}\; b_{vk} \sum_{x'y'}[\alpha'_{x'y'}]_k \langle \psi_{rf} \mid \cos(xx')\cos(yy') \mid \psi_{ri}\rangle \quad (13.101)$$

and for anti-Stokes scattering

$$[\alpha_{xy}] = \sqrt{v_{ki}}\; b_{vk} \sum_{x'y'}[\alpha'_{x'y'}]_k \langle \psi_{rf} \mid \cos(xx')\cos(yy') \mid \psi_{ri}\rangle \quad (13.102)$$

The rotational components depend upon the rotational wave functions and the Euler angles.

The expressions developed for Raman cross-sections are fairly complex, depending upon the nature of the molecule. Here we discuss diatomic molecules exclusively, which does cover most of the molecules of interest in the thermosciences. To find Raman scattering amplitudes, it becomes necessary to find the squares of the polarizability tensor components. Because each rotational energy level has a magnetic degeneracy of $2J+1$, it is necessary to average over all of the m'' magnetic sublevels. As one example

$$\overline{[\alpha_{xy}]^2_{v'J',v''J''}} = \frac{1}{(2J+1)} \sum_{m''}[\alpha_{xy}]^2_{v'J'm',v''J''m''} \quad (13.103)$$

where we have reverted to using $'$ and $''$ to denote the upper and lower energy levels.

Rotational selection rules

Equation (13.99) indicates that the existence of a rotational Raman polarizability depends upon an overlap integral including the rotational wave functions and Euler angles. While rotational (orbital) wave functions have been discussed [see, for example, equation (8.63)], full development of the rotational Raman selection rules is beyond the scope of this chapter. The outcome is actually quite simple. For pure rotational Raman, $\Delta J = \pm 2$ (S and O branches). For ro/vibrational Raman, $\Delta J = 0, \pm 2$. The transition $\Delta J = 0$ for purely rotational Raman is really just Rayleigh scattering.

A common justification for this selection rule states that, for rotational scattering, the dipole observed by the electric field must exhibit

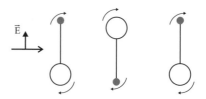

Figure 13.7: One full revolution, including two rotational aspect changes.

an "aspect change". As shown in Figure 13.7, one full rotation includes two aspect changes. The frequency of aspect change is thus twice the frequency of rotation, hence the $\Delta J = \pm 2$.

More on ro/vibrational amplitudes

Following "tedious algebra", Long tabulates the J-dependent coefficients (also called the Placzek-Teller coefficients) that determine the rotational contribution to scattering amplitudes for diatomic molecules.

Raman scattering patterns are dependent upon molecular structure. Even in the case where we induce spontaneous Raman with linearly polarized light and measure the scattering along the polarization coordinate, the signal will be depolarized in a way that is determined by molecular structure. For this reason, detailed descriptions of Raman cross-sections will always be couched in terms of the entire polarizability tensor. Here we give just a few examples, taken from Long's central set of tables (tables H and I in reference [32]).

The dipole moment contribution to the Raman cross-section is determined by the space-averaged polarizability tensor components squared. When $\Delta v_k = 0$ and $\Delta J = 0$, the diagonal elements (α_{ii}, where $ii = xx$, yy, or zz) are

$$\overline{[\alpha_{ii}]^2_{v'J',v''J''}} = (a)^2_\circ + \frac{4}{45}b_{J,J}(\gamma)^2_\circ \tag{13.104}$$

where the various $b_{J,J}$-type terms (the Placzek-Teller coefficients) are defined below, in equation (13.112). The off-diagonal elements under the same transition are given by

$$\overline{[\alpha_{ij}]^2_{v'J',v''J''}} = \frac{1}{15}b_{J,J}(\gamma)^2_\circ \tag{13.105}$$

Equations (13.104) and (13.105) describe a Rayleigh solution, written in terms of the equilibrium mean polarizability and the anisotropy of the equilibrium ellipsoid (not the derived components).

When $\Delta v_k = 0$ and $\Delta J = +2$, the diagonal elements are

$$\overline{[\alpha_{ii}]^2_{v'J',v''J''}} = \frac{4}{45} b_{J+2,J} (\gamma)^2_\circ \qquad (13.106)$$

and the off-diagonal elements under the same transition are given by

$$\overline{[\alpha_{ij}]^2_{v'J',v''J''}} = \frac{1}{15} b_{J+2,J} (\gamma)^2_\circ \qquad (13.107)$$

Purely rotational Raman spectral features are generated by the anisotropy of the equilibrium polarizability $(\gamma)_\circ$. Purely isotropic molecules, therefore, are not rotationally Raman active. This is easy to understand; if the molecular dipole moment is spherically symmetric, then the electric field will not encounter a time-dependent change in polarizability owing to pure rotation. Because purely rotational Raman cross-sections are determined by the equilibrium polarizability, these cross-sections are usually ten times those of ro/vibrational Raman (which are determined by the derived polarizability components).

As a final example, when $\Delta v_k = +1$ and $\Delta J = 0$ (Q branch transitions), the diagonal elements are

$$\overline{[\alpha_{ii}]^2_{v'J',v''J''}} = (v_k + 1)(\alpha')^2_k\, b^2_{vk} \left((a')^2 + \frac{4}{45}\, b_{J,J}\, (\gamma')^2 \right) \qquad (13.108)$$

and the off-diagonal elements under the same transition are given by

$$\overline{[\alpha_{ij}]^2_{v'J',v''J''}} = (v_k + 1)(\alpha')^2_k\, b^2_{vk} \left(\frac{1}{15}\, b_{J,J}\, (\gamma')^2 \right) \qquad (13.109)$$

Conversely, when $\Delta v_k = +1$ and $\Delta J = \pm 2$ (S and O branches), the diagonal elements are

$$\overline{[\alpha_{ii}]^2_{v'J',v''J''}} = (v_k + 1)(\alpha')^2_k\, b^2_{vk} \left(\frac{4}{45}\, b_{J\pm2,J}\, (\gamma')^2 \right) \qquad (13.110)$$

and the off-diagonal elements under the same transition are given by

$$\overline{[\alpha_{ij}]^2_{v'J',v''J''}} = (v_k + 1)(\alpha')^2_k\, b^2_{vk} \left(\frac{1}{15}\, b_{J\pm2,J}\, (\gamma')^2 \right) \qquad (13.111)$$

Note that the Q branch polarizability for the diagonal elements is determined both by the mean and the anisotropy of the derived polarizability, while for the S and O branches it is determined simply by the anisotropy. This would indicate a Q branch response that is much stronger than that of the S and O branches. In practice, the S and O branches are very weak relative to the Q branches. Most commonly, one finds that the Q branches alone are utilized for flowfield research.

The Placzek-Teller coefficients are defined by

$$
\begin{aligned}
b_{J,J} &\equiv \frac{J(J+1)}{(2J-1)(2J+3)} \\
b_{J+2,J} &\equiv \frac{3(J+1)(J+2)}{2(2J+1)(2J+3)} \\
b_{J-2,J} &\equiv \frac{3J(J-1)}{2(2J+1)(2J-1)}
\end{aligned}
\tag{13.112}
$$

In addition to these transition strengths, the expression for the Raman cross-section involves a Boltzmann fraction for the number of molecules in the initial state. Equation (2.85) is modified somewhat to provide this fraction:

$$
\frac{N_{\text{rot}}}{N} = g_I(2J+1)\frac{e^{-\frac{E_J}{k_B T}}}{Q_{\text{rot}}}
$$

Because homonuclear molecules are Raman active, the contribution of g_I will surface in spectrally resolved measurements of homonuclear diatomics. Because g_I depends upon whether J is odd or even in such cases, the rotational band will have a variation from line to line, based simply upon g_I.

The *total* response of a Raman ro/vibration band is the same as the response of the same pure vibrational transition (without rotation). Another way to say this is that the rotational portions (the Boltzmann fraction times the relative rotational strengths) are fractional contributions that must add to 1.

Ro/vibrational Raman spectra

Purely rotational Raman shifts are determined by the rotational molecular spectra (sometimes termed the microwave spectra) described in

Chapter 9. For a rigid rotator, the wave number locations can be adapted from equation (9.28)

$$E_J = \frac{\hbar^2}{2I_e}J(J+1)$$

which can be written in terms of wave numbers as

$$F(J) = \frac{E_J}{hc} = \frac{\hbar}{4\pi c I_e}J(J+1) \tag{13.113}$$

Although we have not included them, vibration/rotation interactions can be included using expressions in Chapter 9. The Raman shift is given by simply inserting the rotational selection rule.

To generate ro/vibrational Raman shifts, we simply add vibrational and rotational shift values.

Raman lineshapes

To measure spontaneous Raman lineshapes is extremely difficult and it is almost never done. This is because the system that disperses the Raman spectrum would need to have better spectral resolution than the lines themselves. In practice, the overall lineshapes observed in spontaneous Raman scattering are set by the spectral response of the measurement device, not Raman scattering. Having said that, Raman scattering does exhibit both homogeneous broadening (due to lifetime phenomena) and inhomogeneous broadening (Doppler again). These can be more easily studied using techniques such as coherent anti-Stokes Raman scattering (see, for example, Farrow and Palmer [95]).

13.5.1 Raman Flowfield Measurements

Unfortunately, the complexity of the development provided above can mask the process by which measurements might actually be made. As mentioned already, the complete details of theoretical Raman cross-sections are not critical for flowfield measurements because the signal is calibrated. Consider, for example, the issue of depolarization in the Raman signal. It will be depolarized, and experimental systems will most likely affect signal polarization (e.g. a grating will pass more of one polarization than another). If one calibrates the measurement, however, and then does not make a change to any device that affects polarization, then this issue is irrelevant.

With this global approach in mind, therefore, we can consolidate the Raman signal expressions into a simple formalism. We begin with equations like 13.89, which describes vibrational Raman scattering from one molecule. Such an expression is multiplied by the number of scatterers in the form $N_{total} f_B$. The global response is then couched in terms of a scattering cross-section as defined in equation (13.93). The outcome is a deceptively simple expression that contains a calibration constant [1, 88]:

$$\boxed{S_R = CI\frac{d\sigma}{d\Omega}\Omega N f(T)l} \qquad (13.114)$$

where S_R is the Raman signal produced by the instrument, C is a factor that includes optics collection efficiency, electronic signal conversion efficiency, depolarization ratios, and any other phenomena that affect signal generation but do not appear explicitly in the equation, I is the input irradiance, $d\sigma/d\Omega$ is the Raman cross-section, Ω is the optics collection solid angle, N is the number density of the scattering species, $f(T)$ combines all of the temperature dependencies, and l is the interaction length. It may be worthwhile discussing the sudden appearance of l in equation (13.114). The signal is generated by a total number of scatterers, not the number density. It is therefore necessary to multiply N by the sample volume. In this case, the sample volume is given by a cross-section $(d\sigma/d\Omega)\Omega$ times interaction length l. Use of the cross-section to define the sample volume is appropriate when the cross-section is defined as it is in equation (13.93). The other terms in equation (13.114) will be discussed in the following paragraphs.

Raman cross-sections are commonly measured and reported [1, 88], typically with vertical polarization (relative to the scattering plane) and measurement at $\theta = \pi/2$. These published measurements are typically used to model a Raman instrument [using equation (13.114)]; to predict signal levels and spectra at various temperatures, for example. It is quite common to extract the frequency dependence from the cross-section, giving an alternative to equation (13.114):

$$S_R = CI(\breve{\nu}_o \pm \breve{\nu}_k)^4\frac{d\sigma}{d\Omega}\Omega N f(T)l \qquad (13.115)$$

where this cross-section is different from the one used in equation (13.114).

If one has performed a calibration in the lab, the cross-section [together with $(d\sigma/d\Omega)\Omega\,l$ etc.] can be folded into a new C, which would then represent the calibration measurement. In subsequent measurements, the calibration is scaled for changes in temperature or laser irradiance.

The temperature correction can be either fairly complicated or fairly simple, depending upon the spectrum detected. The function $f(T)$ includes the Boltzmann fraction terms that describe populations in the states that are detected. Because Raman scattering is nonresonant, the laser couples all of the levels that are thermally populated. If sone of these levels are not detected, then the signal will change with temperature simply because population enters or leaves the collection band that is actually sensed, even when the total number density is constant. In addition, at constant pressure the value of N is linearly dependent upon temperature through the ideal gas law, and this dependence must also be included in $f(T)$. When a Raman calibration is performed at one temperature, but measurements are made at another temperature, $f(T)$ must be used to correct the measurement.

It is often possible to observe an entire vibrational branch structure for a molecule. Even when molecules are anharmonic, the overtones may be located near the fundamental and in some cases the entire band structure can be collected. In such a case, equation (13.92) contains the Boltzmann temperature dependence. Otherwise, it will be necessary to model the temperature dependence of states actually sensed by the instrument. Note that the individual rotational cross-sections depend both upon vibrational and rotational quantum numbers, and these must be incorporated in some way if the system spectrally filters. It is not uncommon to develop fairly detailed numerical simulations for this purpose [96].

It is for reasons such as these that so much detail on Raman cross-sections has been provided. Each researcher must decide whether or not their calibration is properly scaled with changes in important variables such as temperature and laser irradiance.

Most commonly, flowfield researchers observe entire vibrational bands (usually a Q branch, given the strength of that band). This is done for two reasons. First, by sensing a total band the absolute signal level is larger than when individual rotational transitions are measured. This is important, especially for single-pulse measurements. Next, if one observes only part of a band or just one line, one must have very good control of the spectral filtering function in order to model the change in rotational structure with temperature (Boltzmann fraction). While not strongly J-dependent, the Placzek-Teller coefficients require that the experimentalist ensure the instrument is measuring the same set of lines assumed in the analysis of the data. In contrast, to observe and correct entire vibrational band transitions is much simpler. At times it is not possible to do this because there are spectral overlaps. If many calibration cases are performed the overlaps can be corrected using a matrix approach (see, for example, Barlow *et al.* in reference [3]). For multi-species measurements, this matrix approach works quite well.

13.6   Conclusions

Rayleigh and Raman scattering have been reviewed with an intention to explain how various published expressions can be developed from basic principles. With such an understanding in hand, one can use these measurements in flowfields with confidence. Usage typically involves detailed calibration under known conditions.

Chapter 14

THE DENSITY MATRIX EQUATIONS

14.1 Introduction

Equation (10.28) is a perfectly good way to describe the Einstein A coefficient. The assumptions used in the development of equation (10.28) are appropriate for the description of a relatively slow, linear decay of energy from an initially excited state, as A_{nm} is intended. This A_{nm} can then be used to find the other Einstein coefficients via equations (3.95) and (3.96) (by assuming LTE). The Einstein coefficients were originally developed for this kind of rate equation modeling. The formalism is therefore self-consistent.

Under certain circumstances, however, the rate equation approach is inaccurate because it does not account for atomic or molecular phenomena that can occur. One circumstance where the rate equations do not apply is during rapid transient events (short relative to collisional times). Daily [45] has shown that even during the leading edge of nanosecond-pulse laser excitation, the rate equations do not accurately describe the transfer of population between the laser-coupled levels. A laser/atom interaction can easily violate thermodynamic equilibrium, to be restored by collisional events over a relatively long timescale. In addition, the rate equations cannot describe nonlinear optical techniques. For these reasons, the density matrix equations (DME) are becoming more commonplace as a tool used to describe certain details of laser diagnostics.

The DME are required to describe short-pulse interactions accurately. By "short-pulse", we mean shorter than collision times, so that the atomic or molecular coherences (defined below) remain, and therefore must be accounted for in a model (see, for example, Brochinke and Linne in reference [3]). Because the times of the interaction are so short, however, we can usually describe the interaction as one that is isolated to two energy levels and no more. To involve additional levels would require energy transfer to occur during this interaction. Significant transfer usually does not occur on such short time scales. The optical bandwidth of a short pulse is large, and so the pulse may couple multiple sets of levels within a molecule. Usually it remains appropriate to consider this a collection of simple, two-level interactions. It is not an appropriate assumption if the same ground state is coupled to two upper states, but such an occurrence is usually avoidable. In some nonlinear processes this is also an incorrect assumption. Reichardt and Lucht [97], for example, have shown that short-pulse degenerate four-wave mixing (DFWM) relies upon coherent emission from a nonlinearly prepared polarization state. That state is subject to energy transfer. This book deals primarily with linear processes, however, so the treatment of the DME in this paper will assume a two-level system. Such an approach makes an introduction to the density matrix equations straightforward.

The DME are written in a form that provides a quantum description of the electromagnetic interaction. It is a quantum-optical formalism, because the outcome is naturally mated to an electromagnetic description of light. For this reason, the density matrix solutions gravitate towards forms that look similar to wave equation outcomes (e.g. the susceptibility), rather than rate equation outcomes (e.g. the Einstein coefficients). The susceptibility and the Einstein coefficients are ultimately related to each other.

In what follows we will present a development of the density matrix equations, relying primarily upon the developments by Boyd [22] and Verdeyen [26]. Fittingly, Boyd's book is devoted to nonlinear optics. Resonant nonlinear optics can only be described in detail with the DME. The subject of Verdeyen's book is lasers. For linear resonance techniques, primarily for short-time events, we find many useful solutions to the DME in the laser literature. This chapter also borrows heavily from the work of Settersten [98, 42, 46]. For an extension of the following treatment, interested readers are referred to the paper by Lucht *et al.* [79]. In addition to these sources, Shore [73] provides a very detailed treatment for atoms.

Following the development of the DME, we will simplify them, describe several outcomes, very briefly discuss how they can be adapted to multiple levels, and show how they can generate the standard Einstein coefficients in the steady-state limit when the quantum dipole interacts with an externally applied field.

14.2 Development of the DME

In quantum mechanics, one can find perturbation solutions that generate an expression for the linear susceptibility, but they are not appropriate for description of the resonance response because they do not include damping terms. As our treatment of the Lorentz atom demonstrated, resonance lines are broadened in part by damping. If we wish to describe an event that falls within the Lorentzian profile of a resonance, we must therefore include a description for the damping that produces this profile. To account for Doppler broadening (the non-Lorentzian term in gas-phase broadening) it then becomes necessary to sum over velocity classes. It is thus necessary to include classical phenomena in a quantum-mechanical model for resonance response, and this is most easily done through the density matrix equations.

Development of the undamped DME

Following Boyd [22], we begin with a quantum-mechanical system in state s, described by the Schrödinger equation:

$$i\hbar \frac{\partial \Psi_s(\vec{r}, t)}{\partial t} = \hat{H} \Psi_s(\vec{r}, t) \tag{14.1}$$

In this case, $\hat{H}$ contains an external electromagnetic field that will interact with the quantum dipole. This is a departure from the approach used in Chapter 10, which described a passive event—spontaneous emission from a dipole that had somehow become excited. We write the Hamiltonian as an unperturbed, free-atom Hamiltonian $\hat{H}_\circ$ (which is not time dependent) plus an interaction term $\hat{V}$ (which is time dependent):

$$\hat{H} = \hat{H}_\circ + \hat{V}(t) \tag{14.2}$$

This approach is justified in Section 8.3.1. In the case of an unperturbed two-level system, we can assume that the stationary wave functions [written as $u_j(\vec{r})$ here] are known. They are simply the eigenfunctions of

$$\hat{H}_o u_j(\vec{r}) = E_j u_j(\vec{r}) \qquad (14.3)$$

Note that we have chosen to avoid the use of the subscripts n and m. Those subscripts typically denote the excited and ground states of a two-level system respectively. In this development, we wish to remain somewhat general before simplifying to a recognizable two-level system. Here, the indices j and k will denote simply two generally allowed quantum states and nothing more. The full time-dependent wave function appearing in equation (14.1) can then be described by superposition:

$$\Psi_s(\vec{r}, t) = \sum_j c_j^s(t) u_j(\vec{r}) \qquad (14.4)$$

Similar to equation (10.2), the time evolution of the state wave function can be written in terms of the time rate of change of the expansion coefficients $c_j^s(t)$. To do this here, we insert the expression for $\Psi_s(\vec{r}, t)$ given by equation (14.4) into the Schrödinger equation (14.1):

$$i\hbar \sum_j \frac{\partial c_j^s(t)}{\partial t} u_j(\vec{r}) = \sum_j c_j^s(t) \hat{H} u_j(\vec{r}) \qquad (14.5)$$

We then take advantage of the orthogonality property of the eigenfunctions by taking an inner product (written in Dirac notation for compactness):

$$i\hbar \langle u_k | \sum_j \frac{\partial c_j^s(t)}{\partial t} | u_j \rangle = \langle u_k | \sum_j c_j^s(t) \hat{H} | u_j \rangle \qquad (14.6)$$

On the left-hand side this inner product simply filters the kth terms out of the summation, while on the right-hand side we define a new quantity, the matrix element of the Hamiltonian (H_{kj}):

$$i\hbar \frac{\partial c_k^s(t)}{\partial t} = \sum_j H_{kj} c_j^s(t) \qquad (14.7)$$

where:

$$H_{kj} \equiv \langle u_k | \hat{H} | u_j \rangle \qquad (14.8)$$

Equation (14.7) is actually the same as equation (14.1), except that it is written in terms of the probability amplitudes. We will return to equation (14.7) momentarily, but it will be necessary to digress for a few paragraphs.

The expectation value for a Hermitian operator $\hat{A}$ can be written in terms of the usual expansion:

$$\langle A \rangle = \langle \Psi_s | \hat{A} | \Psi_s \rangle = \sum_{kj} \langle u_k | c_k^{s*} \, \hat{A} \, c_j^s | u_j \rangle \tag{14.9}$$

If we then define the matrix element of the operator $\hat{A}$ as

$$A_{kj} \equiv \langle u_k | \hat{A} | u_j \rangle \tag{14.10}$$

the expectation value can be written as

$$\langle A \rangle = \sum_{kj} c_k^{s*} c_j^s A_{kj} \tag{14.11}$$

These collected terms $c_k^{s*} c_j^s$ are a quantum-mechanically pure (and therefore not entirely useful) density matrix. Combined in this way, they represent the probability density for mixed states. As before, since we are considering an ensemble of these mixed quantum states, for a transition between one lower and one upper state, it is a matrix.

As Boyd points out, we will need to specify the states for a large number of atoms or molecules to make this useful, and that becomes impossible. We rely, therefore, on statistical mechanics to provide the probability of state s, and then we rely on quantum mechanics to describe that state in detail. Hence, we define the density matrix as

$$\rho_{jk} \equiv \sum_{s} p_s c_k^{s*} c_j^s \tag{14.12}$$

where p_s is the probability of state s. The diagonal elements of the density matrix, ρ_{jj} or ρ_{kk}, are the probability that the system will be in the eigenstate j or k. The off-diagonal elements, ρ_{kj} for $k \neq j$, are the coherences and they can be related to the induced dipole moment.

The expectation value for operator $\hat{A}$ can then be written in terms of the density matrix via

$$\langle A \rangle = \sum_{s} p(s) \sum_{jk} c_k^{s*} c_j^s A_{kj} = \sum_{jk} \rho_{jk} A_{kj} \tag{14.13}$$

This can be simplified by:

$$\langle A \rangle = \sum_j \left(\sum_k \rho_{jk} A_{kj} \right) = \sum_j (\rho A)_{jj} = tr\left(\hat{\rho}\hat{A}\right) \tag{14.14}$$

Hence, the expectation value of an observable is given by the trace of the product between the density matrix and the matrix elements of the associated operator. Here, again, we use operators and matrices almost interchangeably because the operators are associated with quantized levels that render them as sorted matrices.

We seek to describe the dynamics of the radiation/dipole interaction, and this can be done by taking the time derivative of the density matrix:

$$\dot{\rho}_{jk} = \sum_s \frac{dp(s)}{dt} c_k^{s*} c_j^s + \sum_s p(s) \left(c_k^{s*} \frac{dc_j^s}{dt} + \frac{dc_k^{s*}}{dt} c_j^s \right) \tag{14.15}$$

We can assume that the probability for state s is constant, making the derivative $dp(s)/dt = 0$. The time derivatives of the probability amplitudes can be found using Schrödinger's equation in the form given by equation (14.7). We then find

$$c_k^{s*} \frac{dc_j^s}{dt} = -\frac{i}{\hbar} c_k^{s*} \sum_l H_{jl} c_l^s \tag{14.16}$$

and

$$c_j^s \frac{dc_k^{s*}}{dt} = \frac{i}{\hbar} c_j^s \sum_l H_{kl}^* c_l^{s*} = \frac{i}{\hbar} c_j^s \sum_l H_{lk} c_l^{s*} \tag{14.17}$$

where the last equality follows because H is Hermitian. Then

$$\begin{aligned}
\dot{\rho}_{jk} &= \frac{i}{\hbar} \sum_s p(s) \sum_l \left(c_j^s H_{lk} c_l^{s*} - c_k^{s*} H_{jl} c_l^s \right) \\
&= \frac{i}{\hbar} \sum_l \left(\rho_{jl} H_{lk} - H_{jl} \rho_{lk} \right) \\
&= \frac{i}{\hbar} [\hat{\rho}\hat{H} - \hat{H}\hat{\rho}]
\end{aligned}$$

where equation (7.23) has been applied. Therefore

$$\boxed{\dot{\rho}_{jk} = -\frac{i}{\hbar}[\hat{H}, \hat{\rho}]_{jk}} \tag{14.18}$$

A simple example

Boyd provides a two-level example, and it proves useful so we will do the same. Consider an upper level denoted by n and a lower level denoted by m. Then the state is described by

$$\Psi_s(\vec{r}, t) = c_m^s(t)u_m(\vec{r}) + c_n^s(t)u_n(\vec{r}) \tag{14.19}$$

and the density matrix is given by

$$\rho_{nm} = \begin{pmatrix} \rho_{nn} & \rho_{nm} \\ \rho_{mn} & \rho_{mm} \end{pmatrix} \tag{14.20}$$

The probability densities for the diagonal terms (written either as ρ_{nn} or as ρ_{mm}) represent the probabilities that the system will be in a stationary quantum state (represented by quantum numbers n or m). The off-diagonal terms (written either as ρ_{nm} or as ρ_{mn}) represent the coherences. They are the means by which radiant energy couples one level to another, and they are related to the atomic or molecular transition dipole moment.

The matrix representation for the dipole moment operator is

$$\vec{\mu}_{nm} = \langle u_n|\hat{\mu}|u_m\rangle = -q_e\langle u_n|\hat{r}|u_m\rangle = \begin{pmatrix} 0 & \mu_{nm} \\ \mu_{mn} & 0 \end{pmatrix} \tag{14.21}$$

The terms $\langle u_n|\hat{r}|u_n\rangle$ and $\langle u_m|\hat{r}|u_m\rangle$ are zero because of the symmetry of $\hat{r}$. The expectation value of the dipole moment is then

$$\langle \vec{\mu} \rangle = tr(\hat{\rho}\hat{\mu}) \tag{14.22}$$

where the matrix given by $\hat{\rho}\hat{\mu}$ is

$$\hat{\rho}\hat{\mu} = \begin{pmatrix} \rho_{nn} & \rho_{nm} \\ \rho_{mn} & \rho_{mm} \end{pmatrix} \begin{pmatrix} 0 & \mu_{nm} \\ \mu_{mn} & 0 \end{pmatrix} = \begin{pmatrix} \rho_{nm}\mu_{mn} & \rho_{nn}\mu_{nm} \\ \rho_{mm}\mu_{mn} & \rho_{mn}\mu_{nm} \end{pmatrix} \tag{14.23}$$

The outcome is

$$\langle \vec{\mu} \rangle = \rho_{nm}\mu_{mn} + \rho_{mn}\mu_{nm} \tag{14.24}$$

demonstrating that $\langle \vec{\mu}_{nm} \rangle$, the transition dipole moment, is determined by the off-diagonal elements, which are the coherences.

Development of the damped DME

Equation (14.18) describes the evolution of the density matrix owing exclusively to the interactions that are embedded in the Hamiltonian operator. In order to account for collisional interactions, phenomenological expressions for damping are added to the equations. We continue, therefore, by assuming the same closed two-level system.

Consider the Hamiltonian $\hat{H} = \hat{H}_o + \hat{V}(t)$ in more detail. The stationary portion $\hat{H}_o$ is related to the energy levels of the stationary states via equation (14.3). The matrix representation for $\hat{H}_o$ is thus diagonal, and it is represented by

$$\hat{H}_{o,nm} = E_n \delta_{nm} \tag{14.25}$$

We then assume that the potential $(\hat{V})$ is given by the interaction of the electric field with the dipole moment:

$$\hat{V}(t) = -\hat{\mu}\vec{E}(t) \tag{14.26}$$

In such a simple system $\vec{\mu}$ and $\vec{E}$ act colinearly. Moreover, the diagonal terms are zero owing to "definite parity" of the states. That is, they are symmetric and so $\mu_{nn} = \mu_{mm} = 0$. In the electric dipole approximation, the off-diagonal elements of the interaction Hamiltonian are therefore given by

$$V_{nm}(z,t) = -\mu_{nm}E(z,t), \qquad V_{mn}(z,t) = -\mu_{mn}E(z,t) \tag{14.27}$$

In general, the off-diagonal elements are complex-conjugate. The dipole matrix elements μ_{mn} and μ_{nm} can be made purely real and equal, however, with no loss of generality, if the basis vectors describing the energy eigenstates m and n are judiciously chosen. We assume that condition is met in the following model.

Recall that the Hamiltonian is Hermitian. The state of the system is described by $\hat{\rho}$ as given by equation (14.20), with $\rho_{nm} = \rho_{mn}^*$ [see, for example, equation (14.12)]. The time evolution of $\hat{\rho}$ is given by equation (14.18), but we can break the representation of the Hamiltonian into the time-independent and time-dependent parts [i.e. into $\hat{H}_o$ and $\hat{V}(t)$]:

$$\dot{\rho}_{nm} = -\mathrm{i}\,\omega_{nm}\rho_{nm} - \frac{\mathrm{i}}{\hbar}\sum_v (V_{nv}\rho_{vm} - \rho_{nv}V_{vm}) \tag{14.28}$$

where

$$\omega_{nm} \equiv \frac{E_n - E_m}{\hbar} \tag{14.29}$$

For these two levels, then, the undamped DME are

$$\dot{\rho}_{nm} = -\mathrm{i}\omega_{nm}\rho_{nm} + \frac{\mathrm{i}}{\hbar} V_{nm}(\rho_{nn} - \rho_{mm})$$

$$\dot{\rho}_{nn} = -\frac{\mathrm{i}}{\hbar}(V_{nm}\rho_{mn} - \rho_{nm}V_{mn})$$

$$\dot{\rho}_{mm} = -\frac{\mathrm{i}}{\hbar}(V_{mn}\rho_{nm} - \rho_{mn}V_{nm}) \tag{14.30}$$

We have not written an equation for $\dot{\rho}_{mn}$ because $\rho_{nm} = \rho_{mn}^*$. In addition, because we consider a closed two-level system

$$\dot{\rho}_{nn} + \dot{\rho}_{mm} = 0$$
$$\rho_{nn} + \rho_{mm} = 1 \tag{14.31}$$

Now we add decay terms. State n, for example, may not be in thermal equilibrium, and it will then necessarily decay towards equilibrium (state m). We represent this decay with a phenomenological rate represented by Γ_{nm}. This gamma is the inverse of the characteristic decay time for the population difference, T_1, via

$$\Gamma_{nm} = \frac{1}{T_1} \tag{14.32}$$

We also introduce a phenomenological decay of the coherences, because at equilibrium they must be nonexistent. This is called the total coherence dephasing rate, γ, which is similarly related to the total dephasing time T_2:

$$\gamma = \frac{1}{T_2} \tag{14.33}$$

Adding these decay terms to DME (14.30) produces

$$\boxed{\dot{\rho}_{nm} = -\left(\mathrm{i}\,\omega_{nm} + \frac{1}{T_2}\right)\rho_{nm} + \frac{\mathrm{i}}{\hbar} V_{nm}(\rho_{nn} - \rho_{mm})} \tag{14.34}$$

$$\boxed{\dot{\rho}_{nn} = -\frac{\rho_{nn}}{T_1} - \frac{\mathrm{i}}{\hbar}(V_{nm}\rho_{mn} - \rho_{nm}V_{mn})} \tag{14.35}$$

$$\dot{\rho}_{mm} = \frac{\rho_{nn}}{T_1} + \frac{i}{\hbar}(V_{nm}\rho_{mn} - \rho_{nm}V_{mn}) \qquad (14.36)$$

where as before

$$\dot{\rho}_{nn} + \dot{\rho}_{mm} = 0 \qquad (14.37)$$

At this point it is useful to discuss the physical implications of the various terms in the DME. The contributions of several of the terms in equations (14.34) to (14.36) are clear. The terms ρ_{nn} and ρ_{mm} are the probability densities for states n and m. The population difference decays towards the equilibrium (steady-state) value with a rate $1/T_1$. Furthermore, it is clear that the two states are coupled via the interaction Hamiltonian.

These parameters can be evaluated in somewhat more detail by simplifying the expressions. Imagine that at time $t = 0$, V_{nm} suddenly drops to zero, meaning that the field has been interrupted. A first observation from equations (14.35) and (14.36) is that state n will then clearly lose population to state m with a characteristic time T_1. Moreover, equation (14.34) is then directly integrable to

$$\rho_{nm}(t) = \rho_{nm}(0)e^{-(i\omega_{nm}+\frac{1}{T_2})t} \qquad (14.38)$$

This result demonstrates that the coherence, ρ_{nm}, oscillates at the same frequency as resonant radiation, $\omega_{nm} = 2\pi\nu_{nm}$.

In fact, when radiation is present and in phase with the coherence, it deposits energy into the system (i.e. stimulated absorption), and when it is 180° out of phase it extracts energy from the system (stimulated emission). Moreover, ρ_{nm} has a buildup and decay time constant of T_2. The transition dipole moment $\langle\mu_{nm}\rangle$ then follows, via expression number (14.24).

The total coherence dephasing rate is given by

$$\frac{1}{T_2} = \frac{1}{2T_1} + \gamma_{nm} \qquad (14.39)$$

where γ_{nm} is the dipole collisional dephasing rate, due to "processes that are not associated with transfer of population" [22]. As an example, a collision can disturb an existing coherence between ρ_{nm} and the optical wave, shifting this optical coherence to a slightly different phase over a short period of time, thus introducing some homogeneous broadening to the spectral line.

It may seem surprising that the rate at which the coherences decay is controlled somewhat by the individual state lifetimes (via T_1). As Boyd points out, this dependence can be justified by use of the expansion coefficients c_n and c_m. The lifetime of the upper state is T_n. The probability of population in level n will then decay according to a single exponential:

$$|c_n(t)|^2 = |c_n(0)|^2 e^{-\frac{t}{T_n}} \tag{14.40}$$

The probability amplitude then decays according to

$$c_n(t) = c_n(0)e^{-i\omega_n t}\, e^{-\frac{t}{2T_n}} \tag{14.41}$$

The time dependence of the probability amplitude of population in level m will be similar:

$$c_m(t) = c_m(0)e^{-i\omega_m t}\, e^{-\frac{t}{2T_m}} \tag{14.42}$$

and the coherence must then decay with a time dependence that includes the form:

$$c_n^*(t)c_m(t) = c_n^*(0)c_m(0)e^{-i\omega_{nm} t}\, e^{-\frac{t}{2T_1}} \tag{14.43}$$

where T_1 is the characteristic decay time of the population difference:

$$T_1 = \left(\frac{1}{T_n} + \frac{1}{T_m}\right)^{-1} \tag{14.44}$$

Equation (14.39) is therefore appropriate for description of a decay that includes transfer of population and nontransfer effects.

14.3   Interaction with an Electromagnetic Field

The DME describe the interaction of an electric field with a resonant quantum dipole. Following Settersten *el al.* [98] and Settersten and Linne [46], the electric field can be expressed as the product of a slowly varying envelope function and a rapidly varying oscillation, both of which are functions of the spatial coordinate z and the time t:

$$\vec{E} = \vec{E}_\circ(z,t) \left[e^{i[k\, z - \omega t + \varphi(z,t)]} + c.c. \right] \tag{14.45}$$

Both the angular frequency ω and the propagation constant k are considered constants of the field, whereas the slowly varying phase term

$\varphi(z,t)$ allows for dispersive effects. With no loss of generality, $\varphi(z,t)$ can be incorporated into the envelope function $\vec{E}_\circ$, resulting in a complex envelope function written as $\tilde{E}_\circ$ (with real and imaginary components $\vec{\varepsilon}^R$ and $\vec{\varepsilon}^I$):

$$\vec{E} = \left[\tilde{E}_\circ(z,t)\, \mathrm{e}^{\mathrm{i}(kz-\omega t)} + c.c.\right] \tag{14.46}$$

This transformation allows us to describe the propagation of an optical signal in terms of the propagation of a complex envelope, because the rapidly varying oscillation does not change form. We describe the net induced polarization of the sample in an analogous way:

$$\vec{P} = \left[\tilde{P}_\circ(z,t)\, \mathrm{e}^{\mathrm{i}(kz-\omega t)} + c.c.\right] \tag{14.47}$$

with real and imaginary components $\vec{p}^R$ and $\vec{p}^I$ of $\tilde{P}_\circ$.

When a resonant or nearly resonant optical electric field is applied, the quantum dipole responds at the optical frequency $\omega_{nm} \sim \omega$. The coherence can then be written in terms of the product of a slowly varying complex envelope $\tilde{\sigma}_{nm}$ and the same rapidly varying oscillation used in equations (14.46) and (14.47):

$$\rho_{nm} = \tilde{\sigma}_{nm}\, \mathrm{e}^{\mathrm{i}(kz-\omega t)} \tag{14.48}$$

The envelope function for the coherence has real and imaginary parts σ^R_{nm} and σ^I_{nm} respectively.

Substituting equations (14.48) and (14.27) into the differential equations (14.34) and (14.36), the following two equations result:

$$\frac{\partial \rho_{mm}}{\partial t} = \frac{\rho_{nn}}{T_1} + \frac{\mathrm{i}\mu_{nm}}{\hbar}\left[-\tilde{E}_\circ\tilde{\sigma}^*_{nm}\right.$$

$$\left. +\tilde{E}^*_\circ\tilde{\sigma}_{nm} + \left(\tilde{E}_\circ\tilde{\sigma}_{nm}\, \mathrm{e}^{2\mathrm{i}(kz-\omega t)} - c.c.\right)\right] \tag{14.49}$$

$$\frac{\partial \tilde{\sigma}_{nm}}{\partial t} = -\left[\mathrm{i}(\omega_{nm}-\omega) + \frac{1}{T_2}\right]\tilde{\sigma}_{nm}$$

$$-\frac{\mathrm{i}\mu_{nm}}{\hbar}(\rho_{nn}-\rho_{mm})\left[\tilde{E}_\circ + \tilde{E}^*_\circ\, \mathrm{e}^{-2\mathrm{i}(kz-\omega t)}\right] \tag{14.50}$$

where $1/T$ terms have been used to describe decay mechanisms to avoid confusion between forms of γ. Equation (14.35) (for $\partial\rho_{nn}/\partial t$) is not included for now because it is redundant with equation (14.37). This conservation expression [equation (14.31)] provides an easy conversion between the two.

Equations (14.49) and (14.50) have terms that are slowly varying $(\omega_{nm}-\omega)$, and terms that oscillate at twice the optical frequency. In the rotating-wave approximation, the rapidly varying terms are neglected. This approximation is based upon physical observation. The term that is close to the natural frequency $(\omega_{nm}-\omega)$ is the one that is slow enough to couple states. The rapidly varying term is much too fast, it couples very weakly, and hence is ignored for linear processes. These equations can then be reformulated in terms of the complex and real parts of the envelope functions for the electric field and for the coherences. This results in three linear equations:

$$\frac{\partial \rho_{mm}}{\partial t} = \frac{1}{T_1}(1 - \rho_{mm}) + \frac{2\mu_{nm}}{\hbar}\left(\varepsilon^I \sigma_{nm}^R - \varepsilon^R \sigma_{nm}^I\right) \qquad (14.51)$$

$$\frac{\partial \sigma_{nm}^R}{\partial t} = -\frac{\sigma_{nm}^R}{T_2} + (\omega_{nm} - \omega)\sigma_{nm}^I + \frac{\mu_{nm}}{\hbar}\varepsilon^I (1 - 2\rho_{mm}) \qquad (14.52)$$

$$\frac{\partial \sigma_{nm}^I}{\partial t} = -\frac{\sigma_{nm}^I}{T_2} - (\omega_{nm} - \omega)\sigma_{nm}^R - \frac{\mu_{nm}}{\hbar}\varepsilon^R (1 - 2\rho_{mm}) \qquad (14.53)$$

Given an initial electric field envelope function, these equations can be directly integrated to produce the time evolution of the coherence at a position z.

The induced polarization of an atom, in response to the applied electric field, is given by the trace of the matrix resulting from the density matrix acting on the operator [see equation (14.14)]. As before, the expectation value is simply

$$
\begin{aligned}
\langle \mu \rangle &= \rho_{mn}\mu_{nm} + \rho_{nm}\mu_{mn} \\
&= \mu_{nm}\left\{\tilde{\sigma}_{nm}\exp\left[i(kz - \omega t)\right]\right\} \qquad (14.54)
\end{aligned}
$$

By solving the DME, therefore, one can determine the temporal response of an atom to a laser pulse of any strength, shape, and width (provided it does not violate the slowly varying envelope approximation).

A simple example

For illustrative purposes, we consider a very simple case; one for which we can analytically solve the DME. Imagine that a resonant electric field is turned on at time $t = 0$. With no loss of generality, we can set $\varepsilon^{I}(t) = 0$ and $\varepsilon^{R}(t \geq 0) = \varepsilon$, where ε is a constant. Furthermore, we neglect relaxation effects so that $1/T_1 \to 0$ and $1/T_2 \to 0$, and neglect any atomic motion. Before the field is turned on, all of the population is in the ground state, $\rho_{mm}(t < 0) = 1$, and there is no coherent excitation, so $\sigma^{R}_{nm}(t < 0) = \sigma^{I}_{nm}(t < 0) = 0$.

Consider equation (14.52) for the evolution of the real part of the coherence envelope function. The entire right-hand side of the equation is zero under the conditions outlined above. Therefore σ^{R}_{nm} must be zero for all time.

We introduce a parameter called the Rabi frequency (denoted by Ω):

$$\Omega \equiv \frac{2\mu_{nm}\varepsilon}{\hbar} \tag{14.55}$$

Its significance will be discussed below. With this definition and the simplifying assumptions, the remaining DME become

$$\frac{\partial \rho_{mm}}{\partial t} = -\Omega \sigma^{I}_{nm} \tag{14.56}$$

$$\frac{\partial \sigma^{I}_{nm}}{\partial t} = -\frac{\Omega}{2}(1 - 2\rho_{mm}) \tag{14.57}$$

Enforcing the initial conditions, the solution to this set of differential equations takes the following form:

$$\rho_{mm} = \frac{1}{2}[\cos(\Omega t) + 1] \tag{14.58}$$

$$\sigma^{R}_{nm} = 0 \tag{14.59}$$

$$\sigma^{I}_{nm} = \frac{1}{2}\sin(\Omega t) \tag{14.60}$$

This result is shown in the top half of Figure 14.1. This demonstrates that the population and coherence oscillate 90° out of phase, at a Rabi frequency that is defined by the applied field. The oscillation period is

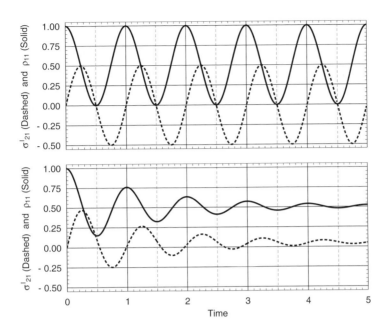

Figure 14.1: Examples of Rabi oscillations in the resonant population ρ_{mm} and coherence σ^I_{nm} for an electric field turned on at time zero. The time axis is in units of the Rabi period. The top set of plots ignores relaxation effects, while the bottom plots show strong damping.

$2\pi/\Omega$. One way to think of Rabi flopping (in the absence of any damping) is to think of an excited state population that is driven up to 100% by an incident field. The only possible action that continued field interaction can generate is subsequent depopulation by stimulated emission. Stimulated emission therefore drives the excited state population back to zero, at which point the field begins to stimulate absorption and drive the excited state population back to 100%. Clearly, the speed at which this occurs would depend upon the field strength.

Including damping terms simply damps these oscillations as shown in the bottom of Figure 14.1. Usually, the Rabi period is much longer than characteristic relaxation times, and these coherent effects are then observed only at low pressure conditions.

Inclusion of Doppler broadening

Equations (14.51), (14.52) and (14.53) include homogeneous broadening mechanisms through the phenomenological decay times T_1 and T_2. They can be modified to include Doppler broadening by including the Doppler shift due to the species velocity component v:

$$\frac{\partial \rho_{mm}}{\partial t} = \frac{1}{T_1}(1 - \rho_{mm}) + \frac{2\mu_{nm}}{\hbar}\left(\varepsilon^{\mathrm{I}}\sigma_{nm}^{\mathrm{R}} - \varepsilon^{\mathrm{R}}\sigma_{nm}^{\mathrm{I}}\right) \tag{14.61}$$

$$\frac{\partial \sigma_{nm}^{\mathrm{R}}}{\partial t} = -\frac{\sigma_{nm}^{\mathrm{R}}}{T_2} + \left(\omega_{nm} - \omega\left(1 - \frac{\mathrm{v}}{c}\right)\right)\sigma_{nm}^{\mathrm{I}}$$
$$+ \frac{\mu_{nm}}{\hbar}\varepsilon^{\mathrm{I}}(1 - 2\rho_{mm}) \tag{14.62}$$

$$\frac{\partial \sigma_{nm}^{\mathrm{I}}}{\partial t} = -\frac{\sigma_{nm}^{\mathrm{I}}}{T_2} - \left(\omega_{nm} - \omega\left(1 - \frac{\mathrm{v}}{c}\right)\right)\sigma_{nm}^{\mathrm{R}}$$
$$- \frac{\mu_{nm}}{\hbar}\varepsilon^{\mathrm{R}}(1 - 2\rho_{mm}) \tag{14.63}$$

These equations are written in the rest frame of the atom or molecule. Atoms or molecules traveling with different velocities along the optical axis (along $\vec{k}$, where $\vec{k} \cdot \vec{v} = k\mathrm{v}$ because we have written the component of $\vec{v}$ along the direction of light propagation as v for notational simplicity) will respond at optical frequencies that are Doppler shifted relative to line center. Note that this shift enters the DME as an effective detuning of the resonance. A frequency detuning can be defined by

$$\delta \equiv \omega_{nm} - \omega\left(1 - \frac{\mathrm{v}}{c}\right) \tag{14.64}$$

and the DME become

$$\frac{\partial \rho_{mm}}{\partial t} = \frac{1}{T_1}(1 - \rho_{mm}) + \frac{2\mu_{nm}}{\hbar}\left(\varepsilon^{\mathrm{I}}\sigma_{nm}^{\mathrm{R}} - \varepsilon^{\mathrm{R}}\sigma_{nm}^{\mathrm{I}}\right) \tag{14.65}$$

$$\frac{\partial \sigma_{nm}^{\mathrm{R}}}{\partial t} = -\frac{\sigma_{nm}^{\mathrm{R}}}{T_2} + \delta\sigma_{nm}^{\mathrm{I}} + \frac{\mu_{nm}}{\hbar}\varepsilon^{\mathrm{I}}(1 - 2\rho_{mm}) \tag{14.66}$$

$$\frac{\partial \sigma_{nm}^{\mathrm{I}}}{\partial t} = -\frac{\sigma_{nm}^{\mathrm{I}}}{T_2} - \delta\sigma_{nm}^{\mathrm{R}} - \frac{\mu_{nm}}{\hbar}\varepsilon^{\mathrm{R}}(1 - 2\rho_{mm}) \tag{14.67}$$

Using the definition of the Rabi frequency given in equation (14.55), we specify the real and imaginary Rabi frequencies as being those obtained by considering the real and imaginary electric field envelope functions respectively:

$$\Omega^R = \frac{2\mu_{nm}\,\varepsilon^R}{\hbar} \tag{14.68}$$

$$\Omega^I = \frac{2\mu_{nm}\,\varepsilon^I}{\hbar} \tag{14.69}$$

The DME can also be cast in these terms:

$$\frac{\partial \rho_{mm}}{\partial t} = \frac{1}{T_1}(1 - \rho_{mm}) + \left(\Omega^I \sigma_{nm}^R - \Omega^R \sigma_{nm}^I\right) \tag{14.70}$$

$$\frac{\partial \sigma_{nm}^R}{\partial t} = -\frac{\sigma_{nm}^R}{T_2} + \delta\sigma_{nm}^I + \Omega^I(1 - 2\rho_{mm}) \tag{14.71}$$

$$\frac{\partial \sigma_{nm}^I}{\partial t} = -\frac{\sigma_{nm}^I}{T_2} - \delta\sigma_{nm}^R - \Omega^R(1 - 2\rho_{mm}) \tag{14.72}$$

Because we have introduced an inhomogeneous detuning (broadening) term, equations (14.61), (14.62) and (14.63) are written for just one class of species (those at velocity v along the optical axis). The response of a collection of atoms or molecules will therefore involve a statistical averaging process. In the density matrix formalism, the expectation value of an observable is given by the trace of the product between the density matrix and the matrix elements of the associated operator, as shown in equation (14.14). At the most fundamental level for a single atom or molecule, the interaction is described by the dipole moment [equation (14.54)]. Note that equation (14.54) relies upon the coherence ($\tilde{\sigma}_{nm}$) to calculate the dipole moment. In such a system, equations (14.61), (14.62), and (14.63) are solved at one value of v, providing a velocity specific value for $\tilde{\sigma}_{nm}$. That value can then be used to find a velocity specific $\langle\mu\rangle$ via equation (14.54); v necessarily enters into $\langle\mu\rangle$ as a parameter.

A linear system of equations

The net polarization induced by the electric field is a macroscopic description of the response of a statistical sample. The polarization P is defined as the volume average of the individual dipole moments of all N atoms in volume V:

$$P(z,t) \equiv \frac{1}{V} \sum_{i=1}^{N} \langle \mu \rangle_i \qquad (14.73)$$

Velocity thus enters into the polarization as a parameter. One can define a partial polarization as the velocity-dependent dipole moment, written as [42]

$$\mathcal{P}(z,t;\mathrm{v}) \equiv \langle \mu(z,t;\mathrm{v}) \rangle \qquad (14.74)$$

Because $\mathcal{P}$ is parameterized by v, the summation in equation (14.73) is treated in statistical mechanics by introduction of the velocity distribution. Normally, the steady-state Maxwellian velocity distribution can be used. Here, the fraction of species with a velocity component v in the interval v→(v+dv) is given by equation (2.45) (modified somewhat for our current purposes):

$$f(\mathrm{v})d\mathrm{v} = \sqrt{\frac{m}{2\pi k_B T}} \exp\left(-\frac{m\mathrm{v}^2}{2k_B T}\right) d\mathrm{v} \qquad (14.75)$$

Equation (14.73) is rewritten as

$$P(z,t) = \mathrm{N} \int_{-\infty}^{+\infty} \mathcal{P}(z,t;\mathrm{v}) f(\mathrm{v}) \, d\mathrm{v} \qquad (14.76)$$

In terms of equations (14.54) and (14.47), this can be written as

$$\tilde{P}(z,t) = \mathrm{N}\mu_{nm} \int_{-\infty}^{+\infty} \tilde{\sigma}(z,t;\mathrm{v}) f(\mathrm{v}) \, d\mathrm{v} \qquad (14.77)$$

As discussed in Chapter 4 [equation (4.69)], the material polarization then enters into the wave equation via

$$\left(\frac{\partial^2}{\partial z^2} - \frac{1}{c^2}\frac{\partial^2}{\partial t^2}\right) E = \frac{1}{\epsilon_o c^2}\frac{\partial^2}{\partial t^2} P(z,t) \qquad (14.78)$$

where we have written a one dimensional equation that would model propagation of one laser beam, for example. A more complicated three dimensional wave equation can be used if necessary. The term $P(z,t)$ can be represented as shown in equation (14.47), and $\tilde{P}_o$ can be broken down into real and imaginary components p^R and p^I. By adopting a retarded time $\tau = t - z/c$ and manipulating the wave equation, Settersten [42] has shown that the second-order wave equation can be reduced to two first order equations that are simpler to solve:

$$\frac{\partial}{\partial z}\varepsilon^R(z,\tau) = -\frac{k}{2\epsilon_\circ}p^I(z,\tau) \tag{14.79}$$

$$\frac{\partial}{\partial z}\varepsilon^I(z,\tau) = \frac{k}{2\epsilon_\circ}p^R(z,\tau) \tag{14.80}$$

These equations can be solved together with the DME (e.g. in the form given by equations (14.65), (14.66) and (14.67), written also in terms of τ) to provide populations of states and coherences as well as the effect of the atoms or molecules on the electric field (and hence the irradiance). It is important to point out, however, that this treatment has been restricted to linear interactions.

14.4 Multiple Levels and Polarization in the DME

Recently, Lucht *et al.* [79] have evaluated the DME for multiple energy levels, including those that are degenerate. They have also developed and evaluated DME expressions that include polarization orientations. Significantly more information on multiple levels is contained in a related article by Reichardt and Lucht [99]. Reference [99] can be used in conjunction with reference [79] to understand multilevel linear interactions.

Equations (14.18) and (14.28) of this chapter actually contain DME expressions for multiple energy levels, but they are missing the phenomenological damping terms. If damping is inserted into them, a multi-energy-level equation set used by Lucht *et al.* emerges:

$$\dot{\rho}_{nm} = -\left(i\,\omega_{nm} + \frac{1}{T_2}\right)\rho_{nm} - \frac{i}{\hbar}\sum_v(V_{nv}\rho_{vm} - \rho_{nv}V_{vm}) \tag{14.81}$$

$$\dot{\rho}_{nn} = -\frac{\rho_{nn}}{T_{1n}} - \frac{i}{\hbar}\sum_v(V_{nv}\rho_{vn} - \rho_{nv}V_{vn}) + \sum_v\frac{\rho_{vv}}{T_{1vn}} \tag{14.82}$$

In a case where polarization orientation matters, equation (14.27) must be written as a dot product:

$$V_{nm} = -\vec{\mu}_{nm}(\vec{r}, t) \cdot \vec{E}(\vec{r}, t) \tag{14.83}$$

where $\vec{\mu}_{nm}$ is the matrix element of the dipole moment defined in equation (10.9). Equation (14.83) now accounts for the relative polarization overlap between the laser and the dipole.

As an example application, equations (14.50), written for two levels, can be written in terms of this same orientational dot product:

$$\frac{\partial \rho_{mm}}{\partial t} = \frac{\rho_{nn}}{T_1} + \frac{i}{\hbar}\left[-\vec{\mu}_{nm} \cdot \tilde{E}_{\circ}\tilde{\sigma}^*_{nm} + \tilde{\sigma}_{nm}\vec{\mu}^*_{nm} \cdot \tilde{E}^*_{\circ} \right] \tag{14.84}$$

$$\frac{\partial \tilde{\sigma}_{nm}}{\partial t} = -\left[i(\omega_{nm} - \omega) + \frac{1}{T_2} \right]\tilde{\sigma}_{nm}$$

$$\qquad\qquad - \frac{i}{\hbar}(\rho_{nn} - \rho_{mm})\vec{\mu}_{nm} \cdot \tilde{E}_{\circ} \tag{14.85}$$

where the rotating wave approximation has been applied but real and imaginary parts have not been extracted. Note that there are several differences between equations (14.84) and (14.85) and the same equations found in Lucht *et al.* [79]. First, we have neglected natural rates of transfer from level m to n in this presentation. Next, Lucht *et al.* use the complex conjugate formalism [equations (4.40) and (4.41)] for the electric field (as does this chapter). Knowing that $E_{\circ}$ is counted twice in this formalism, they have divided by 2. In our work, we assume that each $E_{\circ}$ in equations (4.40) and (4.41) are half as much as they would be if the formalism in equations (4.38) and (4.39) were used. As mentioned in Chapter 4, it is incumbent upon the reader to understand the formalism used by an author and to understand that these differences are not disagreements.

Reichardt and Lucht [99] supply detailed expressions for various state and field interactions (pumping parameters) for multiple levels, written in terms of real and imaginary Rabi frequencies. Both papers also supply detailed expressions for the the matrix element of the dipole moment for coupling to various levels (degenerate and otherwise), including the Zeeman sublevels in each state. These sources provide the necessary details for multi-state calculations.

Reference [79] takes the DME to the steady state limit and finds a classical absorption coefficient. The same approach is used to find Einstein coefficients in the next section of this chapter.

14.5 Two-level DME in the Steady-state Limit

Steady-state expressions can be derived from the DME [as expressed in equations (14.51), (14.52) and (14.53)]. In the treatment that follows, we have chosen to neglect Doppler detuning because we wish to compare a result from the homogeneously broadened DME to a result from the homogeneously broadened ERT. The outcome will remain general because Doppler broadening affects the two formalisms in the same way. We are in fact analyzing just one velocity class, but it can be any velocity class. We consider a two-level model here in order to simplify the presentation.

By definition, at steady state the time derivatives are zero in both systems of equations. A subscript "ss" denotes a steady-state value for a variable. At steady state, therefore, the DME can be rewritten:

$$\frac{\partial \rho_{mm}}{\partial t} = \frac{1}{T_1}(1 - \rho_{mm,\text{ss}})$$
$$+ \frac{2\mu_{nm}}{\hbar}(\varepsilon^I \sigma_{nm,\text{ss}}^R - \varepsilon^R \sigma_{nm,\text{ss}}^I) = 0 \qquad (14.86)$$

$$\frac{\partial \sigma_{nm}^R}{\partial t} = -\frac{\sigma_{nm,\text{ss}}^R}{T_2} + \delta \sigma_{nm,\text{ss}}^I$$
$$+ \frac{\mu_{nm}}{\hbar}\varepsilon^I(1 - 2\rho_{mm,\text{ss}}) = 0 \qquad (14.87)$$

$$\frac{\partial \sigma_{nm}^I}{\partial t} = -\frac{\sigma_{nm,\text{ss}}^I}{T_2} - \delta \sigma_{nm,\text{ss}}^R$$
$$- \frac{\mu_{nm}}{\hbar}\varepsilon^R(1 - 2\rho_{mm,\text{ss}}) = 0 \qquad (14.88)$$

where the frequency detuning is now defined by

$$\delta = (\omega_{nm} - \omega) \qquad (14.89)$$

Although equations (14.51), (14.52) and (14.53) are a common form for the density matrix equations, we now add an equation for the excitation fraction ρ_{nn} by inserting $\rho_{mm} = 1 - \rho_{nn}$ into equation (14.51). This will produce the same outcome that would have resulted if we had carried equation (14.35) all the way through the treatment given just above. We do this because it will be possible to compare the steady-state excitation fraction from the DME with the excitation fraction from the

RE, and thus extract an expression for the Einstein coefficients. The steady-state equation for ρ_{nn} (the excitation fraction) is then

$$\frac{\partial \rho_{nn}}{\partial t} = -\frac{\rho_{nn,ss}}{T_1} - \frac{2\mu_{nm}}{\hbar}(\varepsilon^{I}\sigma^{R}_{nm,ss} - \varepsilon^{R}\sigma^{I}_{nm,ss}) = 0 \qquad (14.90)$$

Using the definitions for real and imaginary Rabi frequency, equations (14.86), (14.87), (14.88), and (14.90) are rearranged to obtain expressions for steady-state values of the population fractions $\rho_{mm,ss}$ and $\rho_{nn,ss}$ and the real and imaginary coherence functions $\sigma^{R}_{nm,ss}$ and $\sigma^{I}_{nm,ss}$ respectively:

$$\rho_{mm,ss} = 1 + T_1 \left(\Omega^{I} \sigma^{R}_{nm,ss} - \Omega^{R} \sigma^{I}_{nm,ss} \right) \qquad (14.91)$$

$$\rho_{nn,ss} = -T_1 \left(\Omega^{I} \sigma^{R}_{nm,ss} - \Omega^{R} \sigma^{I}_{nm,ss} \right) \qquad (14.92)$$

$$\sigma^{R}_{nm,ss} = T_2 \, \delta\sigma^{I}_{nm,ss} + \frac{\Omega^{I} T_2}{2}(2\rho_{nn,ss} - 1) \qquad (14.93)$$

$$\sigma^{I}_{nm,ss} = -T_2 \, \delta\sigma^{R}_{nm,ss} - \frac{\Omega^{R} T_2}{2}(2\rho_{nn,ss} - 1) \qquad (14.94)$$

where $\rho_{mm} = 1 - \rho_{nn}$ was used to find the form of equations (14.93) and (14.94).

Equations (14.93) and (14.94) can be used to eliminate the real and imaginary forms of $\sigma_{nm,ss}$ in equation (14.92). We assume a fixed, real value for the electric field envelope ε_{o}, producing a fixed, real Rabi frequency $\Omega_{o} = 2\mu_{nm}\varepsilon_{o}/\hbar$. The outcome is an expression for $\rho_{nn,ss}$:

$$\rho_{nn,ss} = \frac{1}{2}\frac{\Omega_{o}^{2} T_1}{T_2} \left[\frac{1}{\delta^2 + \frac{1}{T_2^2}(1 + \Omega_{o}^{2} T_1 T_2)} \right] \qquad (14.95)$$

This expression looks somewhat like a resonance response parameter multiplied by a spectral profile. In fact, the term in square braces is a Lorentzian, with a FWHM width of

$$\Delta\omega_{H} = \frac{2}{T_2}\sqrt{1 + \Omega_{o}^{2} T_1 T_2} \qquad (14.96)$$

Equation (14.95) can be compared with the ERT result via manipulation of equation (3.58):

$$\frac{dN_n(t)}{dt} = N_m(t)W_{mn} - N_n(t)[W_{nm} + A_{nm} + Q_{nm}]$$

Here, N_n is the number density of atoms or molecules in state n. For a two-level system, $N_n + N_m = N_{total}$. We can define the number density of each level by dividing by N_{total}:

$$\rho_{nn} = \frac{N_n}{N_{total}}$$

$$\rho_{mm} = \frac{N_m}{N_{total}} \tag{14.97}$$

where we have purposely used ρ to facilitate comparison. Equation (3.58) can now be recast by dividing by N_{total}:

$$\frac{d\rho_{nn}}{dt} = W_{mn} - \rho_{nn}[W_{nm} + W_{mn} + A_{nm} + Q_{nm}] \tag{14.98}$$

where we have also used $\rho_{mm} = 1 - \rho_{nn}$. At steady state the time derivative is eliminated and equation (14.98) produces

$$\rho_{nn,ss} = \frac{W_{mn}}{[W_{nm} + W_{mn} + A_{nm} + Q_{nm}]} \tag{14.99}$$

In order to develop a comparison with the DME result, it is necessary to evaluate W_{nm} and W_{mn}. These are defined in equations (3.56) and (3.57):

$$W_{mn} \equiv B_{mn} \int_\nu \frac{I_\nu}{c} Y_\nu \, d\nu$$

$$W_{nm} \equiv B_{nm} \int_\nu \frac{I_\nu}{c} Y_\nu \, d\nu$$

Because the electric field envelope is a constant, the laser spectrum is a delta function in frequency:

$$I_\nu = 2\epsilon_\circ c \varepsilon_\circ^2 \, \delta(\nu - \nu_{nm}) \tag{14.100}$$

where we have linked the electric field to irradiance using the same complex conjugate expression used in Section 4.6. The lineshape function is given by a Lorentzian [equation (11.42)]:

$$Y_H(\nu - \nu_{nm}) = \frac{\Delta\nu_H}{2\pi} \frac{1}{(\nu - \nu_{nm})^2 + (\frac{\Delta\nu_H}{2})^2}$$

where $\Delta\nu_H$ is found from equation (11.44): $\Delta\nu_H = (\pi T_2)^{-1}$. Using these expressions in equation (3.56), and writing the equations in terms of angular frequency ω, we find

$$W_{mn} = 2B_{mn}\epsilon_o\mathcal{E}_o \frac{1/T_2}{(\omega_{nm} - \omega)^2 + (1/T_2)^2} \qquad (14.101)$$

Finally, the collected terms $A_{nm} + Q_{nm}$ in equation (14.99) describe the total population decay:

$$T_1 = \frac{1}{A_{nm} + Q_{nm}} \qquad (14.102)$$

Substituting equations (14.101) and (14.102) into equation (14.99) produces

$$\rho_{nn,\text{ss}} = \frac{1}{2}8B_{mn}\epsilon_o\mathcal{E}_o^2 \frac{T_1}{T_2}\left[\frac{1}{\delta^2 + \frac{1}{T_2^2}(8B_{mn}\epsilon_o\mathcal{E}_o^2 T_1 T_2)}\right] \qquad (14.103)$$

which is nearly identical to equation (14.95). They are identical if

$$\Omega_o^2 = 8B_{mn}\epsilon_o\mathcal{E}_o^2 \qquad (14.104)$$

Using the definition of Ω and rearranging terms produces

$$\boxed{B_{mn} = \frac{\mu_{nm}^2}{2\hbar^2\epsilon_o}} \qquad (14.105)$$

This expression for the Einstein B_{mn} coefficient can be related to the A_{nm} coefficient using equations (3.95) and (3.96):

$$\frac{A_{nm}}{B_{nm}} = \frac{8\pi h\nu^3}{c^3}$$

$$g_n B_{nm} = g_m B_{mn}$$

Because we have solved for a two-level system, $g_n = g_m$, and therefore $B_{nm} = B_{mn}$. Solving for A_{nm}, we find

$$A_{nm} = \frac{2\omega_{nm}^2\mu_{nm}^2}{h\epsilon_o c^3} \qquad (14.106)$$

Equation (14.106) differs from equation (10.28) (the Einstein coefficient found in Chapter 10) by a factor of $1/3$ (in this two-level approximation,

$g_n = 1$). This difference arose because the DME development assumed perfect alignment between the electric field and the dipole moment. As discussed in Chapter 12, alignment is totally random in the gas phase. Equations (14.106) and (10.28) are thus based upon different assumptions regarding alignment. If we assume random alignment, equation (14.106) should be divided by 3, which brings it into agreement with equation (10.28) (assuming that $g_n = 1$):

$$\boxed{A_{nm} = \frac{2\omega_{nm}^2 \mu_{nm}^2}{3h\epsilon_\circ c^3}} \tag{14.107}$$

It is worth mentioning that Boyd [22] also calculates a susceptibility in the steady-state limit, which can lead to similar results. Electric fields interacting with quantum dipoles are the natural basis for the DME formalism. The link to the ERT just provided is simply that; a link that demonstrates that the DME and ERT are self-consistent. The Einstein coefficients are appropriate for the ERT, while quantum dipoles, electromagnetic waves, susceptibilities and polarization are appropriate for the DME. One can always convert from an electric field to irradiance using equations (4.78) and (4.79).

14.6 Conclusions

This chapter has presented a common version of the density matrix equations. One can find other forms of the same equations. This set of equations is more physical; it is much simpler to interpret the meaning of each term. Moreover, these equations are often simplified with several assumptions and then converted into the optical Bloch equations. This approach has been taken in the past in order to make it possible to develop analytical solutions. They do involve assumptions, however, and again they are not as easy to interpret. Modern computing techniques make it possible to solve the full DME without simplifications. In this section we have also shown how the DME are expressed in a typical problem (suitable especially for numerical solution), and then demonstrated that they produce a standard resonance response when extended to the steady-state limit.

Example applications of the DME to resonant nonlinear optics in combustion diagnostics would include the perturbative solution to the resonant DFWM problem presented by Abrams *et al.* [100], or the non-perturbative numerical solution to the same problem taken by Lucht and co-workers (see, for example, [97, 99]). A linear example is provided by Settersten and Linne [46], who describe short pulse excitation, and by Settersten *et al.* [98] who analyze picosecond pump/probe absorption spectroscopy.

Appendix A

Units

The subject of dimensional units in electromagnetism is discussed in many books. Wangsness [50] includes an entire chapter on the subject (Chapter 23, entitled "Systems of Units—A Guide For The Perplexed"). In addition, Jackson [49] discusses this subject in an appendix. Boyd [22] discusses the impact of this topic on nonlinear optics in a very brief appendix as well. Here we consolidate these and other commentaries on units.

One might ask, justifiably, why dimensional confusion exists at all. Jackson addresses this question by arguing that any system of units is arbitrary, and such a system is normally chosen for the convenience of the group using it. Therein lies the problem. There are various communities that make use of electromagnetic theory, and they do indeed hold to the convenience of their own chosen unit system and conventions. Spectroscopy is an interdisciplinary field that overlaps several of these communities. Spectroscopists therefore encounter various formalisms. In electromagnetism, this can lead to equation sets that seemingly disagree, and it can lead to errors in reported results unless care is taken.

In general, our approach is first to verify that a reported expression is correct (e.g. does not contain typographical errors). At that point, it is often possible to tell whether an expression is in MKSA or Gaussian units. One easy indicator is the presence of an ϵ_o. In Gaussian units, $\epsilon_o = 1$ (see below for an explanation), and so it never appears. If an expression contains an ϵ_o, therefore, it has portions at least that are in MKSA units.

We prefer to work in MKSA, and so we immediately convert an expression if it is not already in MKSA. We then evaluate expressions by inserting numerical values, being careful to include all the dimensions for the physical constants inserted into the equations. Next, we check to see if the units work out. If not, this may be because one of the physical constants simply needs to be converted. If that is not the case, it is likely that we have a combined system (usually Gaussian and MKSA). In this case we choose to adjust the expression to make it entirely MKSA, by performing the equivalent of unit conversions using factors like $4\pi\epsilon_o$.

Having said that, it remains instructive to discuss the systems in use, how differences arise, and to present a formal method for cross-conversion. In what follows, we discuss the foundations of Maxwell's equations and the ways in which an approach can affect units (following most closely the texts by Wangsness and Jackson), we generate Maxwell's equations in the Gaussian unit system (in vacuum for simplicity), and then we present a coherent scheme for converting between MKSA and Gaussian units. It may be helpful to refer back to relevant portions of Chapter 4 while reading the material below.

The following material can be summarized briefly by saying that there are four of Maxwell's equations, which are based upon fundamental physical laws. Maxwell's equations contain four constants of proportionality that must be fixed. The fixing of these constants then determines the dimensional units one must use for the electromagnetic variables (e.g. E, B, ρ, J etc.). It turns out that we do not need to specify four constants because we can also specify two relations among the constants. One of these relationships ensures that the speed of light will appear in the correct location in the wave equation. The other relationship simply relates two of the constants to each other (allowing factors of "1" to appear). We are therefore left to set two of the constants. Those are set by the fundamental relationships underlying Maxwell's equations, Coulomb's law, and the Biot-Savart law. The constants then define the dimensional units for the variables in Maxwell's equations, the wave equation, and so on.

The most basic units used in the physical sciences are: mass m, length l, and time t. In electromagnetism, we also need a dimension either for charge q, or for current I. The commonly accepted absolute current is defined by a force balance between two wires [see equation (4.15)]. Moreover, charge and current are related by $I = \mathrm{d}q/\mathrm{d}t$. The two are also related through charge conservation [equation (4.6)]:

$$\vec{\nabla} \cdot \vec{J} + \frac{\partial \varrho}{\partial t} = 0 \tag{A.1}$$

and we should reasonably expect to resolve the units in a self-consistent fashion within these expressions. Equation (A.1) looks the same in Gaussian and MKSA units, although the dimensional units for current density and charge density are different in the two systems.

We now rewrite Coulomb's law [equation (4.1)], upon which Gauss' electric law is based, in a slightly different form:

$$\vec{F} = k_1 \frac{q_1 q_2}{R^2} \hat{R} \tag{A.2}$$

Here, k_1 is determined either by dimensions within Coulomb's law, if units of charge have already been identified, or it is chosen arbitrarily in order to define charge. At this point, all we know is that the product $k_1 q_1 q_2$ necessarily has the basic units ml^3/t^2. Furthermore, the definition of an electric field [equation (4.3)] could be written: as

$$E = k_1 \frac{q}{R^2} \tag{A.3}$$

This is a defined quantity, and we might be tempted to add another constant of proportionality. Since E is the first defined quantity, however, it is left as shown in equation (A.3).

Gauss' electric law is then written as

$$\vec{\nabla} \cdot \vec{E} = 4\pi k_1 \varrho \tag{A.4}$$

Gauss' magnetic law [equation (4.7)] will not enter into this conversation, because it makes no contribution to the definition of dimensional units.

Next, Faraday's law can be generalized in a dimensional sense as well. Instead of letting the constant of proportionality be simply "1", as found in equation (4.11), we now make it k_2:

$$\vec{\nabla} \times \vec{E} = -k_2 \frac{\partial \vec{B}}{\partial t} \tag{A.5}$$

We will make use of this expression after discussing Ampere's law, but we can state now that both sides of equation (A.5) must have the same units, and this sets the units of k_2.

The Biot-Savart law can be written with a generalized constant k_3 by combining equations (4.13) and (4.14):

$$\frac{\mathrm{d}F}{\mathrm{d}\ell} = 2k_3\frac{I_1 I_2}{L} \tag{A.6}$$

Because current has units of charge per unit time, the constant k_3 has the basic units $ml/t^2 I^2$.

We have now stated enough to draw an important conclusion. Consider the units of equation (A.2):

$$\left(\frac{m\,l}{t^2}\right) = k_1(?)\left(\frac{q^2}{l^2}\right) \tag{A.7}$$

and of equation (A.6):

$$\left(\frac{m}{t^2}\right) = k_3(?)\left(\frac{q^2}{t^2 l}\right) \tag{A.8}$$

The ratio of k_1/k_3 must therefore be

$$\left(\frac{k_1}{k_3}\right) = \left(\frac{l^2}{t^2}\right) \tag{A.9}$$

which are the units of velocity squared. Independent measurements of forces, charge, and current in equations (A.2) and (A.6) place the ratio k_1/k_3 at the value of the speed of light squared, c^2. This value makes sense with respect to the wave equation as well. The relationship

$$\frac{k_1}{k_3} = c^2 \tag{A.10}$$

is an observed constant that electromagnetism must be able to reproduce if it is to model optical wave propagation accurately.

Now consider a generalized Ampere's law [equation (4.13)]:

$$B = 2k_3\alpha\frac{I}{L} \tag{A.11}$$

where we imitate Jackson's variable α for this new constant of proportionality. Another k is not used because it is important to demonstrate that this constant is not independent of the others. We need to maintain a dimensional connection to equation (A.6) in this case.

Consider again the dimensional units of several equations. Equation (A.1) can be used to set the units of J at current per unit area (A/m^2 in MKSA). Equation (A.3) can be used to show that the units for E are

$$E \rightarrow k_1 \left(\frac{q}{l^2} \right) \tag{A.12}$$

and from equation (A.11) [using equation (A.9)]:

$$B \rightarrow 2k_1\alpha \left(\frac{qt}{l^3} \right) \tag{A.13}$$

which gives

$$\frac{E}{B} \rightarrow \frac{l}{\alpha t} \tag{A.14}$$

Equation (A.5) can be rewritten as

$$\vec{\nabla} \times \vec{E} + k_2 \frac{\partial \vec{B}}{\partial t} = 0 \tag{A.15}$$

and this has the units

$$\frac{E}{l} + k_2 \frac{B}{t} = 0 \tag{A.16}$$

Dividing by B then yields

$$\frac{1}{l} \frac{l}{t\alpha} + k_2 \frac{1}{t} = 0 \tag{A.17}$$

The constant k_2 therefore has the dimensions of $1/\alpha$.

Maxwell's equations for vacuum can be generalized to

$$\vec{\nabla} \cdot \vec{E} = 4\pi k_1 \varrho \tag{A.18}$$

$$\vec{\nabla} \cdot \vec{B} = 0 \tag{A.19}$$

$$\vec{\nabla} \times \vec{E} = -k_2 \frac{\partial \vec{B}}{\partial t} \tag{A.20}$$

$$\vec{\nabla} \times \vec{B} = 4\pi k_3 \alpha \vec{J}_{\mathrm{f}} + \frac{k_3 \alpha}{k_1} \frac{\partial \vec{E}}{\partial t} \tag{A.21}$$

If we use these equations to generate a wave equation, we will obtain

$$\nabla^2 E = \frac{k_3 k_2 \alpha}{k_1} \frac{\partial^2 E}{\partial t^2} \tag{A.22}$$

Therefore

$$\frac{k_3 k_2 \alpha}{k_1} = \frac{1}{c^2} \qquad (A.23)$$

We already know that $k_1/k_3 = c^2$, so we can conclude that

$$k_2 = \frac{1}{\alpha} \qquad (A.24)$$

This is perhaps an overlong justification for the fact that, although we have four constants k_1, k_2, k_3, and α, we also have two equations (A.10) and (A.24) that constrain the choices we can make for these four. We also need to ensure that Coulomb's law and the Biot-Savart law are followed, and this sets the other two.

Knowing this, we can rewrite Maxwell's equations as

$$\vec{\nabla} \cdot \vec{E} = 4\pi k_1 \varrho \qquad (A.25)$$

$$\vec{\nabla} \cdot \vec{B} = 0 \qquad (A.26)$$

$$\vec{\nabla} \times \vec{E} = -k_2 \frac{\partial \vec{B}}{\partial t} \qquad (A.27)$$

$$\vec{\nabla} \times \vec{B} = 4\pi \frac{k_3}{k_2} \vec{J}_{\mathrm{f}} + \frac{k_3}{k_1 k_2} \frac{\partial \vec{E}}{\partial t} \qquad (A.28)$$

The two most common unit systems are the MKSA and Gaussian systems. Rationalized MKSA units are used throughout the body of this paper. The Gaussian system is a CGS-related (centimeter, gram, second) combination of electrostatic units ("esu", in pure esu $k_1 = 1$ and $k_3 = 1/c^2$) applied to the electric quantities, and electromagnetic system ("emu", in pure emu $k_1 = c^2$ and $k_3 = 1$) applied to magnetic quantities. Since optical interactions are electrical, the Gaussian system may appear to be the same as the electrostatic system within the domain of optics.

In esu, charge is measured in statcoulombs (stC) (two equal charges of 1 statcoulomb placed 1 cm apart will repel each other with a force of 1 dyne), current is in statamperes (stA) (stC/s), potential is in statvolts (stV) (= 1 erg/stC), capacitance is in statfarads (stF), and resistance is in statohms. Often these units are all simply written "esu" and the reader is expected to know the details. In emu, current is measured in abamperes (two long parallel wires carrying 1 abampere each and

Table A.1: Constants for MKSA and Gaussian systems

System	k_1	k_2	k_3
ESU	1	1	$\frac{1}{c^2}$
EMU	c^2	1	1
Gaussian	1	$\frac{1}{c}$	$\frac{1}{c^2}$
MKSA	$\frac{1}{4\pi\epsilon_o} = 10^{-7} c^2$	1	$\frac{\mu_o}{4\pi} = 10^{-7}$

placed 1 cm apart will experience a force of 2 dynes/cm), charge is then an abcoulomb (1 abampere-s), potential is in abvolts, etc. Often these units are all simply written "emu" and the reader is expected to know the details. In both the esu and emu cases, the writer literally expects the reader to insert appropriate units for variables if necessary. For example, imagine a paper uses both electric charge and capacitance in an expression and ends the expression with "esu", the reader is expected to know that the units for charge are stC and for capacitance they are stF. This may at first seem like a lazy way to work, but that is the beauty of the Gaussian system. It was designed to allow people to focus on issues other than units.

There are other systems of units in electromagnetism (e.g. the Heaviside-Lorentz system), but they are rarely encountered in spectroscopy. Here, we discuss just the two main systems. Details on the others can be found in Wangsness [50] and Jackson [49].

Another way to designate systems is whether they are "rationalized" or not. A rationalized system is arranged to have no factors of 4π within Maxwell's equations. That does not mean the 4π go away, they simply appear elsewhere. The system used in the body of the text is a rationalized MKSA system, and one can see where the 4π are kept. The Gaussian system is not rationalized.

The two main systems then use the values for the k_i constants listed in Table A.1. These values have been chosen to agree with the basic relationships from which they were formed. Note, for example, that k_3 is given a value of 10^{-7} in the MKSA system. This result then agrees with equation (A.6) and the definition of current given in the discussion surrounding equation (4.15). To repeat that definition, when two long wires are placed 1 m apart, and the currents passing through both are equal and set to produce a force of 2×10^{-7} N/m, then that current is defined to be one ampere. Hence the need for 10^{-7}. The result $k_1/k_3 = c^2$, and the fact that we are free to make $k_2 = 1$, then concludes the discussion. In the Gaussian system, the choices for the k_i are made to ensure that the speed of light falls out of the wave equation. Other issues are dealt with by the definition of the dimensional units.

These results can then be used to write Maxwell's equations and the wave equation in both systems (with units). In the MKSA system they are written as shown in equations (A.29) to (A.33). Note that units for individual constants and variables are in parentheses. Collected terms on the right- and left-hand sides are located within {} braces. The overall units for the right- and left-hand sides are contained within square braces.

Maxwell's equations are then:

$$\left\{ \vec{\nabla} \left(\frac{1}{m} \right) \cdot \vec{E} \left(\frac{V}{m} \right) \right\} \left[\frac{V}{m^2} \right] = \left\{ \frac{1}{\epsilon_\circ} \left(\frac{Vm}{C} \right) \varrho \left(\frac{C}{m^3} \right) \right\} \left[\frac{V}{m^2} \right] \qquad (A.29)$$

$$\left\{ \vec{\nabla} \left(\frac{1}{m} \right) \cdot \vec{B} \left(\frac{W}{m^2} \right) \right\} \left[\frac{W}{m^3} \right] = 0 \qquad (A.30)$$

$$\left\{ \vec{\nabla} \left(\frac{1}{m} \right) \times \vec{E} \left(\frac{V}{m} \right) \right\} \left[\frac{V}{m^2} \right] = - \left\{ \frac{\partial \vec{B}}{\partial t} \left(\frac{W}{m^2 s} \right) \right\} \left[\frac{V}{m^2} \right] \qquad (A.31)$$

$$\left\{ \vec{\nabla} \left(\frac{1}{m} \right) \times \vec{B} \left(\frac{W}{m^2} \right) \right\} \left[\frac{W}{m^3} \right]$$

$$= \left\{ \mu_\circ \left(\frac{H}{m} \right) \vec{J}_f \left(\frac{A}{m^2} \right) + \mu_\circ \left(\frac{H}{m} \right) \epsilon_\circ \left(\frac{C}{Vm} \right) \frac{\partial \vec{E}}{\partial t} \left(\frac{V}{ms} \right) \right\} \left[\frac{W}{m^3} \right]$$

and the wave equation is

$$\left\{ \nabla^2 \left(\frac{1}{m^2} \right) E \left(\frac{V}{m} \right) \right\} \left[\frac{V}{m^3} \right] = \left\{ \frac{1}{c^2} \left(\frac{s^2}{m^2} \right) \frac{\partial^2 E}{\partial t^2} \left(\frac{V}{ms^2} \right) \right\} \left[\frac{V}{m^3} \right] \quad (A.33)$$

The base units [length (m), mass (kg), time (s), charge (C), or (A-s) if current is the base unit] for various electromagnetic quantities in the

Table A.2: Basic units for MKSA electromagnetic quantities

Quantity	Basic units
Force	Newton $= \frac{\text{kg m}}{\text{s}^2}$
Capacitance	Farad $= \frac{\text{C}}{\text{V}} = \frac{\text{A}^2\text{s}^4}{\text{kg m}^2}$
Current	Ampere $= \frac{\text{C}}{\text{s}}$
Potential	Volt $= \frac{\text{J}}{\text{C}} = \frac{\text{kgm}^2}{\text{A s}^3}$
Electric field E	$\frac{\text{V}}{\text{m}}$
Charge density ϱ	$\frac{\text{C}}{\text{m}^3}$
Electric permittivity ϵ	$\frac{\text{F}}{\text{m}} = \frac{\text{A}^2\,\text{s}^4}{\text{kg m}^3}$
Magnetic permeability μ	$\frac{\text{H}}{\text{m}} = \frac{\text{N s}^2}{\text{C}^2}$
Magnetic induction B	Tesla $= \frac{\text{Weber}}{\text{m}^2} = \frac{\text{kg}}{\text{A s}^2}$

MKSA system are shown in Table A.2. Note that the unit of charge (C) is considered a basic unit even though it is derived from the definition of current. The use of Table A.2 in equations (A.29) to (A.33) can demonstrate that the units can be resolved.

In the Gaussian system, Maxwell's equations and the wave equation are written as shown in expressions (A.34) to (A.38). On first reading, these equations and the dimensional units may appear to be more contrived. They were contrived in the sense that they were designed to make the system of equations clean within themselves. We will find, for example, that E, D, B, H, P, and M all have the same units (or perhaps the best way to say this is that Maxwell's equations define relationships between the units), which is mathematically convenient. The MKSA system is more closely related to dimensions used elsewhere. It can appear to be less contrived, but it requires the same number of assumptions as the Gaussian system does.

Maxwell's equations are then

$$\left\{ \vec{\nabla}\left(\frac{1}{\text{cm}}\right) \cdot \vec{E}\left(\frac{\text{stV}}{\text{cm}}\right) \right\} = \left\{ 4\pi\varrho\left(\frac{\text{stC}}{\text{cm}^3}\right) \right\} \tag{A.34}$$

$$\left\{ \vec{\nabla}\left(\frac{1}{\text{cm}}\right) \cdot \vec{B}\,(\text{gauss}) \right\} = 0 \tag{A.35}$$

Table A.3: Basic units for Gaussian electromagnetic quantities

Quantity	Basic units
Force	Dyne$=\frac{\text{gm cm}}{\text{s}^2}$
Capacitance	Statfarad$=\frac{\text{stC}}{\text{stV}}=\frac{\text{stA}^2\,\text{s}^4}{\text{g cm}^2}$
Current	Statampere$=\frac{\text{stC}}{\text{s}}$
Potential	Statvolt$=\frac{\text{erg}}{\text{stC}}=\frac{\text{g cm}^2}{\text{stA s}^3}$
Electric field E	$\frac{\text{stV}}{\text{cm}}$
Charge density ϱ	$\frac{\text{stC}}{\text{cm}^3}$
Electric permittivity	(1)
Magnetic permeability	(1)
Magnetic induction B	gauss

$$\left\{\vec{\nabla}\left(\frac{1}{\text{cm}}\right)\times\vec{E}\left(\frac{\text{stV}}{\text{cm}}\right)\right\}=-\left\{\frac{1}{c}\left(\frac{\text{s}}{\text{cm}}\right)\frac{\partial\vec{B}}{\partial t}\left(\frac{\text{gauss}}{\text{s}}\right)\right\} \qquad (A.36)$$

$$\left\{\vec{\nabla}\left(\frac{1}{\text{cm}}\right)\times\vec{B}\,(\text{gauss})\right\}=\left\{4\pi\frac{1}{c}\left(\frac{\text{s}}{\text{cm}}\right)\vec{J}_f\left(\frac{\text{stA}}{\text{cm}^2}\right)+\frac{1}{c}\left(\frac{\text{s}}{\text{cm}}\right)\frac{\partial\vec{E}}{\partial t}\left(\frac{\text{stV}}{\text{cm s}^2}\right)\right\}$$
$$(A.37)$$

and the wave equation is

$$\left\{\nabla^2\left(\frac{1}{\text{cm}^2}\right)E\left(\frac{\text{stV}}{\text{cm}}\right)\right\}=\left\{\frac{1}{c^2}\left(\frac{\text{s}^2}{\text{cm}^2}\right)\frac{\partial^2 E}{\partial t^2}\left(\frac{\text{stV}}{\text{cm s}^2}\right)\right\} \qquad (A.38)$$

Note that equations (A.34) to (A.37) define the need for equality of units between E and B. The base units [length (cm), mass (g), time (s), charge (stC)] for various electromagnetic quantities in the Gaussian system are shown in Table A.3.

We now address the issue of material interactions. The MKSA approach to this topic is presented in detail in the text [see, for example, equations (4.52), (4.53), (4.58), and (4.59), and the discussion around them]. Here, for brevity, we simply explain what is done in the Gaussian system. Note that factors like D, H, χ, and so forth are defined in a completely different fashion, although the same variables are used.

In the Gaussian system, we define

$$\vec{D} \equiv \vec{E} + 4\pi \vec{P} \tag{A.39}$$

$$\vec{H} \equiv \vec{B} - 4\pi \vec{M} \tag{A.40}$$

$$\vec{D} \equiv \epsilon \vec{E} \tag{A.41}$$

$$\vec{H} \equiv \frac{\vec{B}}{\mu} \tag{A.42}$$

$$\vec{P} \equiv \chi_e \vec{E} \tag{A.43}$$

$$\epsilon \equiv 1 + 4\pi \chi_e \tag{A.44}$$

$$\mu \equiv 1 + 4\pi \chi_m \tag{A.45}$$

and finally

$$\vec{S} = \frac{c}{4\pi}(\vec{E} \times \vec{H}) \tag{A.46}$$

From the above, we can conclude the following:

$$(\kappa_e)_{MKSA} = \left(\frac{\epsilon}{\epsilon_o}\right)_{MKSA} = (\epsilon)_{Gaussian} \tag{A.47}$$

$$(\kappa_m)_{MKSA} = \left(\frac{\mu}{\mu_o}\right)_{MKSA} = (\mu)_{Gaussian} \tag{A.48}$$

$$(\chi_{e\ or\ m})_{MKSA} = 4\pi(\chi_{e\ or\ m})_{Gaussian} \tag{A.49}$$

Note that in the Gaussian system there really is no permittivity or permeability of free space. Gaussian units are set up to produce "c" directly in the wave equation without involving fundamental constants, as one finds in MKSA (e.g. $c = (\sqrt{\mu_o \epsilon_o})^{-1}$ in MKSA). What was a dielectric constant in MKSA becomes a material permittivity in Gaussian, and so on.

The results presented above can be compiled into a conversion table (Table A.4, collected from both Wangsness and Jackson). To use the table, the term in the system under use is simply replaced by the term in the system desired. We have included only those terms that apply

Table A.4: Conversion for electromagnetic variables

Quantity	MKSA	Gaussian
Capacitance	C	$4\pi\epsilon_\circ\, C$
Charge	q	$\sqrt{4\pi\epsilon_\circ}\; q$
Charge density	ϱ	$\sqrt{4\pi\epsilon_\circ}\; \varrho$
Dielectric constant	κ_e	ϵ
Dipole moment	$\vec{\mu}_i$	$\sqrt{4\pi\epsilon_\circ}\; \vec{\mu}_i$
Electric displacement	$\vec{D}$	$\sqrt{\frac{\epsilon_\circ}{4\pi}}\; \vec{D}$
Electric field	$\vec{E}$	$\frac{1}{\sqrt{4\pi\epsilon_\circ}}\; \vec{E}$
Magnetic field	$\vec{H}$	$\frac{1}{\sqrt{4\pi\mu_\circ}}\; \vec{H}$
Magnetic induction	$\vec{B}$	$\sqrt{\frac{\mu_\circ}{4\pi}}\; \vec{B}$
Permeability (relative)	κ_m	μ
Polarization	$\vec{P}$	$\sqrt{4\pi\epsilon_\circ}\; \vec{P}$
Light speed	$\frac{1}{\sqrt{\mu_\circ\epsilon_\circ}}$	c
Susceptibility	$\chi_{e\text{ or m}}$	$4\pi\chi_{e\text{ or m}}$

to optical interactions. Significantly more detail is contained in the treatments by Wangsness [50] and Jackson [49].

One warning regarding Table A.4 should be made. If one is applying the table to a Gaussian result worked out for vacuum, $\vec{D} = \vec{E}$ and $\vec{B} = \vec{H}$ in that case. The symbols are sometimes used interchangeably. If an author meant $\vec{E}$ but wrote $\vec{D}$, for example, one should not use the $\vec{D}$ conversion given in the table.

Next, in Table A.5, we provide simple units conversions for numerical values important in optics (note that units of time are the same in both systems).

It is appropriate to mention a few things about conversion of nonlinear coefficients (see Boyd [22] for further details). Recall that equations (4.70) and (4.71) contain alternative descriptions of the nonlinear polarization as used within the MKSA system. They are repeated here for convenience:

$$\vec{P} = \epsilon_\circ \left[\chi^{(1)}\vec{E} + \chi^{(2)}\vec{E}^2 + \chi^{(3)}\vec{E}^3 ... \right]$$
$$\vec{P} = \epsilon_\circ \chi^{(1)}\vec{E} + \chi^{(2)}\vec{E}^2 + \chi^{(3)}\vec{E}^3 ...$$

Table A.5: Conversion for numerical values

Quantity	MKSA	Gaussian
Length	1 meter (m)	10^2 centimeters (cm)
Mass	1 kilogram (kg)	10^3 grams (g)
Force	1 newton (N)	10^5 dyne
Work and energy	1 joule (J)	10^7 erg
Power	1 watt (W)	10^7 erg/s
Capacitance C	1 farad (F)	9×10^{11} statfarad
Charge q	1 Coulomb (C)	3×10^9 statcoulomb
Charge density ϱ	1 C/m^3	3×10^3 statcoulomb/cm^3
Current I	1 Ampere (A)	3×10^9 statampere
Current density $\vec{J}$	1 A/m^2	3×10^5 statampere/cm^2
Potential	1 volt (V)	$\frac{1}{300}$ statvolt
Electric displacement $\vec{D}$	1 C/m^2	$12\pi \times 10^5$ statvolt/cm
Electric ield $\vec{E}$	1 V/m	$\frac{1}{3} \times 10^{-4}$ statvolt/cm
Magnetic field $\vec{H}$	1 A/m	$4\pi \times 10^{-3}$ oersted
Magnetic induction $\vec{B}$	1 tesla=1 weber/m^2	10^4 gauss
Polarization $\vec{P}$	1 C/m^2	3×10^5 statvolt/cm

Clearly, the units of the various $\chi_e^{(i)}$ will be different in the two formalisms.

In the formalism of equation (4.70), since ϵ_o appears in the linear term, the linear susceptibility $\chi_e^{(1)}$ is dimensionless. The dimensions of the higher-order susceptibilities would be

$$\chi_e^{(2)} \quad \rightarrow \quad \frac{\text{m}}{\text{V}}$$

$$\chi_e^{(3)} \quad \rightarrow \quad \frac{\text{m}^2}{\text{V}^2}$$

and so forth.

In the formalism of equation (4.71), the linear susceptibility $\chi_e^{(1)}$ is again dimensionless. Since ϵ_o is missing from the nonlinear terms, dimensions of the higher-order susceptibilities would be

$$\chi_e^{(2)} \quad \rightarrow \quad \frac{C}{V^2}$$

$$\chi_e^{(3)} \quad \rightarrow \quad \frac{Cm}{V^3}$$

and so forth (see Table A.5 for the units of $\vec{E}$ and $\vec{P}$ that determine this result).

The Gaussian system is somewhat different, as one would expect given that the linear χ_e are different. In the Gaussian system we write

$$\vec{P} = \chi_e^{(1)} \vec{E} + \chi_e^{(2)} \vec{E}^2 + \chi_e^{(3)} \vec{E}^3 ... \tag{A.50}$$

As we have already mentioned, the units for $\vec{E}$ and $\vec{P}$ are the same in the Gaussian system, and $\chi_e^{(1)}$ is therefore dimensionless. The dimensions of the higher-order susceptibilities would then be:

$$\chi_e^{(2)} \quad \rightarrow \quad \frac{cm}{statvolt}$$

$$\chi_e^{(3)} \quad \rightarrow \quad \frac{cm^2}{statvolt^2}$$

and so forth.

To convert between Gaussian and MKSA, we have to keep track of which MKSA we mean. Boyd [22] works this out through a simple ratio. The outcome is simply presented here:

$$\chi_e^{(2)} \text{ [MKSA, equation 4.70]} \quad = \quad \frac{4\pi}{3 \times 10^4} \chi_e^{(2)} \text{ (Gaussian)}$$

$$\chi_e^{(2)} \text{ [MKSA, equation 4.71]} \quad = \quad \frac{4\pi\epsilon_\circ}{3 \times 10^4} \chi_e^{(2)} \text{ (Gaussian)}$$

$$\chi_e^{(3)} \text{ [MKSA, equation 4.70]} \quad = \quad \frac{4\pi}{(3 \times 10^4)^2} \chi_e^{(3)} \text{ (Gaussian)}$$

$$\chi_e^{(3)} \text{ [MKSA, equation 4.71]} \quad = \quad \frac{4\pi\epsilon_\circ}{(3 \times 10^4)^2} \chi_e^{(2)} \text{ (Gaussian)}$$

and so forth.

Appendix B

Constants

Throughout this book we have defined various constants and included them in important expressions. Here their values are presented in tabular form.

Table B.1: Useful constants

Constant	Variable	Value	Units
Avagadro's number	N_A	6.022×10^{23}	molecules/mole
Boltzmann's constant	k_B	1.381×10^{-23}	J/(K molecule)
Classical electron radius	r_e	2.818×10^{-15}	m
Electric permittivity of vacuum	$\epsilon_\circ$	8.854×10^{-12}	F/m
Electron rest mass	m_e	0.911×10^{-30}	kg
Elementary charge	e	1.602×10^{-19}	C
Magnetic permeability of vacuum	$\mu_\circ$	$4\pi \times 10^{-7}$	H/m
Planck's constant	h	6.626×10^{-34}	Js
Rydberg's constant	R	1.097×10^{7}	m^{-1}
Speed of light in vacuum	c	2.998×10^{8}	m/s

Bibliography

[1] A. C. Eckbreth. *Laser Diagnostics for Combustion Temperature and Species*. Gordon and Breach, Amsterdam, second edition, 1996.

[2] W. Demtröder. *Laser Spectroscopy, Basic Concepts and Instrumentation*. Springer, Berlin, second edition, 1998.

[3] K. Kohse-Höinghaus and J. Jeffries, editors. *Applied Combustion Diagnostics*. Elsevier, London, 2002.

[4] J.D. Bransford, A. L. Brown, and R. R. Cocking, editors. *How People Learn: Brain, Mind, Experience, and School*. National Academy Press, Washington, D.C., 1999.

[5] A. G. Gaydon. *The Spectroscopy of Flames*. Chapman and Hall, London, 1957.

[6] R. Mavrodineanu and H. Boiteux. *Flame Spectroscopy*. John Wiley and Sons, New York, 1965.

[7] P. R. Griffiths and J. A. deHaseth. *Fourier Transform Infrared Spectrometry*. John Wiley and Sons, New York, 1986.

[8] D. R. Crosley, editor. *Laser Probes for Combustion Chemistry*, ACS Symposium Series 134, Washington, D.C., 1984. American Chemical Society.

[9] K. Kohse-Höinghaus. Laser techniques for the quantitative detection of reactive intermediates in combustion systems. *Progress in Energy and Combustion Science*, 20:203, 1994.

[10] J. Wolfrum. Lasers in combustion: from simple models to real devices. In *Twenty-Seventh (International) Symposium on Combustion*, Pittsburgh, 1998. The Combustion Institute.

[11] R. K. Hanson. Combustion diagnostics: planar imaging techniques. In *Twenty-First (International) Symposium on Combustion*, Pittsburgh, 1986. The Combustion Institute.

[12] R. M. Measures. *Laser Remote Sensing, Fundamentals and Applications*. Krieger Publishing Co., Malabar, 1992.

[13] J. Luque and D. R. Crosley. Lifbase. Technical Report MP 96-001, 1996.

[14] C. N. Banwell and E. M. McCash. *Fundamentals of Molecular Spectroscopy*. Tata McGraw-Hill Publishing Co., Ltd., New Delhi, 1994.

[15] G. Herzberg. *Atomic Spectra and Atomic Structure*. Dover Publications, New York, 1944.

[16] G. Herzberg. *The Spectra and Structure of Simple Free Radicals*. Dover Publications, New York, 1971.

[17] G. Herzberg. *Molecular Spectra and Molecular Structure. I. Spectra of Diatomic Molecules*. Krieger Publishing Co., Malabar, 1989.

[18] G. Herzberg. *Molecular Spectra and Molecular Structure. II. Infrared and Raman Spectra of Polyatomic Molecules*. Krieger Publishing Co., Malabar, 1991.

[19] G. Herzberg. *Molecular Spectra and Molecular Structure. III. Electronic Spectra and Electronic Structure of Polyatomic Molecules*. Krieger Publishing Co., Malabar, 1991.

[20] L. Allen and J. H. Eberly. *Optical Resonance and Two-Level Atoms*. Dover Publications, New York, 1987.

[21] J. I. Steinfeld. *Molecules and Radiation*. MIT Press, Cambridge, 1989.

[22] R. W. Boyd. *Nonlinear Optics*. Academic Press, San Diego, 1992.

[23] J. R. Lalanne, A. Ducasse, and S. Kielich. *Laser Molecule Interaction, Laser Physics and Modern Nonlinear Optics.* John Wiley and Sons, New York, 1996.

[24] S. Mukamel. *Principles of Nonlinear Optical Spectroscopy.* Oxford University Press, New York, 1995.

[25] A. E. Siegman. *Lasers.* University Science Books, Mill Valley, 1986.

[26] J. T. Verdeyen. *Laser Electronics.* Prentice-Hall, Englewood Cliffs, third edition, 1995.

[27] R. C. Hillborn. Einstein coefficients, cross sections, f values, dipole moments and all that. *American Journal of Physics*, 50(11):982, 1982.

[28] R. P. Lucht, R. C. Peterson, and N. M. Laurendeau. Fundamentals of absorption spectroscopy for selected diatomic flame radicals. Report PURDU-CL-78-06, 1978.

[29] W. G. Vincenti and C. H. Kruger. *Introduction to Physical Gasdynamics.* Krieger Publishing Co., Malabar, 1965.

[30] C. Garrod. *Statistical Mechanics and Thermodynamics.* Oxford University Press, New York, 1995.

[31] R. C. Tolman. *The Principles of Statistical Mechanics.* Dover Publications, New York, 1979.

[32] D. A. Long. *Raman Spectroscopy.* McGraw-Hill, New York, 1977.

[33] R. Viskanta and M. P. Mengüç. Radiation heat transfer in combustion systems. *Progress in Energy and Combustion Science*, 13:97, 1987.

[34] H. C. van de Hulst. *Light Scattering by Small Particles.* Dover Publications, New York, 1957.

[35] M. Kerker. *The Scattering of Light and Other Electromagnetic Radiation.* Academic Press, New York, 1969.

[36] C. F. Bohren and D. R. Huffman. *Absorption and Scattering of Light by Small Particles.* John Wiley and Sons, New York, 1983.

[37] F.-Q. Zhao and H. Hiroyasu. The application of laser Rayleigh scattering to combustion Diagnostics. *Progress in Energy and Combustion Science*, 19:447, 1993.

[38] J. W. Daily. Laser induced fluorescence spectroscopy in flames. *Progress in Energy and Combustion Science*, 23:133, 1997.

[39] R. Siegel and J. R. Howell. *Thermal Radiation Heat Transfer*. Hemisphere Publishing Corp., Washington, third edition, 1992.

[40] S. S. Penner. *Quantitative Molecular Spectroscopy and Gas Emissivities*. Addison-Wesley, Reading, 1959.

[41] W. P. Partridge and N. M. Laurendeau. Formulation of a dimensionless overlap fraction to account for spectrally distributed interactions in fluorescence studies. *Applied Optics*, 15:2546, 1995.

[42] T. B. Settersten. *Picosecond Pump-Probe Diagnostics for Combustion*. PhD thesis, Colorado School of Mines, Golden, 1999.

[43] J.W. Daily. Use of rate equations to describe laser excitation in flames. *Applied Optics*, 16:2322, 1977.

[44] R. Kienle, M.P. Lee, and K. Kohse-Höinghaus. A detailed rate equation model for the simulation of energy transfer in OH laser-induced fluorescence. *Applied Physics B*, 62:583, 1996.

[45] J. W. Daily. Saturation of fluorescence with a Gaussian laser beam. *Applied Optics*, 17:225, 1978.

[46] T. Settersten and M. Linne. Modeling pulsed excitation for gas-phase laser diagnostics. *Journal of the Optical Society of America B*, 19(5):954, 2002.

[47] M. Born and E. Wolf. *Principles of Optics, Electromagnetic Theory of Propagation, Interference and Diffraction of Light*. Cambridge University Press, Cambridge, sixth edition, 1980.

[48] E. Hecht. *Optics*. Addison-Wesley, Reading, third edition, 1998.

[49] J. D. Jackson. *Classical Electrodynamics*. John Wiley and Sons, New York, third edition, 1999.

[50] R. K. Wangsness. *Electromagnetic Fields*. John Wiley and Sons, New York, second edition, 1986.

[51] M. V. Klein and T. E. Furtak. *Optics*. John Wiley and Sons, New York, second edition, 1986.

[52] H. M. Schey. *Div, Grad, Curl, and All That, an Informal Text on Vector Calculus*. W.W. Norton and Company, New York, 1973.

[53] J. W. Goodman. *Introduction to Fourier Optics*. McGraw-Hill, New York, second edition, 1988.

[54] R. M. Dickson, D. J. Norris, and W. E. Moerner. Simultaneous Imaging of Individual Molecules Aligned Both Parallel and Perpendicular to the Optic Axis. *Physical Review Letters*, 81(24):5322, 1998.

[55] A. P. Bartko and R. M. Dickson. Imaging three-dimensional single molecule orientations. *Journal of Physical Chemistry*, 103:11237, 1999.

[56] J. B. Marion and S. T. Thornton. *Classical Dynamics of Particles and Systems*. Harcourt Brace Jovanovich, San Diego, fourth edition, 1999.

[57] J. Baggott. *The Meaning of Quantum Theory*. Oxford University Press, New York, 1992.

[58] F. Schwabl. *Quantum Mechanics*. Springer, Berlin, second edition, 1995.

[59] B. L. van der Waerden. *Sources of Quantum Mechanics*. Dover Publications, New York, 1957.

[60] I. N. Levine. *Quantum Chemistry*. Prentice Hall, Englewood Cliffs, fourth edition, 1991.

[61] J. J. Sakurai. *Modern Quantum Mechanics*. Addison-Wesley, Reading, 1995.

[62] S. Brandt and H. D. Dahmen. *The Picture Book of Quantum Mechanics*. Springer-Verlag, New York, 1994.

[63] A. Tonomura, J. Endo, T. Matsuda, and T. Kawasaki. Demonstration of single-electron buildup of an interference pattern. *American Journal of Physics*, 57(2):117, 1989.

[64] M. Karplus and R. N. Porter. *Atoms and Molecules: An Introduction for Students of Physical Chemistry*. W.A. Benjamin, Inc., Menlo Park, CA, 1970.

[65] R. D. Cowan. *The Theory of Atomic Structure and Spectra*. University of California Press, Berkeley, 1981.

[66] R. N. Zare. *Angular Momentum, Understanding Spatial Aspects in Chemistry and Physics*. John Wiley and Sons, New York, 1988.

[67] C. Th. J. Alkemade, Tj. Hollander, W. Snelleman, and P. J. Th. Zeegers. *Metal Vapours in Flames*. Pergamon Press, Oxford, 1982.

[68] I. N. Levine. *Molecular Spectroscopy*. John Wiley and Sons, New York, 1975.

[69] L. S. Rothman, R. R. Gamache, A. Goldman, L. R. Brown, R. A. Toth, H. M. Picket, R. L. Poynter, J.-M. Flaud, C. Camy-Peyret, A. Barbe, N. Husson, C. P. Rinsland, and M. A. H. Smith. The HITRAN Database: 1986 Edition. *Applied Optics*, 26:4058, 1987.

[70] I. Kovacs. *Rotational Structure in the Spectra of Diatomic Molecules*. Hilger, London, 1969.

[71] E. E. Whiting, A. Schadee, J. B. Tatum, J. T. Hougen, and R. W. Nichols. Recommended conventions for defining transition moments and intensity factors in diatomic molecular spectroscopy. *Journal of Molecular Spectroscopy*, 80:249, 1980.

[72] W. Heitler. *The Quantum Theory of Radiation*. Dover Publications, New York, third edition, 1984.

[73] B. W. Shore. *The Theory of Coherent Atomic Excitation*. John Wiley and Sons, New York, 1990.

[74] B. P. Lathi. *Modern Digital and Analog Communication Systems*. Holt, Rinehart and Winston, New York, 1983.

[75] P. P. Feofilov. *The Physical Basis of Polarized Emission*. Consultants Bureau, New York, 1961.

[76] V. D. Kleiman, H. Park, R. J. Gordon, and R. N. Zare. *Companion to Angular Momentum*. John Wiley and Sons, New York, 1998.

[77] R. N. Zare. Molecular level-crossing spectroscopy. *Journal of Chemical Physics*, 45(12):4510, 1966.

[78] P. M. Doherty and D. R. Crosley. Polarization of laser-induced fluorescence in OH in an atmospheric pressure flame. *Applied Optics*, 23:713, 1984.

[79] R. P. Lucht, S. Roy, and T. A. Reichardt. Polarization effects in combustion diagnostics and calculation of radiative transition rates. *Progress in Energy and Combustion Science*, to appear, 2003.

[80] A. Brochinke, W. Kreutner, U. Rahman, K. Kohse-Höinghaus, T. Settersten, and M. Linne. Time, wavelength and polarization-resolved measurements of OH $(A^2\Sigma^+, v' = 1)$ picosecond laser induced fluorescence in atmospheric pressure flames. *Applied Physics B*, 69:713, 1999.

[81] P. Beaud, P. P. Radi, D. Franzke, H.-M. Mischler, A.-P. Tzannis, and T. Gerber. Picosecond investigation of the collisional deactivation of OH $(A^2\Sigma^+, v' = 1, N' = 4, 12)$ in an atmospheric pressure flame. *Applied Optics*, 37(15):3354, 1998.

[82] E. J. McCartney. *Optics of the Atmosphere: Scattering by Molecules and Particles*. John Wiley and Sons, New York, 1976.

[83] S. Bhagavantam. *Scattering of Light and the Raman Effect*. Chemical Publishing Company, Inc., New York, 1942.

[84] D. A. Long. *The Raman Effect: A Unified Treatment of the Theory of Raman Scattering by Molecules*. John Wiley and Sons, New York, 2002.

[85] C. D. Carter. Laser-based Rayleigh and Mie scattering methods. In J. A. Schetz and A. E. Fuhs, editors, *Handbook of Fluid Dy-*

namics and Fluid Machinery. John Wiley and Sons, New York, 1996.

[86] W. M. Pitts and T. Kashiwagi. The application of laser-induced Rayleigh light scattering to the study of turbulent mixing. *Journal of Fluid Mechanics*, 141:391, 1984.

[87] M. Lapp and C. M. Penney. *Laser Raman Gas Diagnostics*. Plenum Press, New York, 1974.

[88] S. Lederman. The use of laser Raman diagnostics in flow fields and combustion. *Progress in Energy and Combustion Science*, 3:1, 1977.

[89] H. Goldstein. *Classical Mechanics*. Addison Wesley, Reading, MA, 1950.

[90] J. A. R. Samson. On the measurement of Rayleigh scattering. *Journal of Quantitative Spectroscopy and Radiative Transfer*, 9:875, 1969.

[91] G. Tenti, C. D. Boley, and R. Desai. On the kinetic model description of Rayleigh-Brillouin scattering from molecular gases. *Canadian Journal of Physics*, 52(4):285, 1974.

[92] C. D. Boley, R. C. Desai, and G. Tenti. Kinetic models and Brillouin scattering in a molecular gas. *Canadian Journal of Physics*, 50:2158, 1972.

[93] A. T. Young and G. W. Kattawar. Rayleigh-scattering line profiles. *Applied Optics*, 22(23):3668, 1983.

[94] N. M. Laurendeau. Temperature measurements by light-scattering methods. *Progress in Energy and Combustion Science*, 14:147, 1988.

[95] R. L. Farrow and R. E. Palmer. Comparison of motionally narrowed coherent anti-Stokes Raman spectroscopy line shapes of H_2 with hard- and soft-collision models. *Optics Letters*, 12(12):984, 1987.

[96] A. Dreizler. RAMSES, software for calculation of Raman spectra. at the time of this writing, RAMSES can be accessed at http://www.tu-darmstadt.de/fb/mb/ekt/main.html.

[97] T. A. Reichardt and R. P. Lucht. Degenerate four-wave mixing spectroscopy using picosecond lasers: theoretical analysis. *Journal of the Optical Society of America B*, 13:2807, 1996.

[98] T. Settersten, M. Linne, J. Gord, and G. Fiechtner. Density matrix and rate equation analyses for picosecond pump/probe combustion diagnostics. *AIAA Journal*, 37(6):723, 1999.

[99] T. A. Reichardt and R. P. Lucht. Resonant degenerate four-wave mixing spectroscopy of transitions with degenerate energy levels: saturation and polarization effects. *Journal of Chemical Physics*, 111(22):10008, 1999.

[100] R. L. Abrams, J. F. Lam, R. C. Lind, D. G. Steel, and P. F. Liao. Phase conjugation and high-resolution spectroscopy by resonant degenerate four-wave mixing. In R. L. Fisher, editor, *Optical Phase Conjugation*. Academic Press, San Diego, 1983.

Index